Mathematical Logic

Mathematical Logic

George Tourlakis

York University
Department of Computer Science and Engineering
Toronto, Ontario, Canada

WILEY

A JOHN WILEY & SONS, INC., PUBLICATION

Published by John Wiley & Sons, Inc., Hoboken, New Jersey.
Published simultaneously in Canada.

For general information on our other products and services or for technical support, please contact our Customer Care Department within the United States at (800) 762-2974, outside the United States at (317) 572-3993 or fax (317) 572-4002.

Wiley also publishes its books in a variety of electronic formats. Some content that appears in print may not be available in electronic format. For information about Wiley products, visit our web site at www.wiley.com.

Library of Congress Cataloging-in-Publication Data:

Tourlakis, George J.
Mathematical logic / George Tourlakis.
   p. cm.
Includes bibliographical references and index.
ISBN 978-0-470-28074-4 (cloth)
1. Logic, Symbolic and mathematical—Textbooks. I. Title.
QA9.T68 2008
511.3—dc22   2008009433

10 9 8 7 6 5 4 3 2

To those who taught me

CONTENTS

Preface

This volume is about the foundation of mathematics, the way it was conceptualized by Russell and Whitehead [56], Hilbert (and Bernays) [22], and Bourbaki[1] [2]: *Mathematical Logic*. This is the discipline that, much later, Gries and Schneider [17] called the "glue" that holds mathematics together.

Mathematical logic, on one hand, *builds* the tools for mathematical reasoning with a view of providing a *formal* methodology—i.e., one that relies on the *form* or *syntax* of mathematical statements rather than on their meaning—that is meant to be *applied* for constructing mathematical arguments that are correct, well documented, and therefore understandable.

On the other hand, it studies the interplay between the written structure (syntax) of mathematical statements and their meaning: Are the theorems that we prove by pure syntactic manipulation true under some reasonable definition of *true*? Are there any true mathematical statements that our tools *cannot* prove? The former question will be answered in the affirmative later in this book, while the latter question, interestingly, has both "no" (Gödel's completeness theorem [15]) and "yes" (Gödel's

---

[1]"Nicolas Bourbaki" is the pen-name of a team of top mathematicians who are responsible for the monumental work, "Élémens de Mathématique", which starts with logic as *the foundation*, or "connecting glue" in the words of [17], and then proceeds to extensively cover fields such as set theory, algebra, topology, analysis, measure, and integration.

(first)*in*completeness theorem [16]) answers![2] Both of these answers are carefully reconstructed in the Appendix to Part II.

Much has been written on logic, which is nowadays a mature mathematical body of knowledge and research. The majority of books written with upper-level undergraduate audiences (and beyond) in mind deal mostly with the *metamathematics* or *metatheory* of mathematical logic; that is, they view logic as a mathematical object and study its abilities and limitations (such as incompleteness), and the theory of models, giving short shrift to the issue of *using* logic as a tool.

There are currently only two books that the author is aware of that chronologically precede this volume and address almost exclusively the interests and needs of the *user* of logic. Both present the subject as a set of tools with which one can do mathematics (or computer science, or philosophy, or anything else that requires reasoning) rigorously, correctly, formally, and with adequate documentation: [2] and [17].

The former tersely introduces logic in its first chapter with a view of applying it as a rigorous tool for *theorem generation* in the numerous (and very advanced) chapter-volumes that follow (from set theory and algebra to topology, and measure and integration).

The latter, a much more recent entry in the literature, is an elementary text (aimed at undergraduate university curricula in computer science) in the same spirit as Bourbaki's, which proposes to use logic, once again, as a tool to prove theorems of interest to computer scientists. Indeed, the second part of [17] is on *discrete mathematics* in the sense that this term, more or less, is understood by most computer science departments today.

Similarly, the volume in your hands aims to thoroughly teach the *use* of logic as a tool for reasoning appropriate for upper-level undergraduate university students in fields of study such as computer science, mathematics, and philosophy. For the first group, this is an introduction to *formal methods*—a subject that is often included in computer science curricula—providing the student with the *tools*, the *methodology*, and a solid grounding on *technique*. As the student advances along the computer science curriculum, this volume's toolbox on formal methods will find serious applications in courses such as design and analysis of algorithms, theory of computation, computational complexity, software specification and design, artificial intelligence, and program verification.

The second group's curriculum, at the targeted level, in addition to a solid course on the use of logic, will normally also require a more ambitious inquiry into the

---

[2]It is not that Gödel was of two minds on the issue. Rather, the question can be made precise in two different ways, and, correspondingly, one gets two different answers. One way is to think of "universal" truth, such as the truth of "$x = x$". Universal truth is completely certifiable by the syntactic tools. The other is to think of truth in the "standard models" of some "rich" theories—rich in what one can formulate and prove in them, that is. Formal (Peano) arithmetic—that is, the axiomatic system that attempts to explain the *set of natural numbers and the arithmetic operations and relations on it*, the standard model— is such a rich theory. Gödel showed the existence of true arithmetical statements in the model that cannot be syntactically proved in the axiom system of Peano arithmetic. One such true statement says, "I am not a theorem."

capabilities and limitations of logic viewed as a mathematical tool. This further trip into the metatheory of logic will traditionally want to delve into two foundational gems beyond those of *soundness* and propositional *completeness*. Both are due to Gödel, namely, his completeness and incompleteness theorems. The Appendix to Part II settles the former in full detail, and also offers a proof of the latter (actually, only of the *first* incompleteness theorem[3]), basing it on an inherent limitation of "general" models of computation, in the process illuminating the connection between the phenomena of *uncomputability* and *unprovability*.

As a side-effect of constructing a self-contained proof of the first incompleteness theorem, we had to develop a fair amount of *computability* theory that will be of direct interest to all the readers, in particular, those in computer science.

The third group of readers, philosophy majors, traditionally require less coverage in a course in logic than what I have presented here; however, philosophy curricula often include a course in symbolic logic at an advanced undergraduate level, and this volume will be an apt companion for such studies.

The book's aim to teach the practice of logic dictates that it must look and feel much like a serious text on programming. In fact, I argue at the very beginning of the first chapter, that learning and practicing logic is a process like that of learning and practicing programming. As a result, the emphasis is on presenting a multitude of tools, and on *using* these tools in many fully written and annotated proofs, an approach that is intended to enhance the reader's effectiveness as a "prover", giving him[4] many examples for emulation.

There are some important differences—despite the superficial similarities that the common end-aims impose—between the approach and content in this volume and that in its similarly aimed predecessors [2] and [17].

Bourbaki provides tools for use by the "practicing mathematician" and does not bother with any semantic issues, presumably on the assumption that the mathematician knows full well how the syntactic and semantic notions interact and relate, and has an already well developed experience and ability to use *semantic methods* toward finding *counterexamples* when needed. He merely introduces and uses the so-called Hilbert style of proofs (cf. 1.4.12) that is most commonly used by mathematicians.

The text of [17] is equally silent about the interplay between syntax and semantics, and about any aspect of the metatheory, and refers to Hilbert-style proofs only tangentially. The authors prefer to exclusively propound the *equational* (or *calculational*) proof style (cf. Section 2.2), originally proposed in [11]. Moreover, unlike [2], they take liberties with their formalism.[5] For example, even though they argue in their introduction in favor of using *formal methods* in practical reasoning, they distance themselves from a true syntactic approach, especially in their Chapter 8, where facts

---

[3]The second incompleteness theorem, that the freedom of contradiction of "rich" axiomatic systems such as Peano arithmetic *cannot* be proved "from within", is beyond the scope of this volume. Indeed, the only complete proofs in print for this result are found in [22], Vol. II, and in [53].

[4]*His, him, he* and related terms that grammatically indicate gender are, by definition, gender neutral in this volume.

[5]A *formalism* in the context of mathematical logic is any particular way logicians structure their formal methods.

outside logic taken from algebra and number and set theory are presented as axioms of predicate logic.

While the approach in this volume is truly formal, just like Bourbaki's, it is not as terse; we are guilty of the opposite tendency! We also believe that, unlike the seasoned practitioner, the undergraduate mathematics, computer science, and philosophy students need some *reassurance* that the form-manipulation proof-writing tools presented here indeed prove (mathematical) "truths", "all truths", and "nothing but truths". This means that we cannot run away from the most basic and fundamental metatheoretical results. After all, every practitioner needs to know a few things about the properties of his tools; this will make him more effective in their use.

Thus I include proofs of the *soundness* (meta)theorems for both propositional and predicate logics (this addresses the "truths", and "nothing but truths" part) and also the two "completeness" results, of propositional and predicate logics (this is the "all truths" part). However, to maintain both the emphasis on the use of logic and an elementary but rigorous flow of exposition I have delegated the much-harder-to-prove completeness metatheorem of predicate logic ([15]) to a sizable appendix at the end of the book.

Why are soundness and completeness relevant to the needs of the user? Completeness of propositional logic, along with its soundness, give us the much-needed—in the interest of user-friendliness—license to mix semantic and syntactic tools in formal proofs *without sacrificing mathematical rigor*. Indeed, this license (to use *propositional* semantic tools) is extended even in predicate logic, and is made possible by the trick of adding and removing quantifiers ("for all" and "for some"). On the other hand, soundness of the two logics allows the user to *disprove* statements by constructing so-called *countermodels*.

There are also quite a few simpler metatheoretical results, beyond soundness and completeness, that we routinely introduce and prove as needed about formulae (e.g., about their syntax) and about proofs (e.g., the validity of principles of proof such as *hypothesis strengthening*, *deduction theorem*, and *generalization*), using the basic tool of induction (essentially on formula and proof lengths).

The Hilbert style of proving theorems is prevalent in the mathematical literature and is prominently displayed and practiced in this volume. On the other hand, the equational-style of displaying proofs has been gaining in popularity especially in computer science curricula. It is a style of proof that seems well adapted to areas in computer science such as software engineering (in particular, in the field of software engineering *requirements*) and program verification.

For the above reason, equational-style proofs receive a thorough exposition in this volume. It is my intention to endow the reader with enough machinery that will make him proficient in *both* styles of proof, but more importantly, will enable him to choose the style that is best suited to writing a proof for any particular theorem.

In terms of prior knowledge (tools) needed to cope with this volume the reader should at least have high school mathematics (but I expect that this includes mathematical induction and some basic algebra). A degree of mathematical maturity, but no specific additional knowledge, of the kind an upper-level undergraduate will normally have will also be handy.

A word on pedagogical approach. I repeatedly taught the material included here to undergraduate computer science students at York University in Toronto, Canada. I think of this book as the record of my lectures. I have endeavored to make these lectures user-friendly, and therefore accessible to readers who do not have the benefit of an instructor's guidance. Devices to that end include anticipation of questions, promptings for the reader to rethink an issue that might be misunderstood if glossed over ("pauses"), numerous remarks and examples that reflect on a preceding definition or theorem.

Using the symbols ☇, and ☇ ☇, I am marking those passages that are very important, and those that can be skipped at first reading, respectively.

My fondness for footnotes is surely evident (a taste acquired long ago, when I was studying Wilder's excellent *Introduction to the Foundations of Mathematics* ([57]).

I give (mostly) *very* detailed proofs, as I know from experience that omitting details normally annoys students. Moreover, I have expectations that students will achieve a certain style, and effectiveness, in proofs. The best way to teach them to do so is by repeatedly giving examples how. In turn, students will have the opportunity to test and further their understanding by doing several exercises, some of which are embedded in the text while others appear at chapters' end (a total of more than 190 exercises).

Book structure. The book is in two approximately equal-length parts, one on Boolean (or propositional) logic and one on predicate logic. A thorough exposition of Boolean logic pedagogically prepares the reader for the much more difficult predicate logic, at the same time endowing him with several tools that are transferable such as the ubiquitous *Post's theorem* (propositional completeness) and *deduction theorem*.

Part I is in three chapters. Chapter 1 starts with the basic formation rules of propositional (Boolean) formulae—the syntax—and introduces "induction on formulae" as a tool via which we can prove facts about syntax. It proceeds with Boolean semantics (truth tables) and then continues with the concept of formal proofs—those effected via purely syntactic manipulation—from axioms and rules of inference. Chapter 2 is a veritable database of proofs and theorems, presenting several proofs and proof techniques, including the *deduction theorem*. Both the equational and Hilbert style of proof layouts are used extensively. Chapter 3 revisits semantics, and proves both the soundness and completeness (Post) theorems, thus demonstrating the full equivalence and interchangeability of the semantic and syntactic proof techniques in Boolean logic. It concludes with an exposition of the technique of *resolution* in Boolean logic.

Part II on predicate logic (or calculus) contains five chapters and a lengthy Appendix. Predicate calculus is introduced as an extension of the logic of Part I, so that every tool that we obtained in Part I is still usable in Part II. This part's first chapter, Chapter 4, is about the syntax of formulae, and introduces the axioms, the rules of inference, and the concept of proof, extending without discarding anything of the corresponding concepts of Part I. Chapter 5 simplifies the metatheoretical arguments by introducing a *simpler-to-talk-about* logic, equivalent to ours; that is, a logic with a simpler metatheory. Chapter 6 proves and extensively uses powerful rules of inference that were not postulated up front: techniques for adding and removing the universal quantifier, powerful Leibniz rules, and techniques for adding and removing the existential quantifier. Our version of predicate calculus, as is common in the literature nowadays, includes equality (=). Chapter 7 advances some basic properties of equality as these flow from the axioms and the rules of inference.

Chapter 8 is a "working" first approximation to Tarski-like semantics and proves (in detailed outline) the soundness theorem for predicate calculus. This is an important tool toward constructing counterexamples, or *countermodels* as we prefer to call them, aiming to show that certain predicate logic formulae are *not* provable.

The Appendix at the very end does several things: It revisits Tarski semantics that were naïvely presented in Chapter 8, proves soundness again, this time totally rigorously, and also proves Gödel's completeness theorem. It then introduces *computability*, that is, the part of logic that makes the concepts of algorithm, computation, and computable function mathematically precise. In this particular approach to computability, I am using the programming language known in the literature as the Shepherdson-Sturgis ([44]) *unbounded register machines* (URMs). The topics included constitute the very foundation of the theory of computation and they will be of interest not only to mathematics readers but also to those in philosophy and, especially, in computer science, who will find ample supplemental material for their theory of computation courses. These include *partial computable functions*, *primitive recursive functions*, a complete characterization in number-theoretic terms of the partial functions computable by URMs, the *normal form theorems*, the "Kleene predicate" and a "universal" URM, computable and *semi-computable* relations and their behavior in connection with Boolean operations and quantification, *computably enumerable relations*, *unsolvability*, *verifiers* and *deciders*, *first-order definability*, and the *arithmetical relations*. This machinery will next allow us to tackle Gödel's first incompleteness theorem. This we prove by basing the proof on the nonexistence of a URM program that solves the following problem (*halting problem*) for any choice of x and y: "Will program x ever terminate if its input is y?"

Suggested coverage. A computer science curriculum in formal logic will probably cover everything but the Appendix. The course MATH1090 at York University, especially designed for computer science majors, does exactly that. However, a hybrid course in logic and computability, often included in computer science curricula, will adjust its pace (e.g., going faster through Part I) to include the computability and Gödel incompleteness topics of the Appendix. A mathematics major will typically see his first course in logic in an upper-undergraduate year. His syllabus will likely require that the book be studied from cover to cover (again, going fast through Part I). A philosophy major's needs in a course in logic are harder to fit to a prescribed template. Advanced students will likely find all of Part I relevant along with chapters 4–6 of Part II. They will also find a high degree of relevance in the computability and Gödel incompleteness topics of the Appendix.

GEORGE TOURLAKIS

Toronto
June 2008

Acknowledgments

The writing of this book would have not been successful without the support and warmth of an understanding family. Thank you.

I also wish to thank my York colleagues, Rick Ganong, Ilijas Farah, Don Pelletier and Stefan Mykytiuk, who have used drafts of this book in their MATH1090 sections and offered me helpful suggestions for improvements. I want to credit all my students in logic courses that I taught over the years, as they have provided me not only with the pleasure of teaching them, but also with the motivation to create a friendly yet mathematically rigorous, thorough, up-to-date, and correct text for their use.

I acknowledge the ever supporting environments at York University and at the University of Crete—where I spent many long-term visits, most recently in the spring term of 2006–07 while on Sabbatical leave from York—which are eminently conducive to projects like this one.

Steve Quigley and Melissa Yanuzzi have been supportive and knowledgeable editors who made all the steps of the project smooth and effective. It has been a pleasure to work with both of them.

I finally wish to record my appreciation of Donald Knuth, Leslie Lamport, and the TUG community (http://www.tug.org) for the typesetting tools TEX and LATEX and the UNIX-based distribution of comprehensive typesetting packages that they have made available to the technical writing community, making the writing of books such as this one fun.

PART I

BOOLEAN LOGIC

CHAPTER 1

THE BEGINNING

Mathematical logic, or as we will simply say, "logic", is the science of mathematical reasoning. Its core consists of the study of the *form*, *meaning*, *use*, and *limitations* of logical deductions, the so-called *proofs*.

This volume, which is aimed at upper-level undergraduate university students who follow a course of study in computer science, mathematics, or philosophy, will emphasize mainly *the use* of proofs—it is written with the interests of the user in mind.

1.0.1 Remark. (Before we Begin) The symbol " ⚠ "[6] goes at least as far back as the writings of Bourbaki. It has been made widely accessible to authors—who like to typeset their writings themselves—through the typesetting system of Donald Knuth (known as "TEX").

I use these "road signs" as follows: A passage enclosed between two single " ⚠ " symbols is purported to be *very noteworthy*, so please heed!

On the other hand, a passage enclosed between two *double* signs (" ⚠ ⚠ ") means two things.

---

[6]This symbol is a stylized typographical version of the "(dangerous) winding-road" road sign.

The *bad news* is that it is rather difficult, or esoteric, or *both*. The *good news* is that you do not *need* to understand (or even read) its contents in order to understand all that follows. It is only there in the interest of the "demanding" reader. Such "doubly dangerous" passages allow me to *digress* without injuring continuity—you can ignore these digressions!   □

Learning to *use* logic, which is what this book is about, is like learning to use a *programming language*.

In the latter case, probably familiar to you from introductory programming courses, one learns the correct syntax of programs, and also learns what the various syntactic constructs do—that is, their semantics. After that, one embarks—for the balance of the programming course—on a set of increasingly challenging programming exercises, so that the student becomes proficient in programming in said language.

We will do an exactly analogous thing in this volume: We will learn to write *proofs*, which are nothing else but *annotated* sequences of formulae and are similar to computer programs in terms of syntactic structure—the annotations playing a role closely similar to that of *comments* in computer programs.

But to do that, we need to know, to begin with, what *are* the *rules* of *correctly writing down* a formula and a proof! We have to start with the *syntax* of these objects—formulae and proofs—precisely as it is done in the case of programming and its related objects, the programs.

Thus, we will begin with learning the syntax of the logical language, that is, what syntactically correct formulae and proofs *look like*. We will also learn what various syntactic constructs "say" (semantics). For example, we will learn that a formula makes a "statement". A proof also makes a statement, that *every formula in it is true in some very intuitively acceptable sense*.

We will learn that correctly written proofs are finite and "checkable" means toward discovering mathematical "truths". We will also learn via a lot of practice how to write a large variety of proofs that certify all sorts of useful truths of mathematics.

The above task, writing proofs—or "programming in logic" if you will—is our main aim. This will equip you with a *toolbox* that you can use to discover or certify truths. It will be handy in your studies in computer science, and in whatever area of study or research you embark upon and where reasoning is required.

However, we will also look at this toolbox, the logic, as *an object of study* and study some of its properties. After all, if you want to take up, say, carpentry, then you need to know *about* tools such as hammers—their properties (e.g., hard and heavy) and limitations (e.g., unfriendly to fingers).

 When *using* the toolbox to prove theorems, you work *within* logic. On the other hand, when *studying* the toolbox, you work in logic's *metatheory* (in metalogic) to talk and reason *about* logic.

People often do this kind of study with programming languages, looking at them as objects of study rather than as instruments to write programs with. For example, in an advanced course on the comparative study of programming languages one looks at several programming languages and compares them for features, suitability for certain programming tasks—for any specific task some are more suitable than others—limitations, etc.

Here is another analogy: In the "real world" that we live in, one builds flight simulators, which we use to simulate flying an airplane, and in the process we learn how to do so. The real world where the simulator is *built* is the simulator's metatheory, where we can, among other things, study the properties and limitations of simulators and compare several simulators for features such as relative "power" (i.e., how effective or realistic they are), etc. Similarly, formal logic is built within "real mathematics", as we will see in the next section. It, too, is a "simulator" employed to write formal proofs that certify the truth of mathematical statements. These proofs imitate the kind of informal proofs one typically employs in informal mathematics but do so within a precisely specified system of notation (called *language*), rules, and assumptions. Thus, *using* formal logic is a means to *learn* how to write proofs—and not only *formal* proofs!—just as using a flight simulator is a means of learning how to fly a real plane. The metatheory of logic—the "real mathematics"—addresses questions among the deepest of which is the question of how far formal logic can go in discovering mathematical truths.

Let us next look more closely at the similarity between *programming languages and programming* on one hand and *logical languages and proving* on the other, and argue that, similar as the two activities may be, the second one is a bit easier!

(1) In programming, you use the syntactic rules to write a program that solves a problem.

(2) In logic, you use the syntactic rules to write a proof that establishes a theorem.

In the latter task you are done as soon as the proof ends. At the end of the proof you have your theorem, exactly as stated.

In the former task, programming, it is not enough to just write a program! You next have to convince your boss, or your instructor, that the program indeed solves the problem; that it is "semantically correct" with respect to the problem's specification.

Note that in proving a theorem you have a purely syntactic task. Once your correctly written proof ends with the theorem you were trying to prove, you are done. There is no messing about with semantics.

There is another reason why programming is harder than proving theorems: Programming has to be painstakingly precise because it involves your writing instructions for a dumb machine to "understand" and follow. You must be absolutely and pedantically clear in your instructions.

On the other hand, you address a proof to a human who knows as much as you do, or more, about the subject. This human will in general accommodate a few shortcuts that you may want to take in your presentation.

In short, proofs are read by "intelligent" humans, while programs are read by "dumb" computers. We need to work really hard to speak at the level of the latter.

Will you ever need to deal with semantics in logic? Yes! Semantics is useful when you want to *disprove* (or *refute*) something, that is, to prove that it is a false statement, a fallacy. We will talk about semantics later—three times: once under Boolean logic, once under predicate logic, and one last time in the Appendix.

There are many methodologies or paradigms (and corresponding programming languages suitable for the task) for writing programs. For example (add the word *programming* after each italicized keyword), *procedural* (Algol, Pascal, Turing), *functional* (LISP), *logic* (Prolog), and *object-oriented* (C++, Java). Most computer science departments will expose their students to many of the above.

Similarly there are several methodologies for writing proofs. For example (add the word *style* after each italicized keyword), *equational* (the one favored by [17]), *Hilbert* (favored by the majority of the mathematics, computer science, and logic literature), *Gentzen's natural deduction*, etc.

My aim is to assist the reader to become an able user of the first two styles: the equational and the Hilbert style of proof.

In both methodologies, an important required component is the systematic annotation of the proof steps. Such annotation explains why we do what we do, and has a function similar to that of comments in a program.

Okay; one can grant that a computer science student needs to learn programming. But logic? You see, the proper understanding of propositional logic is fundamental to the most basic levels of computer programming, while the ability to correctly use variables, scope, and quantifiers is crucial in the use of loops, and subroutines, and in software design. Logic is used in many diverse areas of computer science, including digital design, program verification, databases, artificial intelligence, algorithm analysis, computability, complexity, and software specification. Besides, any science that requires you to reason correctly to reach conclusions uses logic.

When one is learning a programming language, one often starts by learning a small *subset* of the language, just to smooth the learning curve. Analogously, we will first learn—and practice—a subset of the *logical language*. This we will do not due to some theoretical necessity, but due to pedagogical prudence. This particular, "easy" subset of (the "full") logic that we will embark upon learning goes by many names: *Boolean logic*, *propositional logic*, *sentential logic*, *sentential calculus*, and *propositional calculus*.

The "full logic" we will call by any of the names *predicate calculus*, *predicate logic*, or *first-order logic*.

I like the *calculus* qualifier. It connotes that there is a *precise way* to "calculate" within logic. It emphasizes that building proofs is an algorithmic and precise process, just like programming.

Indeed, it turns out that you can write a program, say, in Pascal, that will accept no input, but if it is allowed to run forever it will print all the theorems of logic[7] (and not just those of the Boolean variety)—and never print a non-theorem!—in some order, possibly with some repetitions (cf. A.4.7 on p. 270).

---

[7]We will soon appreciate that there are infinitely many theorems in logic.

Equivalently,[8] we can write a program that is a *theorem verifier*. That is, given as input a theorem, the program will verify that it is so, in a finite number of steps. If the input is a non-theorem, our verifier will decline an answer—it will run forever.

Thus, proving theorems is a *mechanical process*!

Digression: The above assertion is an example of a true assertion *about* the logic, not one that we can prove *using exclusively the tools of* logic as a tool. It is a *metatheorem* of logic as we say, *not* a theorem.

The proof of this metatheorem requires techniques much more powerful than—indeed external to—those that the logic provides. We will prove this metatheorem in the Appendix to Part II (A.4.6).

So metatheorems are truths *about* the logic that we prove with tools external to the logic, while theorems are truths that the logic itself is *capable of proving*.

There is some danger that the above statement, "proving theorems is a mechanical process", may be misinterpreted by some as one advocating that we build proofs by mindlessly shuffling symbols. *Nothing is further from reality.*

The statement must be understood precisely as written. It says that there *is* a "mindless" way, a programmable way, to generate and print *all possible theorems* of logic, and, equivalently, also a programmable way to *verify* all theorems, which, however, refuses to verify any non-theorem by "looping" forever when presented with any such as input.

But it is not a recipe for how *we* ought to behave when we write proofs. This is *not* the way a mathematician, or you or I, go about proving things—mindlessly. In fact, if we do not understand what is going on, we cannot go too far.

Moreover, interesting, even important, as this result (about the existence of theorem verifiers) may be *theoretically*, it is *useless practically*, as we further discuss below.

Our task is different. In general, we are more inquisitive. Given an arbitrary (mathematical) statement, we do not know ahead of time *if it is a theorem or not.* This italicized statement, the so-called *decision problem* of logic, is what we normally are interested in. Thus, our "verifier" is not very helpful, for if the statement that we present it as input is *not* a theorem, then the verifier will run forever, not giving an answer.

Hmm. Can we not write a *decider* for logic? The answer to this is interesting, but also reassuring to mathematicians (and all theorists): Their jobs are secure!

(1) For Boolean logic, we can, since the question "Is this statement a theorem?" translates to "Is this statement a tautology?" (cf. 3.2.1). The latter can be settled algorithmically via truth tables. But there is a catch: Checking a formula (the formal counterpart of a "statement") for tautology status is an *unfeasible* problem.[9] So we can do it *in principle*, but this fact is devoid of any practical value.

---

[8] That this formulation of the claim is equivalent to the preceding one is a standard result of computability. Cf. Appendix to Part II, Remark A.3.91 on p. 262.

[9] The term *unfeasible*—also *intractable*—has a technical connotation in complexity theory: It means a problem for which we know of no algorithm that runs in polynomial time as a function of the input

(2) For predicate logic, the answer is more pleasing to mathematicians.

First, there exists *no decider* for this logic if we expand it minimally so that it can reason about the theory of natural numbers (this is Alonzo Church's theorem, [3, 4]).

Second, even if one were to be satisfied simply with a *verifier* for theorems, then we still would have *no general solution of any practical value* in hand. Indeed, again considering the logic augmented so that it can "do number theory", *any chosen verifier V for this logic* would be extremely slow in providing answers in the following precise sense: For *any choice* of a step-counting function $f(n)$, there is *an infinite subset, S*, of the set of theorems of number theory, such that *each* theorem-member, T, of S that is composed of n symbols requires for its verification more than $f(n)$ steps to be performed by V.[10] This is a result of Hartmanis ([19]).

Let us stop digressing for now. In the next section we begin the study of the sublogic known as *propositional calculus*.

1.1 BOOLEAN FORMULAE

We will continue stressing the algorithmic nature of the discipline of proving, just as it is the case in the discipline of programming.

In particular, just as in serious programming courses the *programming language* is introduced via precise *formation rules* that allow us to write syntactically correct programs, we will be every bit as serious by introducing very precisely the rules for writing syntactically correct (1) formulae and (2) proofs.

Once again, the syntax of the logical language is much simpler to describe than that of any commercially available programming language.

So, how does one build—i.e., what are the rules for writing down correctly—formulae?

Continuing with the programming analogy, you will recall that to define a programming language, i.e., the syntax of its programs, one starts with the list of admissible symbols, the so-called *alphabet*. In some languages, the alphabet includes symbols such as "$3, 4, 0, [, A, B, c, d, E, +, \times, -$" and "keywords"—that is, multiple-character symbols—such as *if, then, else, do, begin*.

Similarly, in Boolean logic, we start with the basic building blocks, which *collectively* form what is called the *alphabet* (for formulae). Namely,

---

length—or worse, we know that such an algorithm does not exist. In this case it is the former. However, there is a connection with the so-called "P vs. NP" open question (see [5]). If a polynomial algorithm that recognizes tautologies does exist, then the open problem is settled as "P = NP", something that the experts in the field consider highly unlikely. The truth table method runs in exponential time.

[10]For example, consider $f(n) = 2^{2^{2^n}}$. If we think of $f(n)$, for each n, as representing picoseconds of run time of the verifier V (1 picosecond is 10^{-12} seconds), then every member of S of length more than 4 symbols will require the verifier V to run for more than 5.70045×10^{288} years!

A1. *Symbols* for *variables*, called the *Boolean* or *propositional* or *sentential* variables. These are p, q, r, with or without primes or subscripts (i.e., p', q_{13}, r_{51}''' are also symbols for variables).

We often need to write down expressions such as "$A[p := B]$", to be defined later (1.3.15), but do not wish to restrict them to the specific variable p. Nor can we say things such as "for any Boolean variable p we consider $A[p := B]\ldots$" as there is only one specific p!

We get around this difficulty by employing so-called *meta*variables or *syntactic variables*—i.e., *symbols outside the alphabet that we can use to refer to or point, generally, to* any *variable*. We adopt the names for those to be the boldface $\mathbf{p}, \mathbf{q}, \mathbf{r}$ with or without primes or subscripts. Thus $\mathbf{p}_{91}''$ names *any* variable p, q, r''', q_{987}'', etc. Rarely if ever in this volume will we need to use more Boolean metavariables than these two: $\mathbf{p}, \mathbf{q}$.

We can now use the expression "for every Boolean variable $\mathbf{p}$ we consider $A[\mathbf{p} := B]\ldots$" referring to what $\mathbf{p}$ *names* rather than to $\mathbf{p}$ itself. Two analogous examples are, from algebra, "for every natural number n" (n is not a natural number!) and, from programming, where we might say about Algol, "for each variable $\mathbf{x}$, the instruction $\mathbf{x} := \mathbf{x} + 1$ means to increase the value of $\mathbf{x}$ by one." Again, $\mathbf{x}$ is not a variable of Algol; $X13, YXZ99$, though, are. But it would be meaningless to offer the general statement "*for each variable $X13$, the instruction $X13 := X13 + 1$, etc.*" since $X13$ is a *specific* variable of the Algol syntax. The programming language metavariable $\mathbf{x}$ allows us to speak of all of Algol's variables collectively!

On the other hand, the expression "for every Boolean *meta*variable" refers to the set of metavariables themselves, $\{\mathbf{p}, \mathbf{q}, \mathbf{r}_{99}''', \ldots\}$ and will be rarely, if ever, used. The expression "for every Boolean *meta*variable $\mathbf{p}$" is as nonsensical as "for every Boolean variable p".

A2. Two *symbols* for Boolean *constants*, namely $\top$ and $\bot$. These are pronounced variously in the literature: *verum* (also *top*, or *symbol "true"*) and *falsum* (also *bottom*, or *symbol "false"*[11]).

A3. Brackets, namely, (and).

A4. "Boolean connectives", namely, the symbols listed below, separated by commas

$$\neg, \wedge, \vee, \rightarrow, \equiv \qquad (i)$$

Let us denote by $\mathcal{V}$ the alphabet consisting of the symbols described in **A1–A4**.

---

[11] Usually, the qualifier *symbol* is dropped and then the context is called upon to distinguish between "true/false" the *symbols* vs. "true/false" the *Boolean values* of the metatheory (introduced in Section 1.3). In particular, cf. Definition 1.3.2 and Remark 1.3.3.

1.1.1 Remark. (1) Even though I say very emphatically that p, q, r, etc., and also $\top$ and $\bot$, are just *symbols*,[12]—the former standing for variables, the latter for constants—yet, I will stop using the qualification *symbols*, and just say *variables* and *constants*. This entails an agreement: I always *mean* to say *symbols*, I just don't say it.

(2) Most variable symbols are formed through the use of "subsymbols"—such as $0, 1, 2, '$ —that are not members of the alphabet $\mathcal{V}$ themselves; e.g., p'''_{110034}. This does not detract from the fact that each variable (name) is a *single* symbol of $\mathcal{V}$, entirely analogously with, say, the keywords of Algol *if*, *then*, *begin*, *for*, etc.

(3) Readers who have done some elementary course in logic, or in the context of a programming course, may have learned that $\neg, \vee$ are the only connectives one really *needs* since the rest can be expressed in terms of these two. Thus we have deliberately introduced redundancy in the adopted set of connectives (i) above. This choice in the end will prove to be user-friendly and will serve our aim to give a prominent role to the connective $\equiv$, in the axioms and in rules of inference (Section 1.4). □

1.1.2 Definition. (Strings or **Expressions; Substrings)** We call a *string* (also *word* or *expression*), *over* a given alphabet, any *ordered* sequence of the alphabet's symbols, written adjacent to each other *without* any visible separators (such as spaces, commas, or the like).

For example, $aabba$ is a string of symbols over the alphabet $\{a, b, c, 0, 1, 2, 3\}$ (note that you don't have to use all the alphabet symbols in any given string, and, moreover, repetitions are allowed). *Ordered* means that *the position of symbols in the string matters*; e.g., $aab \neq aba$.

We denote arbitrary strings over the alphabet **A1–A4** by *string variables*, i.e., names that stand for arbitrary[13] or specific[14] strings. Specific strings, or string constants, are sometimes enclosed in double quotes to avoid ambiguity. For example, if we say

Let A be the string aab.

we need to know whether the period is part of the string or not. If it is *not* we symbolically indicate so by writing

Let A be the string "aab".

If it were part of the string, then we would have written instead

Let A be the string "aab.".

String *variables*—by agreement—will be denoted by uppercase letters A, B, C, D, E, P, Q, R, S, W etc., with or without primes or subscripts. In particular, since Boolean expressions (and theorems) are strings, this naming is valid for this special case, too.

---

[12]Some logicians put it more emphatically: "*meaningless* symbols".
[13]E.g., "let A be any string".
[14]E.g., "let A stand for $(\neg(p \wedge q))$".

The major operation on strings is *concatenation*, that is, juxtaposition. To concatenate the strings (named) A and B, in that order, is to form the string (named) AB that consists of the symbols in A, from left to right, immediately followed by the symbols in B, from left to right. Thus, if A is aab and B is 00110, then AB is $aab00110$.

Clearly, concatenation is an *associative* operation, i.e., $(AB)C = A(BC)$. Hence, when we omit brackets, as we normally do, and simply write ABC, there is no ambiguity since wherever we may insert the brackets that we "forgot" makes no difference!

There is a very special string that we call the *empty string* and denote by ϵ (this being a specific string, a constant, we deviate from the naming convention $A, B, C, \ldots$ above). What is special about it is that it contains no symbols, so that $A\epsilon = \epsilon A = A$.

"B is a *substring* of A" means that for some strings C and D we have $A = CBD$. For example, over the alphabet $\{a, b\}$ we have that a is a substring of aab. Indeed there are two *occurrences* of a in aab as substrings: A first (shown boxed) is justified by noting $aab = \epsilon\,\boxed{a}\,ab$ and a second is justified by noting $aab = a\,\boxed{a}\,b$. $\square$

We can build all sorts of *expressions* over our Boolean alphabet $\mathcal{V}$, such as $pp, pqr, \neg\wedge, (r \rightarrow r)$, and a lot of others.

Some such strings (e.g., the last one above) are *well-formed-formulae* (in short, *wff*), the rest being gibberish.

Hmm. How can we tell? For example, if we asked an unsuspecting (not logically trained) passerby which of the following are "well-formed"

$p \equiv p$

$p \rightarrow$

$((p \vee q) \equiv q) \equiv ((p \wedge q) \equiv p)$

we would have no right to expect any better than lucky guesses from him (we can check, by asking him "why?" in each case).

So, *how* can we tell? The obvious (silly) answer would be, "Why not tabulate *all* formulae? Then we can check any string for formula status by table look-up. If the string is in the table, then it is a formula; otherwise it is not a wff."

Of course this *is* silly, for we cannot write down an infinitely long table such as a table of *all* formulae would be.

 We must find a way to define a set of infinitely many strings (the formulae) *by a finite text*.

Pause. Can we do such a thing?

Absolutely. We will give a precise *process* that every time it is applied builds a formula, and will never build a nonformula. Moreover, it is "general enough" so that if it is applied over and over, for ever, it will build *all* formulae.

We are ready to define *formula-calculation*.

1.1.3 Definition. (Formula-Calculation or **Formula-Parse)** We will call *formula-calculation* (or formula-parse) any finite (ordered) sequence of strings that we may write respecting the following three requirements:

(1) At any step we may write any symbol from **A1** or **A2** of the alphabet (p. 9).

(2) At any step we may write the string $(\neg A)$, *provided* we have already written the string A.

(3) At any step we may write any of the strings $(A \wedge B)$, $(A \vee B)$, $(A \rightarrow B)$, $(A \equiv B)$, *provided* we have already written the strings A and B.   □

1.1.4 Example. In the *first* step of any formula-calculation, only requirement (1) of Definition 1.1.3 is applicable, since the other two require the existence of prior steps. Thus, in the first step, we may write *only* a variable or a constant. In all other steps, *all* the requirements (1)–(3) are applicable.

Here is a calculation (the comma is not part of the calculation, it just separates strings written in various steps):

$$p, \top, (\neg\top), q$$

Verify that the above obeys Definition 1.1.3.

Here is a more interesting one:

$$p, q, (p \vee q), (p \wedge q), ((p \vee q) \equiv q),$$
$$((p \wedge q) \equiv p), (((p \vee q) \equiv q) \equiv ((p \wedge q) \equiv p))$$ □

1.1.5 Definition. (Boolean Expressions or **wff** or **Formulae)** A string A over the alphabet **A1–A4** will be called a *Boolean expression* or a *well-formed-formula* iff[15] it is a string written at some step of some formula-calculation.

The set of Boolean expressions we will denote by **WFF**. A member of **WFF** is often called a wff (a formula).   □

1.1.6 Remark. (1) The idea of presenting the definition of formulae as a "construction" or "calculation" goes at least as far back as [2, 21].

(2) We used, in the interest of user-friendliness, active or *procedural* language in Definition 1.1.3[16] (i.e., that *we* may *do* this or that in each step). A mathematically more austere (hence "colder"(!)) approach that does not call upon anyone to write anything down—and does not speak of "steps"—would say *exactly the same thing as Definition 1.1.3* rephrased as follows:

A *formula-calculation* (or formula-parse) is any finite (ordered) sequence of strings, $A_1, A_2, \ldots, A_n$ such that—for all $i = 1, \ldots, n$—A_i is one of:

(I) Any symbol from **A1** or **A2**

(II) $(\neg A)$, *provided* A is the same string as some A_j, where $1 \le j < i$

---

[15]If and only if.
[16]Exactly as [21] does.

(III) Any of the strings $(A \wedge B)$, $(A \vee B)$, $(A \rightarrow B)$, $(A \equiv B)$, *provided A is the same string as some* A_j, where $1 \leq j < i$ and B is the same string as some A_k, where $1 \leq k < i$ (it *is* allowed to have $j = k$, if needed)

(3) There is an advantage in the procedural formulation 1.1.3. It makes it clear that we build formulae in stages (or steps), each stage being a calculation step.

In each step where we apply requirement (2) or (3) of 1.1.3, *we are building a more complex formula from simpler formulae by adding a Boolean connective.*

Moreover, *we are building a formula from* previously built *formulae.*

These last two remarks are at the heart of the fact that we can prove properties of formulae by induction on the number of steps (stages) it took to build it, or more simply, by induction on its "complexity" (that is, the total numbers of connectives in the formula, counting repetitions; see next section). □

The concluding remark above motivates an "inductive" or "recursive" definition of formulae, which is the favorite definition in the "modern" literature, and we should become familiar with it:

1.1.7 Definition. (Alternative (Recursive) Definition of WFF) The set of all well-formed-formulae is the *smallest* set of strings, **WFF**, that satisfies

(1) *All Boolean variables* are in **WFF**, and so are the symbols $\top$ and $\bot$. We call such formulae *atomic*.

(2) If A and B are any strings in **WFF**, then so are the strings $(\neg A)$, $(A \wedge B)$, $(A \vee B)$, $(A \rightarrow B)$, $(A \equiv B)$. □

1.1.8 Remark. (a) Why "recursive"? Because item (2) in 1.1.7 defines the concept *formula* in terms of (a *smaller, or earlier, instance of*) "itself": It says, in essence, "... $(A \vee B)$[17] is a formula, *provided we know that A and B are formulae.* ..."

In programming terms, confronted with, say, the 3rd subcase of case (2) of the definition, we "call" the definition recursively, twice, to settle the questions "Is A a wff?" and "Is B a wff?" If "yes" for both, then we proclaim that $(A \vee B)$ is a wff.

(b) Part (1) in 1.1.7 defines the most basic, most trivial formulae. This part constitutes what we call the *Basis* of the inductive (recursive) definition, while part (2) is called the *inductive*, or *recursive*, part of the definition.

(c) 1.1.7 and 1.1.5 say the same thing, *looking at it from opposite ends*: Indeed, suppose that we want to establish that a given string D is a formula. If we are using 1.1.5, we will try to build D via a formula-calculation, starting from atomic

---

[17]Each of A and B are substrings (cf. 1.1.2) of $(A \vee B)$, so they are "smaller" than the latter. They are also "earlier" in the sense that we must already *have them*—i.e., *know* that they are formulae—in order to proclaim $(A \vee B)$ a formula.

ingredients and building as we go successively more and more complex formulae until finally, in the last step, we obtain D.

If on the other hand we are using 1.1.7, we are working backwards (and build a formula-calculation in reverse!). Namely, if D is not atomic, we try to guess—from its form—what was the *last* connective applied. Say we think that it was $\to$, that is to say, D is $(A \to B)$ for some strings A and B. *Now we have to verify our guess!* This requires that the strings A and B are formulae. Thus, taking *each* of A and B in turn as (a new, smaller) "D" we repeat this process. And so on. This is a terminating process since the new strings we obtain (for testing) are always smaller than the originals.

Of course, I did not *prove* here that the two definitions define the *same* set **WFF**.[18] But they do!

Technically, the term *smallest* is crucial in 1.1.7 and it corresponds to the similarly emphasized *iff* of 1.1.5. A proof that the two definitions are equivalent is beyond our syllabus. □

1.1.9 Example. Let us verify using 1.1.7 that $((p \lor q) \lor r)$ is a formula.

call #1 We guess that the rightmost "$\lor$" is the last to apply; thus, using (2) in 1.1.7 we must now verify that $(p \lor q)$ and r are formulae. Well, r is by (1) in **WFF**. However, $(p \lor q)$ leads to call #2.

call #2 Again, using (2)—this time we do not need any guessing; there is only one connective—we must verify that p and q are formulae. This is so by (1) in 1.1.7.

When we use Definition 1.1.7 to verify that a string is a formula, we say that we *parse the string top-down*. On the other hand, when we build the formula using Definition 1.1.5, then we are parsing it *bottom-up*.

Can we parse the above string in another top-down way? Obviously, our recursive call to the definition hinges around one of the two "$\lor$" symbols in the string. We must guess which is "the right one" (as the last connective to apply). Why is the leftmost connective not "right"?

Here is where *metamathematical analysis* comes in: The leftmost connective will work as the "last one to apply" iff (according to 1.1.7) "p" and "$q) \lor r$" are both formulae. The metamathematical analysis of formula syntax[19] (see next section)

---

[18] I only hand-waved to that effect, arguing that for any string D in **WFF**, 1.1.5 builds a calculation the normal way, while 1.1.7 builds it backwards. I conveniently swept under the rug the case where D is *not* in **WFF**, i.e., is not correctly formed.

[19] I know that the separation of "mathematics" from "metamathematics" is at first tricky. Think of the hammer analogy: You do "theory" (or "mathematics" or "logic") when you *use* the hammer. On the other hand, *when articulating a principle* such as "It is inevitable that I will hit my finger with the hammer", then you are doing an analysis of *the hammer's behavior*. You are doing "metatheory" (or "metamathematics" or "metalogic").

Similarly, you do logic when you generate a formula according to 1.1.5, or backwards according to 1.1.7. However, articulating principles such as "There is only one way to parse a formula", or "Every formula has balanced brackets", *studies*, does not *build*, formulae and thus lies within the metalogic.

tells us that every formula must have a balance of left and right brackets. So none of these two is a formula, and therefore the leftmost $\vee$ *cannot* be the last connective to apply. □

In the course of a formula-calculation (1.1.3), we write some formulae down without looking back (step of type (1)). Some others we write down by combining via one of the connectives $\wedge, \vee, \rightarrow, \equiv$ *two* formulae A and B already written, or by prefixing *one* already written formula, C, by $\neg$.

In terms of the construction by stages, the formula built in this last stage had as *immediate predecessors* A and B in the first case, or just C in the second case.

One can put this elegantly via the following definition:

1.1.10 Definition. (Immediate Predecessors) None among the constants $\top$ and $\bot$, or among the variables, have any immediate predecessors.

Any of the formulae $(A \wedge B), (A \vee B), (A \rightarrow B), (A \equiv B)$ have A *and* B as immediate predecessors. A is an immediate predecessor of $(\neg A)$.

Sometimes we use the acronym *i.p.* for *immediate predecessor*. □

 It turns out that a formula uniquely determines its i.p. We give a proof later (1.2.5).

1.1.11 Remark. (Priorities) In practice, too many brackets make it hard to read complicated formulae. Thus, texts (and other writings in logic) often come up with *an agreement on how to be sloppy, but get away with it.*

This agreement tells us what brackets are *redundant*—and hence can be removed—from a formula written according to Definitions 1.1.5 and 1.1.7, still allowing the formula to "say" the same thing as before:

(1) Outermost brackets are redundant.

   For the remaining two cases, it is easiest to think of the process in reverse: How to reinsert correctly (as per Definition 1.1.5) any omitted brackets.

(2) Any other pair of brackets is redundant, if its presence (as dictated by 1.1.5) can be understood from the *priority*, or *precedence*, of the connectives. Higher-priority connectives bind *before* lower-priority ones. That is, if we have a situation where a subformula[20] A of a formula has already been reconstructed as per 1.1.5, and is claimed by two distinct connectives $\circ$ and $\diamond$, among those in ($*$) below, as in "$\ldots \circ A \diamond \ldots$", then the higher-priority connective "glues" first. This means that the implied brackets are (reinserted as) "$\ldots \circ A) \diamond \ldots$" or "$\ldots \circ (A \diamond \ldots$" according as $\circ$ or $\diamond$ has the higher priority, respectively.

The order of priorities (decreasing from left to right) is *agreed to be*:[21]

$$\neg, \wedge, \vee, \rightarrow, \equiv \qquad (*)$$

---

[20] A *subformula* of a formula B is a substring of B that is a formula.

[21] Other agreements for priorities are possible. I offered the one that most people use. But remember: It is only an agreement, which means (1) we *must* stick to it, and (2) it is neither "more right" nor "more wrong" than any alternative agreement.

(3) In a situation like "...$\diamond A \diamond$..."—where A has already been reconstructed as in 1.1.5, and $\diamond$ is any connective listed in ($*$) above, other than $\neg$—the right $\diamond$ acts before the left. Thus the implied bracketing is "...$\diamond (A \diamond$...".

Similarly, in $\neg\neg A$ is short for $(\neg(\neg A))$.

We say that *all connectives are right associative*.

It is important to emphasize:

(a) This "agreement" results in a shorthand notation. Most of the strings depicted by this notation are *not* correctly written formulae, but this is fine: Our agreement allows us to decipher the shorthand and *uniquely recover* the correctly written formula we had in mind.

(b) I gave above the convention that is followed by 99.9% of writings in logic, and in (almost) all programming language definitions (when it comes to "Boolean expressions" or "conditions").

(c) The agreement on removing brackets is a *syntactic* agreement.

In particular, right associativity says simply that, e.g., $p \vee q \vee r$ *is shorthand for* $(p \vee (q \vee r))$ *rather than* $((p \vee q) \vee r)$.

However, *no claim is either made or implied* that $(p \vee (q \vee r))$ and $((p \vee q) \vee r)$ *mean* different things. At this point *meaning* (Boolean values) has not yet been introduced. When it is later on, we will easily see that $(p \vee (q \vee r))$ and $((p \vee q) \vee r)$ mean the same thing. $\square$

1.1.12 Example. p stands for p.

$\neg p$ stands for $(\neg p)$.

$p \to q \to r$ stands for $(p \to (q \to r))$.

If I want to simplify $((p \to q) \to r)$, then $(p \to q) \to r$ is as simple as I can get it to be.

$\neg p \wedge q \vee r$ is short for $(((\neg p) \wedge q) \vee r)$.

If in the previous I wanted to have $\neg$ act last, and $\vee$ to act first, then the minimal set of brackets necessary is: $\neg(p \wedge (q \vee r))$. $\square$

A connection with things to come in a degree program in computer science. Any set of rules that tell us how to correctly write down strings constitutes a so-called *grammar*. *Formal language* theory studies grammars, the sets of strings that grammars define (called *formal languages*), and the procedures (or "machines") that are appropriate to parse these strings.

In an introductory course on "automata and formal language theory", a student learns about formal languages. Such a student would quickly realize that Definition 1.1.7 is, in effect, a definition of a grammar for the "language" (i.e., set of strings) **WFF**. He would utilize a neat notation,[22] such as

$$E ::= A \,\Big|\, (E \wedge E) \,\Big|\, (E \vee E) \,\Big|\, (E \to E) \,\Big|\, (E \equiv E) \,\Big|\, (\neg E)$$

$$A ::= \top \,\Big|\, \bot \,\Big|\, p \,\Big|\, q \,\Big|\, r \,\Big|\, p' \,\Big|\, \ldots$$

where E stands for (Boolean) **E**xpression, A for **A**tom, "::=" is read as "is defined to be" and "$|$" is read as "or", separating alternatives in an "is defined to be"-list.

Thus, the first line says, in English, "A (Boolean) expression *is defined to be* an atom, *or* '(' followed by an expression, followed by '$\wedge$' followed by an expression followed by ')', *or*, etc."

The second line defines *atom* as any of the constants or the variables (note the separating *or*'s).

1.2   INDUCTION ON THE COMPLEXITY OF WFF: SOME EASY PROPERTIES OF WFF

Suppose now that we want to prove that every $A \in$ **WFF**[23] has a "property" $\mathscr{P}$.

The technique is to associate a natural number with each member of **WFF** and prove the property by induction on numbers. The most obvious number one may associate with a formula A is the formula's complexity:

1.2.1 Definition. (Complexity of a Formula) The complexity of a formula is the number of connectives—counting repetitions—occurring in the formula.   □

1.2.2 Example. Note that we can read the complexity accurately even if we write formulae in least parenthesized notation.

Every atomic formula has complexity 0. The complexities of $p \to q \to p'$, $\neg p \vee q \vee s$, and $\neg p \wedge p' \vee p'' \to (p''' \equiv q)$ are 2, 3, and 5 respectively.

Brackets do not contribute to complexity.   □

 A crash course on induction. First off, let us recall what we call *strong* or *course-of-values* induction on the natural numbers (also known as *complete induction*):

Suppose that $\mathscr{P}(n)$ is a property of the natural number n. To prove that $\mathscr{P}(n)$ holds *for all* $n \in \mathbb{N}$[24] it suffices, *in principle*, to prove *for the arbitrary n* that $\mathscr{P}(n)$ holds.

What we mean by "arbitrary" is that we do not offer the proof of $\mathscr{P}(n)$ for some "biased" n such as $n = 42$, or n even, or n with 105 digits, etc. If the proof indeed

---

[22] Known as *BNF notation*, or Backus-Naur-Form notation.

[23] "$x \in y$" is shorthand for the claim "x is a member of the set y", also pronounced "x belongs to y" or "x is in y".

[24] $\mathbb{N}$ denotes the set of all natural numbers $\{0, 1, 2, 3, \ldots\}$. Thus "*for all* $n \in \mathbb{N}$" is elegant notation that says "for $n = 0, 1, 2, 3, \ldots$".

has not cheated by using some property of n beyond "$n \in \mathbb{N}$", then our proof is *equally valid for any $n \in \mathbb{N}$*; we have succeeded in effect to prove $\mathscr{P}(n)$, *for all $n \in \mathbb{N}$.*

> *Now the above endeavor is not always easy. It would probably come as a surprise to the uninitiated that* we can pull an extra assumption out of the blue *and use it toward proving $\mathscr{P}(n)$, and that when all is said and done* this process is as good as if we proved $\mathscr{P}(n)$ without the extra assumption!

This out-of-the-blue assumption is that

$$\mathscr{P}(k) \text{ holds for all } k < n \qquad\qquad (I)$$

or, another way of putting it, that *the history* or *course-of-values* of $\mathscr{P}(n)$,

$$\mathscr{P}(0), \mathscr{P}(1), \dots, \mathscr{P}(n-1) \qquad\qquad (II)$$

holds—that is, it is a sequence of valid statements. It goes by the name *induction hypothesis* (I.H.), and the technique is that of "proof by strong induction".

A couple of comments:

(1) As before, we still have to prove $\mathscr{P}(n)$ for the *arbitrary n*, although now we have the I.H. as extra help.

(2) We note that the history, (II), of $\mathscr{P}(n)$ is empty if $n = 0$. Thus every proof by strong induction has two cases to consider: the one where the history helps, *because it exists*, i.e., when we have $n > 0$, and the one where the history *does not help*, because it simply does not exist, i.e., when $n = 0$.

In summary, strong induction proofs have two cases:

I.S. Where $n > 0$ and we are helped by the I.H. ((I) or (II) above). *I.S.* is an abbreviation for *induction step*.

Basis. Where $n = 0$ and we are on our own! The proof for $n = 0$ is called the *basis* step of the induction.

Since on occasion we will also employ "simple" induction in this book, let me remind the reader that in this kind of induction the I.H. is not the assumption of validity of the entire history, but that of just $\mathscr{P}(n-1)$. As before, simple induction is carried out for the arbitrary n, so we need to work out two cases: when the I.H. is really there ($n > 0$) and when it is not ($n = 0$). The case of proving $\mathscr{P}(0)$ directly is still called the *basis* of the (simple) induction.

Tradition has it that in performing simple induction the majority of users in the literature take as I.H. $\mathscr{P}(n)$ while the I.S. involves proving $\mathscr{P}(n+1)$.

Correspondingly, we organize proofs of properties of formulae X, $\mathscr{P}(X)$, into two *main* cases (rather three, in practice; see below)—essentially carrying out a strong induction with the complexity n of X as a "proxy". However, the complexity n of X is well hidden in the background of the argument and we do not mention it:

(i) Case of atomic formulae (these are the only ones with complexity $n = 0$) where the proof is direct, without the benefit of the I.H.

(ii) Case of nonatomic formulae (corresponding to a complexity (of X) $n > 0$) where we will benefit from the I.H. that $\mathscr{P}(A)$ *holds for all formulae A that are less complex than X (i.e., they have complexity $k < n$)*

In case (ii), if A is any formula less complex than X, we will often say that "the I.H. applies to A", meaning precisely that "by the I.H., $\mathscr{P}(A)$ holds".

Let us apply (i)–(ii) to obtain a framework of proofs *by induction on the set of formulae* or, as we say more simply, *by induction on formulae*.

Now, since every proper[25] subformula A of X has a lesser complexity than X, the I.H. applies on A. In particular, the I.H. applies on all the i.p. of X (the definition of *i.p.* was given in Definition 1.1.10).

Thus, in practice, (i)–(ii) translate into the following simple *framework for proofs by induction on formulae*:

(a) X is atomic: Give a direct proof.

(b) X has the form $(\neg A)$. Give a proof on the assumption (I.H.) that $\mathscr{P}(A)$ holds.

(c) X has the form $(A \circ B)$—where $\circ \in \{\wedge, \vee, \equiv, \rightarrow\}$. Give a proof for each case of $\circ$ on the assumption (I.H.) that $\mathscr{P}(A)$ and $\mathscr{P}(B)$ hold.

Let us now prove a few properties of formulae *by induction on formulae*.

All the "theorems" (and their corollaries) of this section are *about* formulae and their syntax. They are not theorems of logic, but are metatheorems.

1.2.3 Theorem. *Every Boolean formula A has the same number of left and right brackets.*

Proof. The theorem is about formulae written properly, as per Definition 1.1.5, that is, before our agreements to simplify bracketing are applied.

We prove the property by induction on formulae, A.

(1) *Basis*: A is atomic. Each atomic formula has 0 left and 0 right brackets. We are Okay.

(2) A has the form $(\neg B)$. The I.H. applies on the less complex B. So let B have m left and m right brackets. Then A—i.e., $(\neg B)$—has $m + 1$ of each.

(3) A has one of the forms $(B \wedge C)$, $(B \vee C)$, $(B \rightarrow C)$ and $(B \equiv C)$.

The I.H. applies to the less complex subformulae B and C. So let them have m left/right and r left/right brackets respectively. Thus A has $m + r + 1$ left, and as many right, brackets. $\square$

---

[25]That is, not the same string as X.

Note. A string B is a *prefix* of a string A iff there is a string C such that $A = BC$. The prefix is *empty* iff it is the empty string (i.e., it has no symbols in it; it has length 0). It is *proper* iff $A \neq B$.

1.2.4 Corollary. *Any nonempty proper prefix of a Boolean expression A has more left than right brackets.*[26]

Proof. Induction on A.

Since A denotes an arbitrary formula, "induction on formulae" can be rephrased as "induction on A". Compare with "induction on natural numbers" and "induction on n".

(1) A is atomic. Note that none of the atomic formulae has any nonempty proper prefixes, so we are done without lifting a finger.

 This is an instance of a statement being "vacuously true": The statement has a typical instance that says, "All nonempty proper prefixes of **p** have more left than right brackets." Is this true? Absolutely! If you think otherwise, then show me *just one* nonempty proper prefix of **p** that does *not* have an excess of left brackets. You cannot, because there *are* no nonempty proper prefixes of **p**. (The only nonempty prefix of **p** is **p**, but this is improper.)

(2) A has the form $(\neg B)$. The I.H. applies to B. Well, let's check the nonempty proper prefixes of $(\neg B)$. These are (quotes not included, of course):

(a) "(". Okay, by inspection.

(b) "$(\neg$". Ditto.

(c) "$(\neg C$", where C is a nonempty proper prefix of B. By I.H. if m is the number of left and n the number of right brackets in C, then $m > n$. But the number of left brackets of "$(\neg C$" is $m + 1$. Since $m + 1 > n$, we are done.

(d) "$(\neg B$". By 1.2.3, B has, say, k left and k right brackets. We are okay, since $k + 1 > k$.[27]

(3) A has the form $(B \circ C)$—where "$\circ$" is any of $\wedge, \vee, \rightarrow, \equiv$. The I.H. applies on B *and* C. Well, let's check the nonempty proper prefixes of $(B \circ C)$. These are (quotes not included, of course):

(i) "(". Okay, by inspection.

---

[26]*Corollary* is jargon that mathematicians, logicians, computer scientists, philosophers—and other reasoning people—use to characterize a statement that needs proof, but whose proof follows easily from another proved statement, or from the latter's proof. One then speaks of "A is a corollary of B", meaning that A easily follows from B (and/or B's proof).

[27]The I.H. was not needed in this step.

(ii) "$(D$", where D is a nonempty proper prefix of B. By I.H., if m is the number of left and n the number of right brackets in D, then $m > n$. But the number of left brackets of "$(D$" is $m + 1$. Okay.

(iii) "$(B$". By 1.2.3, B has, say, k left and k right brackets. We are okay, since $k + 1 > k$.

(iv) "$(B \circ$ ". The accounting exercise is exactly as in (iii). Okay.

(v) "$(B \circ D$", where D is a nonempty proper prefix of C. By 1.2.3, B has, say, k left and k right brackets. By I.H., D has, say, m left and r right brackets, where $m > r$. Thus, "$(B \circ D$" has $1 + k + m$ left and $k + r$ right brackets. Okay!

(vi) "$(B \circ C$". Easy. □

The following tells us that once a formula has been written down correctly, there is a unique way to understand the order in which connectives apply.

1.2.5 Theorem. (Unique Readability) For any formula A, its immediate predecessors are uniquely determined.

Proof. Obviously, if A is atomic, then we are okay (nothing to prove, for such instances of A have *no* i.p.). Moreover, *no* A can be seen (written) as both atomic and nonatomic.[28] The former do *not* start with a bracket; the latter do (cf. 1.2.6).

Suppose that A is not atomic. Is it possible *to build* this string *as a formula* in more than one way?

Can A have *two different sets of i.p.*, as listed below? (Below, when I say "we are okay" I mean that the answer is "no", as the theorem claims.)

(1) $(\neg C)$ and $(\neg D)$? Well, if so, C is the same string as D (why?); so in this case we are okay.

(2) $(\neg C)$ and $(D \circ E)$, where $\circ$ is any of $\wedge, \vee, \rightarrow, \equiv$? Well, no (which means we are okay in this case too). Why "no"? For if $(\neg C)$ and $(D \circ E)$ are identical strings (they are, supposedly, two ways to read A, remember?), then "$\neg$" must be the same symbol as the *first symbol* of D. Now the first symbol of D is one of "(" or an atomic symbol. None matches "$\neg$".

(3) $(C \circ D)$ and $(E \diamond G)$, where $\circ$ and $\diamond$ are any of $\wedge, \vee, \rightarrow, \equiv$ (possibly the same symbol) and either C and E are different strings, or D and G are different strings, or both? Well, no!

(i) If C and E are different, then, say, C is a proper prefix of E (of course, C is nonempty as well (why?)). By 1.2.4, C has more left brackets than right ones, but—being also a formula—it has the *same* number of left and right

---

[28] So we cannot be so hopelessly confused as to think at one time that A has no i.p. and at another time that it does.

brackets (by 1.2.3). Impossible! The other case, E being a proper prefix of C instead, is equally impossible.

(ii) If C and E match, then $\circ$ and $\diamond$ match. This forces D and G to be the same string, since the strings $(C \circ D)$ and $(E \diamond G)$ are the same—okay again.

Having answered "no" in all cases, we are done. □

1.2.6 Exercise. Prove that the first symbol of any formula A is one of
(1) a variable
(2) ⊤
(3) ⊥
(4) a left bracket
Hint. Induction on formulae, or directly from an analysis of formula-calculations (1.1.3). □

 1.2.7 Exercise. In footnote 20 of p. 15 we defined the concept of *subformula*, saying: "A subformula of a formula A is a substring of A that is also a formula."

This definition does not offer itself toward showing rigorously that, e.g., *"If all the occurrences of a subformula B of A are replaced by the same Boolean variable, say p, then the string so obtained is a formula."*

Do the following:

(1) Try to contradict me (I said, "This definition does not offer itself toward showing rigorously that, e.g., ...")

(2) Regardless of how you did in (1), *give an inductive definition* of the concept *subformula*.

(3) Now use (2) to prove by induction on A that *"If all the occurrences of a subformula B of A are replaced by the same Boolean variable, say p, then the string so obtained is a formula."* □

1.3 INDUCTIVE DEFINITIONS ON FORMULAE

Now that we know (by 1.2.5) that we can decompose a formula uniquely into its constituent parts, we are comfortable with defining functions (more generally "concepts") on formulae *by induction—or recursion—on formula complexity*, or as we rather say, "by induction—or recursion—on formulae".[29]

This recursion will define the concept as follows:

- (Basis) If A is atomic, we define the concept (or function) directly, depending on what we are trying to achieve.

- If A is $(\neg B)$, then we "call" the definition recursively to define the concept for B. Depending on the nature of the concept we then, taking the presence of $\neg$ into account, extend the concept to the entire A.

---

[29]Some people prefer to use the term *induction* for proofs, and *recursion* for constructions. Others do not mind using the term *induction* for either.

- If A is $(B \circ C)$, then we "call" the definition recursively for B and C. Depending on the nature of the concept, taking the presence of $\circ \in \{\wedge, \vee, \equiv, \rightarrow\}$ into account, we extend the concept to the entire A.

I will merely state that the above process is feasible because of 1.2.5, which allows us to have uniquely determined "components" of A, its i.p., on which we perform the "recursive calls".

A rigorous proof that indeed the process works, that we *can* effect recursive definitions on sets such as **WFF**, which themselves have been inductively defined (1.1.7), is beyond our aims. This subject is fully considered in [54].

1.3.1 Example. We consider here a simple example that shows what can happen if we attempt a recursive definition on a set of formulae that were defined in a manner that the uniqueness of i.p. was *not* guaranteed.

This time we define simple arithmetic formulae, without variables. As an alphabet we take

$$\{1, 2, 3, +, \times\} \tag{1}$$

Inductively, we define the set "arithmetic formulae", **AR**:

> **AR** *is the smallest possible set of strings over the alphabet (1) that contains the strings of unit length, 1, 2 and 3, and, moreover, if the strings X and Y are in* **AR**, *then so are $X + Y$ and $X \times Y$.*
> *The strings 1, 2, and 3 are the atomic formulae of* **AR**.

The concept i.p. is defined on the formulae of **AR** in the obvious manner: The atomic formulae do not have any i.p., and $X + Y$ and $X \times Y$ have, each, X and Y as i.p.

$1 + 2 \times 3$ is an example of a formula in **AR** that does not have a unique i.p.

Indeed, according to the definition, we have two sets of i.p. here: $\{1, 2 \times 3\}$ and $\{1 + 2, 3\}$.

> *Both i.p. sets are correct. Remember that any agreement on the priority of the connectives—and we entered into no such agreement—is not part of the rigorous definition of formula syntax for* **AR**; *thus, let us not assume that any such agreement is implied here!*

But why do we fear the multiplicity of i.p. sets for $1 + 2 \times 3$?

Let us attempt to define an "evaluation" function, inductively, on the set **AR**. We will call it EV. Here is the "natural" definition EV:

$$EV(1) = 1$$
$$EV(2) = 2$$
$$EV(3) = 3$$
$$EV(X + Y) = EV(X) + EV(Y)$$
$$EV(X \times Y) = EV(X) \times EV(Y)$$

Now, what is $EV(1 + 2 \times 3)$?

To answer this, we need to decompose $1 + 2 \times 3$ into a set of i.p. so that we can next do our recursive calls of EV (see the two last cases in the definition of EV).

Unfortunately, we obtain two distinct answers!

First we compute according to the decomposition $\{1, 2 \times 3\}$:

$$
\begin{aligned}
EV(1 + 2 \times 3) &= EV(1) + EV(2 \times 3) \\
&= EV(1) + \Big(EV(2) \times EV(3) \Big) \\
&= 1 + (2 \times 3) \\
&= 7
\end{aligned}
$$

Next, let us do so according to the other decomposition, $\{1 + 2, 3\}$:

$$
\begin{aligned}
EV(1 + 2 \times 3) &= EV(1 + 2) \times EV(3) \\
&= \Big(EV(1) + EV(2) \Big) \times EV(3) \\
&= (1 + 2) \times 3 \\
&= 9
\end{aligned}
$$
□

"Natural", or "only"? I said earlier: "Here is the "natural" definition EV."

But is there any other? Yes, infinitely many! We must get used to the idea that once we define the syntax of a set of strings—of a formal language, as we say in the theory of computation, the language here being **AR**—*we do not* have anything more than the syntax, i.e., the knowledge of the "shape" of such strings. All strings in **AR** are "meaningless", and their semantics (or interpretation) is totally up to us. The variety of such interpretations at our disposal is infinite.

I wanted the above example to be immediately relevant to our existing knowledge, to be "natural". That is why I gave the meaning to all the meaningless symbols of alphabet (1) that anyone would likely expect.

However, a "meaningless" symbol such as "1" may stand for infinitely many different objects of mathematics. Staying in algebra, I will mention the following (infinitely many) interpretations: The symbol may be interpreted, as here, to be the *number* "one", but also as the unit "2×2" matrix

$$
\begin{pmatrix} 1 & 0 \\ 0 & 1 \end{pmatrix}
$$

or the unit "3×3" matrix

$$
\begin{pmatrix} 1 & 0 & 0 \\ 0 & 1 & 0 \\ 0 & 0 & 1 \end{pmatrix}
$$

or . . .

But "unit" of some sort or another is not an intrinsic meaning of "1". The symbol could stand for the number 0, or 42, or for some other mathematical object.

Similarly, the meaningless symbol "$+$" can be interpreted as "plus" on numbers, but also as "plus" on $n \times n$ matrices (for various n), and also as concatenation of strings, union of so-called *regular expressions*, etc. Similar comments hold for all the other symbols of the alphabet (1).

There will be two main examples of recursive definitions in this section. The first follows the definition of the concept of *state* and is the inductive definition of *value* of a formula *in a state*. This leads to Boolean formula *semantics*.

The other will be the inductive definition of *substitution* of a formulae into a variable, an operation that is central in the use of the "Leibniz rule" in proofs.

But first, let us introduce states and Boolean semantics.

As we said early on (p. 6), Boolean logic is a subset of the (full) logic on first order languages and is, mainly, a pedagogical tool,[30] since it is "easy" and therefore its study painlessly trains us and prepares us for the study of predicate logic.

Does it have any other use? Yes. We can imagine that the Boolean variables of propositional logic are "abstractions" of statements in mathematics, computer science, philosophy, etc. By the term *abstraction* of a statement I mean the assignment of a name—a Boolean variable—to it, *purposely forgetting* the intrinsic semantic content (i.e., what it says) of the original statement.

As an example, we can decide the logical correctness or not of the statement

$$x = \aleph_0 \rightarrow x = \aleph_0 \vee y > \aleph_1 \tag{$*$}$$

within Boolean logic by the method of abstraction, not bothering with the fact that most of the symbols above—i.e., $x, y, =, >, \aleph_0, \aleph_1$—are not even in the alphabet $\mathcal{V}$ of propositional logic!

Indeed, we abstract the elementary statements[31] "$x = \aleph_0$" and "$y > \aleph_1$", naming them, say, p and q. Since they are two distinct statements, I used two distinct Boolean variables. Thus $(*)$ becomes

$$p \rightarrow p \vee q \tag{$**$}$$

and, as we will soon see, it holds *independently of what hides under the names p and q.*

Two useful observations are motivated from this example on abstraction and are noteworthy:

- The "object" of propositional logic is *not* the study of the elementary "statements" (or "propositions"), that is, of the Boolean variables and their intrinsic "semantic content". After all, variables *have no intrinsic* semantic content. Once we name through such a variable an "elementary"—i.e., connective-free—substatement, we turn around and forget the meaning of the original!

  To put it positively, the "object" of study is, exclusively, the Boolean connectives and their behavior.

---

[30] E.g., advanced texts such as [45, 53] introduce predicate calculus directly and do not cover propositional logic.

[31] What makes them "elementary" is that they do not involve Boolean connectives.

- The statements that we can abstract *are not* restricted to those mathematical ones (or other) that are variable-free, like "$3 = 9 \rightarrow 7 > 101$". E.g., the elementary statement "$x = \aleph_0$" above depends on the variable x and therefore whether it is intrinsically true or false is *indeterminate* (depending on the value of x). But this did not hinder our abstraction in any manner!

Here is why: In the process of abstraction—to which we come back in detail in 4.1.25—the Boolean variables that we use as names *do not inherit* the semantic content of the statements they name. Thus we do not care whether the truth or falsehood of what a Boolean variable names can be determined or not!

The only thing that matters is the Boolean structure of the original statement, that is, how the original statement is put together where the connectives act as "glue". In the previous example, all we needed to know was that the statement had the Boolean structure $(**)$.

The semantics of Boolean formulae is defined—in the metatheory of propositional logic—through a process that allows us to *calculate* whether a formula is *true* or *false*, and this *under certain conditions*.

Our aim below is to make precise what we mean by "conditions", and to give the process according to which we calculate the "truth value" of a formula under any conditions.

As you probably know from programming courses, only two values are possible for a formula in "classical" Aristotelian logic as well as in its descendant, mathematical logic, which we study here. These two values are *true* and *false*—which we collectively call *truth values*.

Thus we need a set of two distinct objects, which we will find outside the alphabet $\mathcal{V}$, in logic's metatheory. We freeze, i.e., reserve, these two values in this volume and they will serve, to the last page, as our truth values.

Our choice is the set $\{t, f\}$ of truth values. We will pronounce t as *true* and f as *false*.

Some programming languages, but even books on logic, use different sets of truth values, such as $\{0, 1\}$. At the end of the day, neither how we write them down nor how we pronounce the truth values matters, as long as we have exactly two distinct ones!

1.3.2 Definition. A state v^{32} is a function that *assigns the value* f or t to *each* Boolean variable, while it assigns *necessarily* the value f to the constant $\bot$ and *necessarily* the value t to the constant $\top$.

We pronounce f and t "false" and "true" respectively. On the chalkboard one usually denotes them by $\underline{f}$ and $\underline{t}$ respectively.

If, say, the value f is assigned to q'', then we write $v(q'') = f$. $\square$

---

[32]"v" for *value*. Alternative letter is "s" for *state*.

1.3.3 Remark. (1) A state v is one of the infinitely many possible "conditions" where we are interested in finding the truth value of a formula and where it is possible to compute such a value.

(2) A function v is, of course, a table of input/output values such as

in	out
$\bot$	**f**
$\top$	**t**
p	**t**
q	**f**
$\vdots$	$\vdots$

where *no two rows contain the same input*. Disobeying this condition would result in ambiguity, assigning to a variable both values **f** and **t**.

By definition, a state is an infinite table, so we cannot fit it on a page. Mathematically, it is an infinite set of input/output ordered pairs.

Since the truth values **f** and **t** lie outside the alphabet $\mathcal{V}$ (p. 9) of our logic, they are symbols that, *despite the similarity of their pronunciation with that of the names of the "meaningless" formal $\bot$ and $\top$ respectively, are different from the latter.*

In particular, neither the metasymbol **f** nor the metasymbol **t** may appear in a formula!

But why the fuss of *assigning* the values **f** and **t** to the (formal) variables and constants? How does this process give a "meaning" to the variables and constants?

Why do we need the **f** and **t**? Where do they come from? Aren't these two *symbols*, well, just symbols? If so, what do we gain by their introduction and why are the $\bot$, $\top$ "meaningless" while the **f** and **t** are "meaningful"? What is their understood meaning?

These are good questions! Here are some answers:

- The symbols $\bot$, $\top$ (but also $\neg, \vee, \wedge, \equiv, \rightarrow$) are "meaningless" in the sense that we know nothing of them besides what the axioms will lead us to know: Their behavior and properties are determined only by the axioms and the rules of writing proofs in Boolean logic.

  On the other hand, the symbols **f** and **t** (as well as the counterparts of $\neg, \vee, \wedge$, $\equiv, \rightarrow$, namely, the $F_\neg, F_\vee$ etc.—see the table in 1.3.4) are directly given via tables in the metatheory, as part of the elementary "Boolean algebra" that we learn in computer programming. It turns out the properties of the latter faithfully track the properties of the former, and thus in a natural way provide a "concrete" interpretation of the former.

- An analogy may shed more light on the above discussion: Think of axiomatic Euclidean geometry. There we learn the properties of, and interrelationships between, the "meaningless" concepts of *point*, *line* and *plane* by rigorous proofs that are based on the axioms and proof-writing rules, but on *nothing else*. Yet, there is another, "naïve", kind of geometry, *analytic geometry*. In this geometry, a "point" is not something abstract whose properties we wait to learn from axioms; rather, it is a concrete, well-understood object of mathematics: an ordered pair of real numbers, (x, y). Similarly, a (planar) "line" is an algebraic expression on two variables x and y and real coefficients a, b, c: $ax + by = c$. One finds that properties of the abstract (or "meaningless") points and lines of the axiomatic version are faithfully tracked by those of the corresponding concepts of the "naïve" versions. In this sense, the points and lines of the latter provide a "concrete" interpretation of the points and lines of the former.

- But why interpret? Because even if we are determined to write proofs solely based on axioms and rigid rules of logic, it aids our motivation and ability to formulate such proofs if we have a "concrete" counterpart in mind. For example, another interpretation of axiomatic geometry is that of geometric drawings, "figures" as we say. You will recall from your high school years how helpful these figures were in your formulation of proofs in geometry. The geometer M. Pasch once wrote a totally figureless monograph on axiomatic geometry, presumably to emphasize that the figures we draw in geometry are *only* intuitive visual aids that are theoretically redundant. Yet, it seems that the human brain does well with some sort of assistance, be it visual or some other well-understood concrete representation of abstract objects, toward understanding and formulating abstract arguments.

- Analogously with the above, we are often motivated and aided by our knowledge of informal Boolean algebra—that is, the set $\{f, t\}$ *and* the various operations on it as introduced in 1.3.4 below—as we construct proofs in axiomatic logic.

- Interpretations of abstract concepts by concrete ones provide a powerful tool toward building counterexamples: The "faithful" nature of such interpretations means that if I can prove a statement involving abstract concepts, let us call it A, then its concrete counterpart, let us call it A', is also verifiable.

Turning this around I get a very useful observation: If I believe that A is *not* provable, I can offer indisputable evidence for this provided I can show, in the concrete domain, that A' is false. This comment will make more sense later, in 3.1.5.

- *Hmm*. It appears that this discussion builds too strong a case for the import of the "naïve" or "concrete" approach, even though early on (p. 4) I said that the abstract (syntactic) approach will be favored in this book. I quote:

> We will learn that correctly written proofs are finite and "checkable" means toward discovering mathematical "truths". We will also learn via a lot of practice how to write a large variety of proofs that certify all sorts of useful truths of mathematics.
>
> The above task, writing proofs—or "programming in logic" if you will—is our main aim.

Yet, we noted that the concrete methods track the abstract (axiomatic) approach faithfully. They provide motivation and aid the construction of proofs. They provide definitive evidence regarding the falsehood of mathematical statements. Then why bother with the axiomatic approach? When it comes to logic, would it not be best to work exclusively with Boolean algebra instead?

There are a number of reasons why the axiomatic approach has attained a prominent status, even in the undergraduate curricula:

(a) There is more to logic (predicate logic of Part II) that cannot be tracked by Boolean algebra. For predicate logic the concrete counterparts are in general infinitary, i.e., deal with infinite sets and operations with infinitely many arguments, such as searching an infinite set to determine if an object belongs to it.

By contrast, syntactic proofs of the axiomatic method continue to be *finite processes*. Where the advantage lies is clear!

(b) The axiomatic method was introduced as a mind-focusing device: Focus on what matters, via axioms and rules, and discard all that is extraneous to our assumptions. The approach literally saved mathematics from the paradoxes that the purely concrete, or naïve, set theory of Cantor introduced.

(c) The axiomatic method makes logic—and any theory that we build upon logic, e.g., modern set theory, Peano number theory—a mathematical object, just as a programming language is a mathematical object. This allows us to use mathematical tools to *study* logic—and any mathematical theories built upon it—as to its power, limitations, freedom from contradiction, etc. $\square$

1.3.4 Definition. (Truth Tables) There are five operations or functions, the *Boolean functions*, that take as inputs only values from the set $\{\mathbf{f}, \mathbf{t}\}$ and produce as outputs only values in the same set. The symbols we choose for these functions, one symbol for each Boolean connective, are

$$F_\neg(x), F_\vee(x, y), F_\wedge(x, y), F_\to(x, y), F_\equiv(x, y)$$

and their behavior is fully described by the following table, known as a *truth table*.

x	y	$F_\neg(x)$	$F_\vee(x,y)$	$F_\wedge(x,y)$	$F_\to(x,y)$	$F_\equiv(x,y)$
f	f	t	f	f	t	t
f	t	t	t	f	t	f
t	f	f	t	f	f	f
t	t	f	t	t	t	t

$\square$

The following definition *extends* a state v so that it can give a Boolean value, hence *meaning*, to all formulae. Note that originally v gave a value *only to atomic formulae*.

In essence, the definition *gives meaning to the Boolean connectives*, as it is clear from the fact that there is a case for each connective of how to compute the value.

Pedantry requires that the extension of v below be denoted by a different symbol, say $\overline{v}$, since, after all, the extension is a much bigger table. While the original had a row only for every atomic formula, the extension has a row for *every* formula. But now that you know about this quibble, we feel safe to use the same symbol, "v", for both.

1.3.5 Definition. (Value of a Formula in a State v) Below I use the metavariable "**p**", so that the "first" equation actually represents infinitely many, one for each variable in the alphabet $\mathcal{V}$:

$$v(\mathbf{p}) = \text{whatever we originally assigned to } \mathbf{p}; \mathbf{t} \text{ or } \mathbf{f}$$
$$v(\top) = \mathbf{t}$$
$$v(\bot) = \mathbf{f}$$
$$v\big((\neg A)\big) = F_\neg\big(v(A)\big)$$
$$v\big((A \wedge B)\big) = F_\wedge\big(v(A), v(B)\big)$$
$$v\big((A \vee B)\big) = F_\vee\big(v(A), v(B)\big)$$
$$v\big((A \to B)\big) = F_\to\big(v(A), v(B)\big)$$
$$v\big((A \equiv B)\big) = F_\equiv\big(v(A), v(B)\big) \qquad \square$$

 The symbol "$=$" above is the "equals" sign of the metatheory, and means that the left-hand side and right-hand side values are the same (equal). It is *not* a formal symbol for (at least) two reasons:

(1) Our $\mathcal{V}$ does *not* include "$=$"

(2) The definition above compares informal (metatheoretical) values (**t** and **f**).

Why the above definition works is clear at the intuitive level: Lack of ambiguity in decomposing a nonatomic formula C, i.e., *uniquely*, as one of $(\neg A), (A \wedge B), (A \vee B), (A \to B), (A \equiv B)$ allow us to know how to compute *a unique* answer. The why at the technical level is beyond our reach (the demanding reader can find a proof in [54]).

The convenience of truth tables can be extended to rephrase the recursive equations in the above definition (from the 4th equation onward). For example, the 5th equation is represented in table form as

A	B	$A \wedge B$
f	f	f
f	t	f
t	f	f
t	t	t

The way we read the above is that for all possible values of the *not necessarily atomic* formulae A and B we have listed the correct value of $A \wedge B$. We do *not* care how A and B actually *obtained* their values. Indeed, we do not care how A and B are built; they can be as complex as they like.

Here is an example of a definition of a "concept" regarding formulae, by induction on formulae.

1.3.6 Definition. (Occurrence of a Variable) We define "p occurs in A" and "p does not occur in A" simultaneously.

Occ1. (Atomic) p *occurs* in p. It does *not* occur in any of q, ⊤, ⊥—where q is a variable distinct from p.

Occ2. p occurs in $(\neg A)$ iff it occurs in A.

Occ3. p occurs in $(A \circ B)$—where $\circ$ is one of $\wedge, \vee, \rightarrow, \equiv$—iff it occurs in A or B or both.[33]   □

1.3.7 Remark. We wanted to be user-friendly (which often means "sloppy") in the first instance, and said that the above defines a "concept": "Occurs"/"does not occur". In reality, all such "concepts" that we may define by recursion on formulae are just functions.

For example, this "concept" can be captured by the function "$occurs(\mathbf{p}, A)$", where $occurs(\mathbf{p}, A) = 0$ means "p occurs in A", and $occurs(\mathbf{p}, A) = 1$ means "p does *not* occur in A".   □

1.3.8 Remark. (Finite "Appropriate" States) A state v is by definition an infinite table. Intuitively, the value of a formula A in any state v should depend only *on the values of the variables that occur in A and on no others*. Thus, for any one A, the state could be truncated into a *finite table* "appropriate" just for A—defined on all the variables of A but undefined elsewhere—without altering the Boolean value of A. Such a table would have one row for each variable that occurs in A, plus the two rows for ⊥ and ⊤—which in the end can be omitted as they offer no surprises.

Our intuition is correct. Here is a proof by induction on A of the relevant statement:

If v and v' are two states that agree on the variables of A, then $v(A) = v'(A)$, where "=" is metamathematical equality on the set $\{\mathbf{f}, \mathbf{t}\}$.

The proof, as it must, goes back and forth among Definitions 1.3.6 and 1.3.5 while, tacitly, it is also mindful at all times of how formulae are formed, going from less to more complex ones (Definitions 1.1.5 and 1.1.10).

Basis. If A is atomic, then either

(1) It is a constant, and hence $v(A) = v'(A)$ by Definition 1.3.5, equations two and three,

or

---

[33]Needless to say, by the "iff", it does *not* occur exactly when we have both: It does not occur in A *and* it does not occur in B.

(2) It is **p**. By assumption, $v(\mathbf{p}) = v'(\mathbf{p})$. Okay!

Complex formulae have two main shapes:

Case where A is $(\neg C)$: The I.H.[34] applies to C. We use it as follows:

(1) Since (1.3.6) C and $\neg C$ have precisely the same variable occurrences, it is that v and v' agree on the variables of C.

(2) By I.H. we get $v(C) = v'(C)$, and so $v(\neg C) = v'(\neg C)$ by 1.3.5.

Case where A is $C \circ D$: By Definition 1.3.6 (**Occ3**), each of C and D have all their variables also occur in $C \circ D$; therefore, by assumption, v and v' agree on all the variables that occur in C and D. By I.H. $v(C) = v'(C)$ and $v(D) = v'(D)$. By 1.3.5, $v(C \circ D) = v'(C \circ D)$. This concludes the proof. $\qquad\qquad\square$

1.3.9 Definition. (Tautologies) Boolean logic is primarily interested in those formulae that are true (**t**) in *all possible states*. Such formulae are called *tautologies* and because of their "shape", i.e., the way they are put together from atomic formulae, brackets, and connectives, are "always" true. We use the shorthand notation $\models_{\text{taut}} A$ to indicate that A is a tautology. $\qquad\qquad\square$

In view of 1.3.8, when checking a formula A for tautology status, we need to check it only on all finite states *appropriate for A*. If and only if we find that its value in *all* those states is **t**, then it is a tautology. Thus there is a finite process to do this, however, one that at the present state of the art is ridiculously inefficient: To so check an A that has n Boolean variables (occurring in it) we need a truth table of 2^n rows. Current research on the "P = NP?" question of Cook ([5]) converges toward the opinion that *it is highly unlikely that we will ever have a way to check tautologyhood, deterministically, in a manner appreciably more efficient than constructing a truth table*.

This is one additional reason why we are interested in discovering tautologies in a different way, *nondeterministically*, one that allows shortcuts in the calculation by allowing the prover to guess the correct next step from a set of candidates, whenever such a choice is offered, thus avoiding having to check all possible avenues that offer themselves. These guesses, when possible,[35] are informed by experience and human intuition and ingenuity and shorten the process of tautology verification. Enter (syntactic) "proofs" of the next section.

1.3.10 Example. (Some tautologies) $\top$ and $p \to p$ are tautologies. The latter follows from $v(p \to p) = F_\to(v(p), v(p))$ and the truth table on p. 29.

How about $p \to q \to p$? First, remember that this is sloppy for $(p \to (q \to p))$. Thus the *last connective* to act is the leftmost.

---

[34]The I.H. is invariably "assume the claim for all formulae less complex than A". As such, it deserves no explicit mention.

[35]It is not known whether there is a fast nondeterministic algorithm that verifies *every* tautology. But even if we discover one such, it is unlikely in view of what I said above that we can eliminate the nondeterminism *without significant speed loss*.

Here's the general technique: The table below has two components. The state-part consists of the first two columns. Each row of these columns is a finite state appropriate for the formula.

The value of the formula in each state is found beneath the last-to-act connective in the process of 1.1.3. To compute this value, we use the values $v(p)$ and that of $q \to p$. The values of the latter are aligned under the relevant connective, the "$\to$" of the subformula $q \to p$. In general, in the value part we align the values that we compute *under the connective that acted* last *in the subformula we are evaluating*.

p	q	p	$\to$	q	$\to$	p
			(1)		(3)	(2)
f	f		t		t	
f	t		t		f	
t	f		t		t	
t	t		t		t	

The number headings (1)–(3) above indicate the order in which the columns are built. By the way, we have just verified that $\models_{\text{taut}} p \to q \to p$.

What about $A \to B \to A$ where A, B are arbitrary formulae? Can we settle this question without knowing the *particular* ways that A, B are put together from variables, connectives, and brackets?

Yes! No matter how they are put together, in any state v we have that each of A, B attain one of two values **t** or **f**. Thus the exact same table as the above, but this time using A, B and $A \to B \to A$ as column headings,

A	B	A	$\to$	B	$\to$	A
			(1)		(3)	(2)
f	f		t		t	
f	t		t		f	
t	f		t		t	
t	t		t		t	

settles the question: We have $\models_{\text{taut}} A \to B \to A$.

We must be sure to read the above table correctly. We are *not* saying that we are assigning values to A and B (we can assign values only to variables and constants). We *are* saying that the possible pairs of values of A and B—in that order—no matter what state we are in, will be among the four listed in the table. Then using Definition 1.3.5 we fill in the last two columns of the table.   □

There are three more concepts related to tautologies that we want to introduce, but first some notation:

We use the metavariables $\mathbf{p}, \mathbf{q}, \mathbf{r}, \mathbf{q}_6'', \ldots$ for variables, and A, B, E, Q for formulae. What shall we use for *sets of formulae*?

Convention: We denote *sets* of formulae by certain capital Greek letters, as a rule, by those that *cannot* be confused with Latin letters. Thus $\Gamma, \Delta, \Sigma, \Theta$ will always stand for sets of formulae (such sets may have zero, one, two, three, one million, or infinitely many members). Of course, we use these letters to denote such sets if either

(i) We do not care what are the members of a set of formulae

or

(ii) We do care, but we are going to refer to that set over and over again in an argument; thus rather than, say, writing $\{\bot, p, p \rightarrow q, p \rightarrow \neg q\}$ over and over again, we may give it a name saying, "Let Σ stand for $\{\bot, p, p \rightarrow q, p \rightarrow \neg q\}$."[36]

By the way, we must *not* confuse the *set* $\{A\}$ with its member A. These are different. Think in terms of *types* (as in programming languages): $\{A\}$ is an object of type *set* while A is an object of type *formula*.

1.3.11 Definition. A formula A is *satisfiable* iff *there is at least one state* v *where* $v(A) = \mathbf{t}$. A *set* of formulae Γ is satisfiable iff *there is at least one state* v *where for every formula* A *in* Γ, $v(A) = \mathbf{t}$. We say that v *satisfies* Γ.

We say that Γ *tautologically implies* A—and write $\Gamma \models_{\text{taut}} A$— iff *for every state* v *that satisfies* Γ *we must have* $v(A) = \mathbf{t}$. We call Γ the *hypotheses* (plural, in general) or *premises* of the implication, while A is the *conclusion*.

We say that a formula A is *unsatisfiable* or a *contradiction* iff *for every state* v, we have $v(A) = \mathbf{f}$. We say that a *set* Γ is unsatisfiable iff *for every state* v *there is at least one* A *in* Γ *such that* $v(A) = \mathbf{f}$. $\qquad\square$

By *convention*, logicians write the simpler $A \models_{\text{taut}} B$ for the correct $\{A\} \models_{\text{taut}} B$, and more generally, prefer to write $A_1, \ldots, A_n \models_{\text{taut}} B$ rather than the correct (but pedantic) $\{A_1, \ldots, A_n\} \models_{\text{taut}} B$.

Note that intuitively, a tautology A is true, no questions asked, and *must* be accepted as such. On the other hand, the *conclusion* of a tautological implication is only *relatively true*, relative to the premises, that is: *If* we accept the premises as true, then we *must* also accept the conclusion as true.

This relativity of truth is at the heart of mathematics. For example, if we accept Euclid's "5th postulate"[37] as true, then we *must* accept that the sum of the angles of *any* triangle equals 180 degrees. This is a relative truth,[38] since Euclid's 5th postulate is not an absolute truth. Accepting any one of its possible negations[39] leads to a totally different (relative) truth regarding the sum of angles in a triangle.

---

[36] We may also say, "Let $\Sigma = \{\bot, p, p \rightarrow q, p \rightarrow \neg q\}$."

[37] It states that through a point that lies outside a given line it is always possible to draw a unique line parallel to the given one.

[38] Strictly speaking, this example lies beyond Boolean logic, since we need predicate logic in order to "speak" and do Euclidean geometry. Nevertheless it illustrates the phenomenon of relativity of truth in a familiar branch of mathematics.

[39] One possible negation is that you can draw two parallels (Lobachevsky). Another is that you can draw no parallels at all (Riemann).

1.3.12 Example. We can in principle verify $A_1, \ldots, A_n \models_{\text{taut}} B$ via a truth table

p	q	$\cdots$	A_1	$\cdots$	A_n	B
	$\cdots$			$\cdots$		$\cdots$
	a state			all **t**		must be **t** too
	$\cdots$			$\cdots$		$\cdots$

We look *only* at those rows that have **t** everywhere between the two "‖" vertical dividers. We must ensure that these rows have a **t** under B as well. By the way, the A_i and B being, in general, complicated formulae, we need the "$p, q, \ldots$" columns to the left of the leftmost ‖ in order to compute the values of the A_i and B in all states.

In general, when checking $A_1, \ldots, A_n \models_{\text{taut}} B$, one cannot avoid building the whole truth table (or doing some equivalently laborious task) for the following reasons:

(1) As we have said, at the present (and foreseeable) state of the art there is *no way* to check whether A is a tautology any more efficiently than it takes to build the whole truth table, in terms of the variables $p, q, \ldots$, of A.

(2) If we had a substantially faster algorithm for tautological implication, we could use it to also establish tautologyhood fast, since $\models_{\text{taut}} A$ iff $\top \models_{\text{taut}} A$.

However, in many cases of *practical interest*, we can do better since we *need to only check the rows that have exclusively* **t***'s between the two* ‖ *dividers, and ignore all the other rows*. Sometimes we can identify these rows without building the whole table.

For example, we can quickly see that $A, B \models_{\text{taut}} A \wedge B$ since whenever $v(A) = v(B) = \mathbf{t}$ we have $v(A \wedge B) = \mathbf{t}$ (cf. 1.3.5). We did not compute the other three rows. A more substantial example is

$$A \vee B, \neg A \vee C \models_{\text{taut}} B \vee C$$

If this is done by the full table method we need 8 rows to compute the values of $A \vee B, \neg A \vee C, B \vee C$ for all possible ordered triples of values $v(A), v(B), v(C)$.

But instead, let us be clever: Okay; we merely need to compute the value $v(B \vee C)$ on the condition that

$$v(A \vee B) = v(\neg A \vee C) = \mathbf{t} \tag{1}$$

We analyze (1) according to two cases:

(i) $v(A) = \mathbf{f}$: By (1) and 1.3.5, $v(B) = \mathbf{t}$. But then $v(B \vee C) = \mathbf{t}$ by 1.3.5.

(ii) $v(A) = \mathbf{t}$: By (1) and 1.3.5, $v(C) = \mathbf{t}$. But then $v(B \vee C) = \mathbf{t}$ by 1.3.5. □

1.3.13 Example. Here is a really important example: $\perp \models_{\text{taut}} A$. Well, I need to ensure that for every v, where $v(\perp) = \mathbf{t}$, I also get $v(A) = \mathbf{t}$.

Since no v satisfies $v(\perp) = \mathbf{t}$—i.e., there are no rows to check—I am done.

We have seen such situations before. The statement is vacuously true. Like before, one can explain this by saying, "The only way to *refute* $\perp \models_{\text{taut}} A$ is to find a v where $v(\perp) = \mathbf{t}$ but $v(A) = \mathbf{f}$. But such a task must fail as no v satisfies $v(\perp) = \mathbf{t}$." □

Here is a simple but nice exercise that you *must* try:

1.3.14 Exercise. (1) Show that $\models_{\text{taut}} A$ iff $\emptyset \models_{\text{taut}} A$.

(2) Show that $A_1, \ldots, A_n \models_{\text{taut}} B$ iff $\models_{\text{taut}} A_1 \to A_2 \to \ldots \to A_n \to B$.

(3) Show that $\Gamma \models_{\text{taut}} B$ iff $\Gamma \cup \{\neg B\}$ is unsatisfiable. We note here that for two sets Γ and Σ the notation $\Gamma \cup \Sigma$—called the *union* of Γ and Σ— denotes the set that contains every member of Γ *and* every member of Σ. $\qquad\qquad\square$

We conclude this section with the definition of *substitution* in a Boolean expression, and with the proof of two important properties of substitution.

Intuitively, the symbol "$A[\mathbf{p} := B]$" is shorthand that means—i.e., expands into— the string we get if we replace *all occurrences* of the variable $\mathbf{p}$ in A by the formula B. We may think of this operation as defining a function from formulae to strings:

Input: $A, \mathbf{p}, B$; output: the string denoted by $A[\mathbf{p} := B]$.

In order to read the following definition of substitution correctly, we emphasize that the "operation" $[\mathbf{p} := B]$ takes place in the metatheory and has the highest priority against all other "formal operations", i.e., the Boolean connectives $\neg, \wedge, \vee, \to, \equiv$. For example, $\neg A[\mathbf{p} := B]$ means $\neg\{A[\mathbf{p} := B]\}$, where the symbols "$\{, \}$" are here meta-brackets inserted to indicate the order of application of "operations". Naturally, because of its placement, this operation is *left*-associative, so that $A[\mathbf{p} := B][\mathbf{q} := C]$ means $\{A[\mathbf{p} := B]\}[\mathbf{q} := C]$.

1.3.15 Definition. (Substitution in Formulae) *In what follows, "$=$" denotes equality in the metatheory, here between strings.* The definition states the obvious: (1) It handles the basis cases in the trivial manner, and (2) when A is actually built from i.p. (cf. 1.1.10), it says that we substitute into *each* i.p. first, and then apply the connective.

As we usually do in definitions we are careful *not* to use formula abbreviations. All brackets are present. *Note the use of metavariables.*

$$A[\mathbf{p} := B] = \begin{cases} B & \text{if } A = \mathbf{p} \\ A & \text{if } A = \mathbf{q} \text{ (where } \mathbf{p} \neq \mathbf{q}), \text{ or} \\ & \quad A = \top, \text{ or } A = \bot \\ (\neg C[\mathbf{p} := B]) & \text{if } A = (\neg C) \\ (C[\mathbf{p} := B] \circ D[\mathbf{p} := B]) & \text{if } A = (C \circ D) \end{cases}$$

where $\circ$ is one of $\wedge, \vee, \to, \equiv$. $\qquad\qquad\square$

We state and prove two easy and hardly unexpected properties of substitution:

1.3.16 Proposition. *For any formulae A and B and variable $\mathbf{p}$, $A[\mathbf{p} := B]$ is a (well-formed!) formula.*[40]

---

[40] A *proposition* is a theorem—here, metatheorem—that did not quite make it to be called that. You see, people reserve the term *theorem*, or *metatheorem*, for the "important" or earth-shattering stuff that

Proof. Induction on A, keeping an eye on Definition 1.3.15.

Basis. If A is atomic, then we get either A or B formula; okay in either case.

Complex formulae have two main shapes:

A is $(\neg C)$: The I.H. applies to C, thus $C[\mathbf{p} := B]$ is a formula, Then so is $(\neg C[\mathbf{p} := B])$ by 1.1.7.

A is $(C \circ D)$: The I.H. applies to C and D, thus $C[\mathbf{p} := B]$ and $D[\mathbf{p} := B]$ both are formulae. Then so is $(C[\mathbf{p} := B] \circ D[\mathbf{p} := B])$, again by 1.1.7. $\square$

1.3.17 Proposition. *If $\mathbf{p}$ does not occur in A, then $A[\mathbf{p} := B] = A$, where for* convenience we once more use "$=$" as metamathematical equality of strings.

Proof. Again, by induction on formulae A:

Basis. A can only be one of $\mathbf{q}$ ($\mathbf{q} \neq \mathbf{p}$), $\top$, or $\bot$, by Definition 1.3.6. Then, by Definition 1.3.15 (2nd case), $A[\mathbf{p} := B] = A$. Okay, so far.

Complex formulae have two main shapes:

A is $(\neg C)$: Does $\mathbf{p}$ occur in C? No, by assumption (it does not occur in A) and definition of occurrence (1.3.6). Now, the I.H. applies to C, thus $C[\mathbf{p} := B] = C$. We are done by 1.3.15, case 3.

A is $(C \circ D)$: By 1.3.6 (**Occ3**) $\mathbf{p}$ occurs neither in C nor in D. The I.H. applies to both these subformulae; thus $C[\mathbf{p} := B] = C$ *and* $D[\mathbf{p} := B] = D$, and we are done by 1.3.15, case 4. $\square$

1.4 PROOFS AND THEOREMS

We are ready to develop a *calculus* that we may use to write down theorems. We will learn to "calculate theorems" just as we have learned to "calculate" (= "parse") formulae.

Boolean logic is a (crude) vehicle through which we formulate and establish mathematical truth. This truth is captured absolutely (tautologies) or relatively to certain *premises* (tautological implications). Thus, when we do Boolean logic, our main task is to discover and verify tautologies, and more generally, to discover and verify tautological implications.

We have already remarked that the presently known mechanical ways to check for tautology status as well as to verify tautological implications are hopelessly inefficient, and that there is every indication that an efficient tautology (or tautological implication) checker will never be discovered.

One turns to utilizing human ingenuity and experience—in other words, utilizing (educated) *guessing*—toward effecting serious shortcuts in the process of certifying tautologies and tautological implications. This guessing (or "nondeterministic")

---

we prove. All else that we prove are just *propositions* (or *lemmata*—singular: *lemma*) if they have just "auxiliary status", just like FORTRAN subroutines; or they are *corollaries*, if they follow trivially—more or less—from earlier results.

process of certifying tautologies and tautological implications is *syntactic* rather than truth table driven[41] (semantic) and is called *theorem proving*.

The theorems that we will learn to prove with this new syntactic technique will be either absolute truths (tautologies) or truths that are relative (conclusions of tautological implications) to certain assumptions that we have accepted.

Our major concern as we are founding this syntactic proof calculus will be to ensure that whatever tools we utilize are capable of certifying *all* absolute and relative truths, and *only those*. That is, these tools will never "certify" a falsehood as a "theorem". Our degree of success in implementing these requirements will be assessed later (Section 3.1 will assess the promise for *only those*, while 3.2 will assess the one for *all*).

As I said before, I kind of like the terms *calculate theorems* and *theorem calculus* as they remind us that *proving* theorems is a precise *syntactic* (synonym for *formal*, i.e., depending only on form) algorithmic process. This observation is the origin of the alternative name of *equational logic* of [11]: *calculational logic*. However, most logicians and mathematicians would proclaim that they "proved" (rather than "calculated") a theorem and would rather call a theorem-calculation a *proof*.

First off, a theorem-calculation or *proof* is a finite sequence of formulae, entirely analogous to formula-calculations. *Each formula occurring in a proof will be called a* theorem.

The specifications of formula-calculations are the following two:

(1) A formula-calculation must start with the *simplest possible kind* of formula: one that is "primitive", i.e., atomic.

(2) Every *operation that extends* the formula-calculation must *preserve* the property "formula": either it is the trivial operation of writing down an atomic formula, or it is an operation that acts on previously written formulae *and produces* a formula.

Entirely analogously, as it flows from the preceding discussion in this section, the specifications of theorem-calculations are the following two:

(1′) A theorem calculation must start with the writing down of a formula that is among the simplest possible theorems—a "primitive theorem" for which the validation process is simply to write it down! We call such a formula an *axiom*.

(2′) Every *operation that extends* the theorem-calculation must preserve the property "theorem"; thus, it is either the trivial act of writing down another axiom, or it is applied to already-written theorems, resulting into a new theorem. Since a theorem calculation must *certify truth*, these nontrivial operations, or rules, must preserve truth. Technically, whenever they are applied to formulae $A_1, \ldots, A_n$ and yield a formula B as a result, it is *necessary* that they obey $A_1, \ldots, A_n \models_{\text{taut}} B$.

We have two types of axioms: The *logical axioms* are certain well-chosen[42] absolute truths; therefore, they are certain tautologies. The other type we will call *special* axioms, but also *assumptions* or *hypotheses*. These are not fixed outright, but

---

[41] Later there will be a weakening of this somewhat dogmatic statement.
[42] The qualifier *well-chosen* will be revisited later.

may change from discussion to discussion.[43] They are not deliberately chosen[44] to be absolute truths, but are formulae that we "accept as true" (cf. discussion on relativity on p. 34) simply because we are interested to explore what sort of (tautological) conclusions we may draw from them.

In intuitive terms—since the quest for theorems is the quest for "mathematical truths", absolute or relative—the axioms are our *initial truths*. Our rules of reasoning will allow us to derive further truths from, and relative to, the axioms.

The nontrivial operations that lengthen proofs (cf. (2′) above) are called *rules of inference*. To achieve the purely syntactic character of proofs, the rules of inference are applied in a manner that the input/output relation is purely syntactic. For example, one of the two *primitive*[45] rules, soon to be introduced, applies to any formulae of the forms A and $A \equiv B$ and "outputs" B. The rule does not care about which specific formulae A or B stand for, nor about the semantics of any of $A, B, A \equiv B$. The only thing that the rule cares about is that it "sees" as input an equivalence on one hand, and the first formula of this equivalence on the other. It immediately "knows" that it must "output" the second formula of the equivalence.

In order to describe the rules of inference, we need *formula schemata*[46] (or, simply, *schemata*).

A *schema* is a string in the metatheory (i.e., outside logic) over the *augmented* alphabet that along with $\mathcal{V}$ (p. 9) includes the symbols "[", ":=", and "]", and all the *syntactic variables* for formulae and Boolean variables.

The syntactic structure of a schema is, by definition, such that if we replace *all* the syntactic variables that occur in it by *any* specific formulae and variables,. as appropriate, then the result names a formula of **WFF**.

1.4.1 Definition. (Schema Instance) An *instance* of a schema is the formula we obtain if we replace *all* its metavariables with specific objects (formulae/Boolean variables) as appropriate.   □

Here are some examples of schemata:

(1) A: It is a formula-metavariable; if we replace the letter "A" by some formula, well, we get that formula as the result!

(2) $(A \equiv B)$: This schema has two formula-metavariables, A and B. Whatever formulae we may replace A and B with, we get a formula by 1.1.7.

(3) $A[\mathbf{p} := B]$: This schema has two formula-metavariables, A and B, and a Boolean metavariable, $\mathbf{p}$. Whatever formulae we replace A and B with, and whichever Boolean variable replaces $\mathbf{p}$, we get a formula by 1.3.16.

---

[43]For this reason, I suppose, some people call them *temporary* assumptions. However, *temporary* is not a technical term. For example, Euclid's 5th postulate has no expiry date. Yet it is not an absolute truth.

[44]The qualification *chosen* is picked purposely: I do not want to rule out as special axioms any that, either by accident or on purpose, have been chosen to be tautologies. Analogously, when we defined the notation $\Sigma \models_{\text{taut}} A$, quite correctly we did not forbid the case where some formulae of Σ may be tautologies.

[45]We will encounter many *derived* rules. These are not "given" or postulated up front, but we prove their validity.

[46]*Schema*, plural *schemata* (and, incorrectly but often, *schemas*), is Greek for *form* and *figure*.

We often write the rules of inference as "fractions", like

$$\frac{P_1, P_2, \ldots, P_n}{Q} \tag{R}$$

where all of $P_1, \ldots, P_n, Q$ are formula schemata. We call the "numerator" the *premise* (case $n = 1$) or *premises* (case $n > 1$) and the "denominator" the *conclusion* or *result* of the rule. Instead of *premises* we also say *hypotheses* or *assumptions*.

The P_i and Q are *syntactically related* so that one can *mechanically check* this input/output relation by simply looking at the *form* of the P_i and Q.

We have already noted an example of this mechanical applicability of a rule such as

$$\frac{A, A \equiv B}{B}$$

The input/output relation of a rule need not be "functional"; that is, the result of a rule need not be uniquely determined by the hypotheses (cf. Leibniz rule below).

It is obvious why a rule is expressed in terms of schemata rather than specific formulae. Schemata allow a rule to be applicable to infinitely many formulae. If conclusion and premises were specific formulae, then there would be just one case where the rule would be applicable, which would hardly qualify it to be called a *rule*, a term that creates the expectation of applicability to a vast number of cases. Here is an analogy: The input/output relation on numbers "in:3 / out:9" is not a rule, but "in:x / out:x^2" is.

How a rule like (R) above is applied will be clear in Definition 1.4.5.

Let us finally introduce the two primitive rules of Boolean logic that we will adopt in this volume.

1.4.2 Definition. (Rules of Inference) The following two are our *primitive* or *primary* rules of inference, given with the help of the syntactic variables $A, B, \mathbf{p}, C$:

Inf1

$$\frac{A \equiv B}{C[\mathbf{p} := A] \equiv C[\mathbf{p} := B]} \tag{$Leibniz$}$$

Inf2

$$\frac{A, A \equiv B}{B} \tag{$Equanimity$}$$

An instance of a rule of inference is obtained by replacing all the letters A, B, C and $\mathbf{p}$ by specific formulae and a specific variable respectively.   □

The rule names conform to those in [17].

1.4.3 Remark. (1) Why *primary*? Do we also have "secondary rules"? Yes. We will soon learn that we can apply additional rules in a theorem-calculation, which are not mentioned in the definition of proof below (1.4.5) because they are not theoretically necessary toward defining *proof* and *theorem*. Such *additional* rules are *not* to be arbitrarily added to our toolbox without question. Instead, we will show before adding them, via a rigorous mathematical argument, that we *are* allowed to use them. This "allowed" means that there is nothing we can prove using these additional rules that we cannot prove without them. We call such additional rules *derived rules*, or *secondary rules*.

Compare, once again, with programming languages. Some general-purpose procedural languages were designed not to contain the instruction **goto** because it was considered "harmful" by some influential computer scientists in the 1970s (cf. [10], which, arguably, started it all)—but not by all (cf. [28]). Nevertheless, one *can* prove that **goto** *can* be simulated by the originally given instructions. It is a "derived" kind of instruction in those **goto**-less languages.

Harmful or not, one will agree that adding one more tool does add to convenience, in general.

(2) Other than the "restriction" that the A, B, C and **p** are metavariables of the agreed-upon kind, *there is no other restriction on the letters*. In particular, we have *no* assumption on whether **p** actually occurs in C of the Leibniz rule. Either way, it is all right.

(3) We have already discussed the mechanical nature of **Inf2**. That also **Inf1** is mechanically applicable is clear: Once we have written "$A \equiv B$", we can pick any formula C whatsoever and any variable **p** and construct the output, first effecting two substitutions and then connecting the results with the connective "$\equiv$" in the indicated order.

Note that the Leibniz rule is not functional: Infinitely many different outputs are possible for a given input $A \equiv B$. $\qquad\qquad\square$

We now turn to our choice of axioms.

 Schemata; again. We noted already that the axioms will be "initial truths", and as such they will be *selected tautologies*.

Suppose then, for the sake of discussion, that one of the tautologies of choice to attain axiom status is $p \equiv p$.

But how about $q \equiv q$? Or how about $p \vee p' \equiv p \vee p'$ and, indeed, how about $(p \vee p' \equiv p \vee p') \equiv (p \vee p' \equiv p \vee p')$?

All these have the same *form*, namely $A \equiv A$, where A stands for an arbitrary formula. Naturally, if we want to include $p \equiv p$, then we will want to include all those tautologies that have the same "shape", $A \equiv A$, as well, since surely they all state the same (absolute) principle: "Every statement is equivalent to itself." If this is a "truth" worth postulating, then it would be absurd to do so just for one of its *special cases*, just for the variable p.

There are two main ways to achieve this generality:

(1) The "modern" and rather obvious way: Rather than saying, "include $p \equiv p$ and $q \equiv q$ and $p \vee p' \equiv p \vee p'$ and $(p \vee p' \equiv p \vee p') \equiv (p \vee p' \equiv p \vee p')$ and ...", we say, "include the *schema* $A \equiv A$."

This means *include all instances of* $A \equiv A$ (cf. 1.4.1).

(2) The "old" way: "Include just $p \equiv p$; however, add a *new primary* rule of inference called *substitution.*" This rule

$$\frac{A}{A[\mathbf{p} := B]} \qquad (Sub)$$

when applied to the formula "$p \equiv p$" (that is, take A to be $p \equiv p$ and $\mathbf{p}$ to be p) will be able to generate, by successive applications, all possible tautologies of the form $B \equiv B$.

There is a catch: Once you add rule (Sub) as a *primary* rule, you have to awkwardly hedge when writing proofs. The rule *cannot* be used in a theorem-calculation unless you *know* that the variable $\mathbf{p}$ does *not* occur in any formulae that are *special* axioms. This restriction in turn makes a very useful tool that we will soon obtain and learn to use—the *deduction theorem*—hard to state and even harder to apply.

Thus, the old way is rightfully abandoned. In particular, we will *not* have any use for (Sub).[47]

We next present the list of *logical* axioms for Boolean logic. The list is infinite, but because of the use of schemata it can be presented by a finite table. That is, there are only finitely many different *forms* of tautologies that we need to take as our starting point. These are largely the ones in [17] with a few adjustments.

What the axioms do is to codify the most basic properties of the connectives. The following list both presents (in partially parenthesized notation[48]) and *names* the axioms.

· **1.4.4 Definition. (Logical Axioms of Boolean Logic)** In what follows, A, B, C denote arbitrary formulae:

Properties of $\equiv$

Associativity of $\equiv$	$((A \equiv B) \equiv C) \equiv (A \equiv (B \equiv C))$	(1)
Symmetry of $\equiv$	$(A \equiv B) \equiv (B \equiv A)$	(2)

Properties of $\bot, \top$

$\top$ vs. $\bot$	$\top \equiv \bot \equiv \bot$	(3)

Properties of $\neg$

Introduction of $\neg$	$\neg A \equiv A \equiv \bot$	(4)

Properties of $\vee$

---

[47]The current edition of [17] uses the old approach in Boolean logic (their Chapter 3) but switches to the modern approach for predicate logic (Chapters 8 and 9). One may assume that by the time the authors decided that the approach with schemata is better it was too late in terms of publication deadlines to rewrite Chapter 3 with schemata.

[48]Recall that brackets associate from right to left.

Associativity of $\vee$	$(A \vee B) \vee C \equiv A \vee (B \vee C)$	(5)
Symmetry of $\vee$	$A \vee B \equiv B \vee A$	(6)
Idempotency of $\vee$	$A \vee A \equiv A$	(7)
Distributivity of $\vee$ **Over** $\equiv$	$A \vee (B \equiv C) \equiv A \vee B \equiv A \vee C$	(8)
Excluded Middle	$A \vee \neg A$	(9)

Properties of $\wedge$

Golden Rule	$A \wedge B \equiv A \equiv B \equiv A \vee B$	(10)

Properties of $\rightarrow$

Implication	$A \rightarrow B \equiv A \vee B \equiv B$	(11)

We will reserve the capital Greek letter "lambda", Λ, to denote the set of all *logical* axioms. This set is, of course, infinite. $\square$

The axioms of Λ, here, however, formulated in their schemata edition, are those that are customary in the "equational" (or "calculational") approach to doing logic, as presented, e.g., in [17] and [32], but with some minor differences—besides the essential one that [17] does not use schemata. For example, [17] uses $\top \equiv A \equiv A$ instead of (3). Moreover, our choice for (4) is natural, and hence easy to remember: It *intuitively* says "negating A is tantamount to saying that it is false" ($A \equiv \bot$). But [17] adopts instead a different axiom that eventually implies our (4): I quote it, but in schema form, "$\neg(A \equiv B) \equiv \neg A \equiv B$". This being intuitively less clear is also less memorable than the rest.

We are ready to calculate! (Compare with Definition 1.1.3 and Remark 1.1.6(2).)

1.4.5 Definition. (Theorem-Calculations—or Proofs) Let Γ be an arbitrary, given set of formulae.

A *theorem-calculation* (or *proof*) *from* Γ is any finite (ordered) sequence of formulae that we may write respecting the following two requirements:

In any stage we may write down

Pr1 Any member of Λ or Γ

Pr2 Any formula that appears in the denominator of *an instance* of a rule **Inf1–Inf2** as long as *all the formulae* in the numerator of the same instance of the (same) rule have already been written down at an earlier stage

We may call a *proof from* Γ by the alternative name Γ-proof. $\square$

1.4.6 Remark. (1) *By definition*, a Γ-proof is a *purely form-manipulation construction* without any reference to semantic concepts, such as **t**, **f**, etc.

(2) We will call Γ the set of *special* axioms (Λ contains the "general" axioms). Special axioms are also called *hypotheses* or *assumptions*. Clearly, while Λ is reserved and "frozen" in the first part of this book, Γ can vary from subject to subject.

(3) Any member of Γ *that is not also in* Λ we will call a *nonlogical* axiom.[49]

(4) Since any theorem-calculation from some Γ is a finite sequence of formulae, only a finite part of Γ and Λ may appear in such a calculation. This is entirely analogous with what happens in a formula-calculation: Each uses only a finite number of formal variables even though we have an infinite supply of those. ☐

1.4.7 Definition. (Theorems) Any formula A that appears in a Γ-proof is called a *Γ-theorem*. We write $\Gamma \vdash A$ to indicate this. If Γ is empty ($\Gamma := \emptyset$)—i.e., we have no special assumptions—then we simply write $\vdash A$ and call A just "a theorem".
Caution! We may also do this out of laziness and call a Γ-theorem just "a theorem" if the context makes clear which $\Gamma \neq \emptyset$ we have in mind.

We say that A is *an absolute*, or *logical* theorem whenever Γ is empty. ☐

(1) Clearly, the symbol "$\vdash$" of the metatheory formulates the predicate "is a theorem".
(2) Definition 1.4.7 says that a formula is a theorem *on one and only one condition*: It occurs in some Γ-proof. We say that such a proof *proves A from Γ, or from the hypotheses (of)* Γ.

In the common parlance of mathematics we may also say that "Γ derives A".

(3) Note how in the symbol "$\vdash A$" we take Λ for granted and do not mention it to the left of $\vdash$.

1.4.8 Remark. Thus, a Γ-proof of a formula A is a sequence of formulae

$$B_1, \ldots, B_n, A, C_1, \ldots, C_m$$

obeying the requirements stated in 1.4.5. It is trivial that if we discard the "tail" part "$C_1, \ldots, C_m$" of the sequence, then

$$B_1, \ldots, B_n, A \tag{1}$$

is still a proof. The reason is that every formula in a proof is either legitimized outright—without reference to any other formulae—or is legitimized by reference to formulae to its *left*. Thus (1) also proves A, since A occurs in it. This technicality allows us to stop a proof as soon as we write down the formula that we wanted to prove. ☐

So, 1.4.7 tells us what kind of theorems we have:

1. Anything in $\Lambda \cup \Gamma$.[50]

2. For any formula C and variable $\mathbf{p}$, the formula $C[\mathbf{p} := A] \equiv C[\mathbf{p} := B]$, *provided* $A \equiv B$ was written down already, and therefore is a (Γ-) theorem.

---

[49]That is, it does not speak *about* logic itself; logic does not need this axiom in order to function properly.
[50]We explained the notation "$\Lambda \cup \Gamma$" in 1.3.14 on p. 36, item (3).

3. B (any B), *provided $A \equiv B$ and A were* written down already and therefore are both (Γ-) theorems.

Hey! The above is a recursive definition of (Γ-) theorems, and is worth recording (compare with Definition 1.1.7).

1.4.9 Definition. (Theorems, Inductively) A formula E is a Γ-theorem iff E fulfills one of **Th1–Th3** below:

Th1 E is in $\Lambda \cup \Gamma$.

Th2 For some formula C and variable $\mathbf{p}$, E is $C[\mathbf{p} := A] \equiv C[\mathbf{p} := B]$, *and* (we know that) $A \equiv B$ is a (Γ-) theorem.

Th3 (We know that) $A \equiv E$ *and* A are (Γ-) theorems. $\square$

1.4.10 Remark. (Theorem vs. Metatheorem) In the expression $\Gamma \vdash A$, "A" is the theorem, *not* "$\Gamma \vdash A$". After all, a theorem by definition has to be a *single formula* that appears somewhere in some proof (1.4.7).

So what is "$\Gamma \vdash A$" then? It is a *meta*theorem. It is a statement that we are making *about* our logic, about what the logic can do, *if* we take all the formulae in Γ as assumptions. It says, "there is a proof, which from assumptions Γ proves A."

But how does one establish the validity of such a meta-result as quoted immediately above?

By *proving within Boolean logic*—according to Definition 1.4.5, that is— the theorem A using assumptions Γ.

This action does two things at once: It *proves A* (from Γ) and it also *metaproves* the statement $\Gamma \vdash A$.

Nevertheless, people are mostly shy of such distinctions and it has become acceptable practice to say (by abuse of language) things like "I proved $\Gamma \vdash A$" all the time. See also the start-up comment in the next chapter. $\square$

1.4.11 Exercise. So that you can check your understanding of the concepts *proof* and *theorem*, show (i.e., present proofs) that

(1) $A \vdash A$, for any A.

(2) A more general form of (1): If A is a member of Σ—also written "$A \in \Sigma$"— then $\Sigma \vdash A$.

(3) $\vdash B$, for any axiom B. $\square$

1.4.12 Remark. (Hilbert Proofs) A Γ-proof is also called a *Hilbert proof* after the name of the great mathematician David Hilbert, who essentially was the first serious proponent of the idea to *use logic to do mathematics*. Hilbert also helped to found modern logic in an axiomatic and rigorous setting, and defined the concept of proof, essentially as above.[51]

---

[51]There are some inessential differences. Hilbert used a different set of logical axioms, and a single rule of inference.

In practice we write a Hilbert proof vertically on the page, i.e., one formula on top of the other, numbering every formula. It is imperative that we provide annotations to explain what we are doing at every step and why. The numbering assists us in referring to previous formulae.

All proofs, whether they are in the Hilbert style or in the equational style (the latter style will be introduced shortly), *must* be annotated.   □

1.4.13 Example. (Some Very Simple Annotated Hilbert Proofs)
   (a) We will verify that "$A, A \equiv B \vdash B$" for any formulae A and B. Thus *we need to write a formal proof of B using A and A $\equiv$ B as hypotheses* (cf. 1.4.10). The part "for any formulae A and B" makes this result applicable to infinitely many instances, one for each specific choice of A and B. It is therefore a *metatheorem schema*. We could have said instead, "prove the schema $A, A \equiv B \vdash B$", a formulation where the part "for any formulae A and B" is redundant.

The same comment applies to any theorems (and metatheorems) that we will prove where we have letters to *stand* for arbitrary formulae.

Let us establish (a) now. Make sure you memorize the style! This is how you will write your own Hilbert proofs. *Every* line must be numbered *and* annotated!

$$(1) \quad A \quad \langle\text{hypothesis}\rangle$$
$$(2) \quad A \equiv B \ \langle\text{hypothesis}\rangle$$
$$(3) \quad B \quad \langle(1) \text{ and } (2) \text{ and Equanimity}\rangle$$

 Worth stating. It is clear from Definition 1.4.5 that assumptions can be scrambled. So we have also established $A \equiv B, A \vdash B$ by the very same proof.

Can we also swap A and B in $A \equiv B$? It turns out that we can. *But* the above proof does *not* address this question: after all, $B \equiv A$ is never mentioned.

It is a fatal error to say, "but is not '$\equiv$' symmetric? I can see that it is so from the truth table of p. 29."

No; you see, we have *not* connected our syntactic proofs with semantics yet. Until such time, the only things that we *may* assume that we *can* do *must* directly follow from 1.4.5 in connection with our axioms and rules of inference.

(b) We next (meta)prove $A \equiv B \vdash C[\mathbf{p} := A] \equiv C[\mathbf{p} := B]$.

$$(1) \quad A \triangleq B \qquad\qquad\qquad \langle\text{hypothesis}\rangle$$
$$(2) \quad C[\mathbf{p} := A] \equiv C[\mathbf{p} := B] \ \langle(1) \text{ and Leibniz}\rangle$$

Hmm. Is there a pattern here? Indeed there is! Any rule like (R) of p. 40 leads to the statement of provability

$$P_1, P_2, \ldots, P_n \vdash Q \qquad\qquad\qquad (R')$$

as the *proof*

$$P_1, P_2, \ldots, P_n, Q$$

establishes. This *once* we have written a Hilbert proof horizontally and without annotation, in order to expedite the obvious.

That this is a $\{P_1, P_2, \ldots, P_n\}$-proof of Q is clear: Referring to Definition 1.4.5 we see that writing $P_1, P_2, \ldots, P_n$ (indeed doing so in any order if we wish) is legitimate by **Pr1**. Then following this by writing Q is also legitimate by an application of rule (R) on the previously written formulae.

This is why in some of the literature (e.g., [43]) rules of inference are written as in (R') above rather than in "fraction form". Tradition has it that *derived* rules of inference are always written in the style (R'). After all, such a rule is a (meta)provable principle and says that from certain assumptions we can prove a certain conclusion.

(c) We (meta) prove a derived rule of inference, called *transitivity*. It is

$$A \equiv B, B \equiv C \vdash A \equiv C \qquad\qquad (Transitivity)$$

Here it goes:

(1) $A \equiv B$ ⟨hypothesis⟩

(2) $B \equiv C$ ⟨hypothesis⟩

(3) $(A \equiv B) \equiv (A \equiv C)$ ⟨(2) and Leibniz, denom. "$A \equiv \mathbf{p}$" where $\mathbf{p}$ is fresh⟩

(4) $A \equiv C$ ⟨(1) and (3) and equanimity⟩

What's this about "fresh"? This means that $\mathbf{p}$ does not occur in any of A, B, C. Actually, all I need here is that it just does not occur in A, but it takes less space to say it is fresh in the annotation, so I can fit it in one line!

I want $\mathbf{p}$ not to occur in A so that when I do "$(A \equiv \mathbf{p})[\mathbf{p} := B] \equiv (A \equiv \mathbf{p})[\mathbf{p} := C]$" I am guaranteed that I get line (3). If $\mathbf{p}$ does occur in A then the substitutions will *change* A and I will not get line (3)! See also 1.3.17.

Pause. But what if I cannot find a fresh $\mathbf{p}$? Actually, I always can, since I have an *infinite supply* of variables, while only finitely many appear in A, B, C.

(d) We next *prove* the theorem (schema) $A \equiv A$. Note that I mentioned *no* assumptions. This is an absolute result.

Another way to say this is "metaprove $\vdash A \equiv A$", or if you hate to say "meta", say instead, "establish that $\vdash A \equiv A$" or "show that $\vdash A \equiv A$".

(1) $A \vee A \equiv A$ ⟨axiom⟩

(2) $A \equiv A$ ⟨(1) and Leib: $A[\mathbf{p} := A \vee A] \equiv A[\mathbf{p} := A]$ where $\mathbf{p}$ is fresh⟩

A few remarks: (i) We may use *any logical axiom* of the form "$\ldots \equiv \cdots$" in place of $A \vee A \equiv A$ in the above proof. By the way, this is our first proof where we used a logical axiom.

(ii) For logical axioms our annotation will just be "axiom". For special axioms, our annotation will always be "assumption" or "hypothesis".

(iii) We do not need to name the axioms (idempotent, etc.) in our annotations, and certainly I do *not* want you to memorize their numbers in the list! (What if I scramble the list?!)

But we must be truthful. Writing, say, "$A \equiv B$" and annotating "axiom" will not get us anywhere.

(iv) Rules *must* be named, and we must annotate how they are applied! In what follows we will make a habit of abbreviating rule names. For example, "Leibniz" will be *Leib* and "Equanimity" will be *Eqn*. Transitivity will be *Trans*.

In the next chapter we systematically prove several theorems (in almost all cases schemata) and metatheorems to enrich our toolbox and enhance our familiarity with the methodology. We also introduce the equational style of proof. $\square$

1.5   ADDITIONAL EXERCISES

1. Which of the following are Boolean formulae? Why? (Do not use any general principles; just try to give a good reason based on the definition of formula-calculation (1.1.3)).

    - p
    - (p)
    - $\top$
    - $\vee$
    - $p \rightarrow q$
    - $(p \rightarrow q)$

2. Are the following string sequences formula-calculations? Why?

    - $p, \top, (p \vee \bot), \bot$
    - $p, \bot, (p \vee \bot), \top$

3. Give a formula-calculation for $\left(\neg \left((p \vee q) \rightarrow \bot \right) \right)$

4. Prove either by analyzing formula-calculations or by induction on formulae that the string () is not a formula.

5. Prove either by analyzing formula-calculations or by induction on formulae that the string $((\neg\bot))$ is not a formula.

6. True or false, and why? "If A is a formula, then so is (A)."

7. Show by induction on formulae, or by analyzing formula-calculations, that every Boolean formula must contain at least one Boolean variable or one Boolean constant.

8. Prove that the complexity of a Boolean formula—correctly written as required by 1.1.5—equals the number of its left brackets.

9. (a) Prove that the *last* symbol of a Boolean formula is never $\wedge$.

   (b) Prove that the string $\wedge\vee$ never occurs as part of a Boolean formula.

   The proof of each part must be either by *induction on the complexity of formulae*, or by analyzing *formula-calculations*. In part (b), you may use the result of part (a).

10. Which of the following schemata are tautologies? Show all work and remember that to show that a schema is *not* a tautology we must identify an instance of it that is not a tautology.

    I am not using all the brackets required by 1.1.5.

    - $((A \to B) \to A) \to A$
    - $A \wedge B \to A \vee B$
    - $A \vee B \to A \wedge B$
    - $A \to B \equiv \neg B \to \neg A$
    - $A \wedge (B \equiv C) \equiv A \wedge B \equiv A \wedge C$
    - $A \vee (B \equiv C) \equiv A \vee B \equiv A \vee C$

11. Using truth tables or truth table shortcuts, determine the validity of the following. Show all your work. Again, a schema is *not* a tautological implication iff some instance of it is not.

    - $p \models_{\text{taut}} p \wedge q$
    - $A, B \models_{\text{taut}} A \wedge B$
    - $A, A \to B \models_{\text{taut}} B$
    - $B, A \to B \models_{\text{taut}} A$
    - $p \wedge q \models_{\text{taut}} p$

12. Use a truth table, or a shortcut of one, to show that

$$\models_{\text{taut}} \left(A \wedge B \wedge C \to D\right) \equiv \left(A \to (B \to (C \to D))\right)$$

13. Use a truth table, or a shortcut of one, to show that

$$C, A \to (B \equiv C) \models_{\text{taut}} A \to B$$

14. Which of the following sets is satisfiable?

    - $\{A, A \to B, A \to \neg B\}$
    - $\{A \vee B, \neg A \vee C, \neg B, \neg C\}$

- $\{A \vee B, \neg A \vee C, B \vee C\}$

15. Calculate the following (show/explain all work!).

    NB. The first bullet below must be done using Definition 1.3.15, step by step. For the rest you are free to rely on the intuitive definition of substitution/replacement. Some of the replacements I ask you to do may be illegal. If so, explain *precisely* *why* they are illegal and don't do them!

     Review priorities!

    - $p \vee (q \to p)[p := r]$
    - $(p \vee q)[p := \mathbf{t}]$
    - $(p \vee q)[p := \top]$
    - $p \vee q \wedge r[q := A]$ (where A is some formula, we don't care which)
    - $p \vee (q \wedge r)[q := A]$ (where A is some formula, we don't care which)

16. If $A \models_{\text{taut}} B$ and also $B \models_{\text{taut}} A$, then we say that A and B are *tautologically equivalent*. Prove that every formula is tautologically equivalent to one that contains no constants ($\bot, \top$) and moreover, the only connectives in it are $\neg$ and $\vee$.

17. Prove that every formula is tautologically equivalent to one that contains no constants ($\bot, \top$) and moreover, the only connectives in it are $\neg$ and $\wedge$.

18. Prove that every formula is tautologically equivalent to one that does not contain the constant $\top$ and moreover, the only connective in it is $\to$.

19. Let us introduce a new Boolean connective $\downarrow$ by "$A \downarrow B$ means $\neg(A \vee B)$". Prove that every formula is tautologically equivalent to one that contains no constants ($\bot, \top$) and moreover, the only connective in it is $\downarrow$.

20. Let us introduce a new Boolean connective $\uparrow$ by "$A \uparrow B$ means $\neg(A \wedge B)$". Prove that every formula is tautologically equivalent to one that contains no constants ($\bot, \top$) and moreover, the only connective in it is $\uparrow$.

CHAPTER 2

THEOREMS AND METATHEOREMS

2.1 MORE HILBERT-STYLE PROOFS

Before we begin. A word on the use of the headings "theorem" and "metatheorem". In what follows we will prove several theorems and a few metatheorems. Some of the theorems will be absolute (no assumptions beyond logical axioms used) and some will be relative. These will be tersely stated with headings like "theorem" and the text of the result will have the format "⊢ *A*" or "Γ ⊢ *A*" respectively. The *theorem* in each case, announced by the heading "theorem", is just "*A*" (cf. the discussion in Remark 1.4.10); thus the heading is purposely abusing terminology—but this conforms with normal practice.

On occasion we will establish statements of the form "if Γ ⊢ *A*, then also Σ ⊢ *B*". This type of result will be a metatheorem that will appear with such a heading.

2.1.1 Metatheorem. (Hypothesis Strengthening) *If* Γ ⊢ *A and* Γ ⊆ Δ, *then also* Δ ⊢ *A*.

Note. Γ ⊆ Δ means that every formula of Γ also occurs inside Δ.

Proof. Any Γ-proof of A is also a Δ-proof, for whenever the legitimacy of writing down a formula B in the proof is by virtue of B being in Γ, then this precise step would also be legitimate if we were composing a Δ-proof, since B is also Δ. The other two legitimate reasons for writing a formula down are independent of our choice of assumptions (cf. 1.4.5). $\square$

2.1.2 Remark. In particular, if $\vdash A$, then also $\Gamma \vdash A$ for any set of formulae Γ. This is because $\emptyset \subseteq \Gamma$ vacuously.[52] $\square$

2.1.3 Exercise. Prove the content of the above remark not as a corollary of 2.1.1 but directly from 1.4.7 or 1.4.9. $\square$

2.1.4 Metatheorem. (Transitivity of $\vdash$) *Suppose that we have* $\Gamma \vdash B_1, \Gamma \vdash B_2, \ldots,$ $\Gamma \vdash B_n$. *Suppose, moreover, that we also have* $B_1, \ldots, B_n \vdash A$. *Then* $\Gamma \vdash A$.

Proof. By assumption we have Γ-proofs (cf. 1.4.8)

$$\boxed{\ldots, B_1}^{53} \tag{1}$$

$$\boxed{\ldots, B_2} \tag{2}$$

$$\vdots$$

$$\boxed{\ldots, B_n} \tag{n}$$

We also have a $B_1, \ldots, B_n$-proof

$$\boxed{\ldots, A} \tag{$n+1$}$$

We now concatenate all proofs (1)–(n) (in any order will be fine) and append to the end of the result proof $(n+1)$ to form the sequence

$$\boxed{\ldots, B_1}, \boxed{\ldots, B_2}, \ldots, \boxed{\ldots, B_i}, \ldots, \boxed{\ldots, B_n}, \boxed{\ldots, A} \tag{$*$}$$

In $(*)$ every formula C satisfies either Case 1 or Case 2:

Case1: Is in a sequence among (1)–(n). Thus C is either written outright (because it is in $\Gamma \cup \Lambda$) or because it follows from a previous formula, via Eqn or Leib. This "previous" consideration is localized in the sequence $((i), i = 1, \ldots, n)$ where the formula belongs.

Case2: Is in the sequence $(n+1)$. Thus C is either written outright (because it is in Λ or is one of the B_i) or because it follows from a previous formula,

---

[52] I trust that you cannot think of any member of $\emptyset$ that is not in Γ.
[53] The boxes are inserted to improve readability.

localized in the sequence $(n+1)$, via Eqn or Leib. We would like to say that $(*)$ is a Γ-proof and rest the case. The only part above, while checking the sequence for legitimacy, that may bother us momentarily is the underlined part in **Case 2** above: Writing B_i outright is fine for a $B_1, \ldots, B_n$-proof but may not be for a Γ-proof since we have no guarantee that B_i is *in* Γ.

No problem: Any B_i that is written down in the $\boxed{\ldots, A}$ (last) segment of the sequence $(*)$ *is* legitimate in the Γ-proof context. Indeed, it was already legitimized as the last formula of the $\boxed{\ldots, B_i}$ segment. So $(*)$ *is a Γ-proof*, and thus A is a Γ-theorem. $\square$

 2.1.5 Remark. The above metatheorem makes derived rules of inference usable. While primary rules, *by definition (1.4.5)*, are applicable to *any formulae* that appear in a proof, it was not *a priori* clear that this applies to derived rules such as "$A \equiv B, B \equiv C \vdash A \equiv C$". Now it is. Any proof that looks like

$$\ldots, A \equiv B, \ldots, B \equiv C, \ldots$$

or like

$$\ldots, B \equiv C, \ldots, A \equiv B, \ldots$$

can be, if it fits our purposes, continued as

$$\ldots, A \equiv B, \ldots, B \equiv C, \ldots, A \equiv C$$

or

$$\ldots, B \equiv C, \ldots, A \equiv B, \ldots, A \equiv C$$

respectively. $\square$

Metatheorem 2.1.4 has a very important corollary:

 2.1.6 Corollary. *If* $\Gamma \cup \{A\} \vdash B$ *and also* $\Gamma \vdash A$, *then* $\Gamma \vdash B$.

The notation $\Gamma \cup \{A\}$ was introduced in 1.3.14(3). A shorter form of it often used in writings in logic is "$\Gamma + A$".

Proof. The proof of B from $\Gamma \cup \{A\}$ utilizes, besides A, only a finite number of formulae from Γ—because every proof has a finite number of steps, so it can use only finitely many formulae. Say the formulae used from Γ are

$$C_1, C_2, C_3, \ldots, C_n$$

Thus the proof of B is from $\{C_1, C_2, C_3, \ldots, C_n, A\}$, that is,

$$C_1, C_2, C_3, \ldots, C_n, A \vdash B \tag{1}$$

Since for any D in Γ we have $\Gamma \vdash D$ (cf. 1.4.11), it follows that

$$\Gamma \vdash C_1, \Gamma \vdash C_2, \Gamma \vdash C_3, \ldots, \Gamma \vdash C_n \tag{2}$$

Statements (1), (2), and the given $\Gamma \vdash A$ jointly satisfy the hypotheses of 2.1.4. Thus we have at once that $\Gamma \vdash B$.   □

2.1.7 Corollary. *If* $\Gamma \cup \{A\} \vdash B$ *and also* $\vdash A$, *then* $\Gamma \vdash B$.

Proof. By 2.1.1, the hypothesis $\vdash A$ can be replaced by $\Gamma \vdash A$. Corollary 2.1.6 concludes the argument.   □

Corollary 2.1.6 essentially says that in a proof (from Γ) we are allowed to write down, not only (1)–(3), that is, (1) any axiom, (2) any member of Γ, (3) any result of an inference rule applied to already-written-down formulae, *but also* we may write down (4) any Γ-theorem (like A above).

In other words, the corollary justifies the legitimacy of *quoting* or *using* in proofs already-proved theorems *without having to prove them all over again every time we need to use them!*

2.1.8 Exercise. Show that $\Gamma \cup \{A\} \vdash B$ and also $\Delta \vdash A$, then $\Gamma \cup \Delta \vdash B$.   □

All the preceding metatheorems in this section, and the exercise above, are independent of the choice of logical axioms and rules of inference as is clear from their proofs.

We next turn to theorems and metatheorems relating to "$\equiv$".

2.1.9 Theorem. $\vdash (A \equiv (B \equiv C)) \equiv ((A \equiv B) \equiv C)$

Note. This is the mirror image of axiom schema (1).
Proof.

$$(1) \quad ((A \equiv B) \equiv C) \equiv (A \equiv (B \equiv C)) \qquad \langle\text{axiom}\rangle$$

$$(2) \quad \Big(((A \equiv B) \equiv C) \equiv (A \equiv (B \equiv C))\Big)$$
$$\equiv \Big((A \equiv (B \equiv C)) \equiv ((A \equiv B) \equiv C)\Big) \ \langle\text{axiom}\rangle$$

$$(3) \quad (A \equiv (B \equiv C)) \equiv ((A \equiv B) \equiv C) \qquad \langle(1,2)+\text{Eqn}\rangle \ \square$$

2.1.10 Remark. The above theorem/axiom (1) pair allows us to insert brackets in any way we please in a chain of $\equiv$ signs, or not insert them at all,[54] since it does not matter one way or another.

If you believed, as you should, what I have just said above, then you may skip this part. Strictly speaking, the pair of 2.1.9 and axiom (1) deal with a chain of *two* $\equiv$ signs. The general case can be dealt with by (strong) induction on the length (i.e.,

---

[54]Recall that "$A \equiv B \equiv C$" is the least-parenthesized notation for "$(A \equiv (B \equiv C))$" according to 1.1.11.

number of A_i) in the chain. The basis ($n = 3$) being settled as noted, we turn to the case of $n > 3$, where we want to show that

$$\vdash A_1 \equiv A_2 \equiv \cdots \equiv A_n \equiv \left(A_1, A_2, \ldots, A_n \right) \tag{0}$$

The notation "$\left(A_1, A_2, \ldots, A_n \right)$" indicates, *solely for the purposes of the exposition here*, an $\equiv$-chain of A_i *in the order given*, where brackets (other than the indicated outermost pair) are arbitrarily placed (we do not care how). Of course, the left-hand side implies that brackets *are* present and inserted from right to left (cf. 1.1.11). We have two cases pertaining to the right-hand side of (0), and these depend on where the *last* $\equiv$ was inserted (*last* in a formula-calculation of $\left(A_1, A_2, \ldots, A_n \right)$, that is):

Case 1:[55] The right-hand side of (0) is (omitting outermost brackets)

$$A_1 \equiv D \tag{1}$$

where D denotes the $\equiv$-chain $\left(A_2, \ldots, A_n \right)$. By the I.H.

$$\vdash A_2 \equiv \cdots \equiv A_n \equiv D \tag{2}$$

By 1.4.9, using the above in an application of Leibniz with "denominator" $A_1 \equiv \mathbf{p}$ ($\mathbf{p}$ fresh) I obtain

$$\vdash \left(A_1 \equiv \left(A_2 \equiv \cdots \equiv A_n \right) \right) \equiv (A_1 \equiv D)$$

which, by right associativity of $\equiv$, is (0).

Case 2:[56] The right-hand side of (0) is (omitting outermost brackets)

$$\left(A_1, \ldots, A_k \right) \equiv E \tag{3}$$

where $k > 1$ and E denotes $\left(A_{k+1}, \ldots, A_n \right)$. By the comment that led to the two cases, E is nonempty, i.e., $k < n$. By the I.H.

$$\vdash A_1 \equiv (A_2 \equiv \cdots \equiv A_k) \equiv \left(A_1, \ldots, A_k \right)^{57} \tag{4}$$

By 1.4.9, an application of Leibniz with "denominator" $\mathbf{p} \equiv E$ ($\mathbf{p}$ fresh) yields from ("numerator") (4):

$$\vdash \left((A_1 \equiv (A_2 \equiv \cdots \equiv A_k)) \equiv E \right) \equiv \left((A_1, \ldots, A_k) \equiv E \right) \tag{5}$$

By Theorem 2.1.9,

$$\vdash \left(A_1 \equiv ((A_2 \equiv \cdots \equiv A_k) \equiv E) \right) \equiv \left((A_1 \equiv (A_2 \equiv \cdots \equiv A_k)) \equiv E \right) \tag{6}$$

---

[55]The last inserted $\equiv$ is the leftmost.

[56]The last inserted $\equiv$ is *not* the leftmost.

[57]Same as "$\vdash A_1 \equiv \cdots \equiv A_k \equiv (A_1, \ldots, A_k)$". The brackets around $A_2 \equiv \cdots \equiv A_k$ are inserted only for emphasis; cf. 1.1.11.

Remembering 2.1.5, and using 1.4.9 with an application of the derived rule $(Trans)$ (p. 47), (5) and (6) yield

$$\vdash \Big(A_1 \equiv ((A_2 \equiv \cdots \equiv A_k) \equiv E)\Big) \equiv \Big((A_1, \ldots, A_k) \equiv E\Big) \qquad (7)$$

By **Case 1**,

$$\vdash A_1 \equiv \cdots \equiv A_k \cdots \equiv A_{n-1} \equiv A_n \equiv \Big(A_1 \equiv ((\cdots \equiv A_k) \equiv E)\Big) \qquad (8)$$

thus, by an application of 1.4.9 and $(Trans)$ on (7) and (8), we get (0).

Thus, in a chain of any number of $\equiv$ signs, we insert brackets merely to visually suggest what we have in mind, but for no other reason, as they have been shown to be redundant. □

2.1.11 Theorem. (The Other Equanimity) $B, A \equiv B \vdash A$

Proof.

(1)	B	⟨hypothesis⟩
(2)	$A \equiv B$	⟨hypothesis⟩
(3)	$(A \equiv B) \equiv (B \equiv A)$	⟨axiom⟩
(4)	$B \equiv A$	⟨(2, 3) + Eqn⟩
(5)	A	⟨(1, 4) + Eqn⟩ □

The original primary "Eqn" says that if I assume an equivalence and the left formula of the equivalence, then I can conclude the right. This derived rule says that if I assume an equivalence and the right formula of the equivalence, then I can conclude the left.

*In all that follows we will call "Eqn" either the primary (**Inf2**) or the derived one without notice.*

2.1.12 Theorem. $\vdash A \equiv A$

Proof. We have already proved this schema in 1.4.13. We have stated it here again because it is important. □

2.1.13 Exercise. We proved the above using Leib and the trick that $A[\mathbf{p} := B]$ expands to A if $\mathbf{p}$ does not occur in A. This time, prove the result without the trick, but be careful not to introduce any circularities in your proof! □

2.1.14 Corollary. $\vdash \bot \equiv \bot$

Proof. This is a specific instance of the theorem schema above. It is one of our very few examples of "theorems" as opposed to (the majority that are) "theorem schemata". □

2.1.15 Corollary. $\vdash \top$

Proof.

$$
\begin{array}{lll}
(1) & \top \equiv \bot \equiv \bot & \langle \text{axiom} \rangle \\
(2) & \bot \equiv \bot & \langle \text{absolute theorem, cf. 2.1.6} \rangle \\
(3) & \top & \langle (1) \text{ and } (2) \text{ and Eqn} \rangle \qquad \square
\end{array}
$$

Worth Repeating: (Γ-) theorems can be inserted in any (Γ-) proof just like axioms are. This is due to Corollary 2.1.6 and has been employed above. In the above case $\Gamma = \emptyset$.

2.1.16 Theorem. (Eqn + Leib Merged) $C[\mathbf{p} := A], A \equiv B \vdash C[\mathbf{p} := B]$

Proof.

$$
\begin{array}{lll}
(1) & C[\mathbf{p} := A] & \langle \text{hypothesis} \rangle \\
(2) & A \equiv B & \langle \text{hypothesis} \rangle \\
(3) & C[\mathbf{p} := A] \equiv C[\mathbf{p} := B] & \langle (2) + \text{Leib} \rangle \\
(4) & C[\mathbf{p} := B] & \langle (1, 3) + \text{Eqn} \rangle \qquad \square
\end{array}
$$

In our annotations we will call the above derived rule, whenever used, "Eqn/Leib" or "Leib/Eqn".

2.1.17 Remark. (Equivalent Logics) A Boolean logic is determined by its language, its logical axioms and its rules of inference. Fixing the language, two different choices of the remaining tools lead to two "different logics" as we say. These are *equivalent* if and only if they have exactly the same theorems. That is, one can do with the first set of tools precisely what one can do with the second.

Compare with programming languages: There is a theorem (one can encounter this theorem in a variety of courses, including possibly data structures, or theory of computation) that the pair of instructions "**if** ... **then** ... **else** ..." and "**goto** ..." have exactly the same power as the pair "**if** ... **then** ... **else** ..." and "**while** ... **do** ...". That is, whatever one can do (i.e., program) using one, one can do using the other.

How does one prove such results? By *simulation*. One proves that each set can simulate or "implement" the other.

Back to logic, we just saw that Eqn and Leib can simulate "Eqn/Leib", so whatever we can do (i.e., prove) using the Eqn/Leib we can also prove using Eqn and Leib—i.e., our original logic.

It turns out that, conversely, Eqn/Leib can simulate each of Eqn and Leib. Thus if we were to make an about face and drop **Inf1** and **Inf2** and adopt instead Eqn/Leib as our only primary rule, keeping the same logical axioms, we would get precisely the same theorems in the new logic as in our current logic. The new logic would be *equivalent* to the original ([32] uses Eqn/Leib as primary).

While we will not do that, and will continue as planned with **Inf1** and **Inf2** as *the* primary rules—where Eqn/Leib is just a derived rule—it is nevertheless instructive to see the truth of my claim.

(I) **Simulation of Eqn by Eqn/Leib:**

(1) A ⟨hypothesis⟩
(2) $A \equiv B$ ⟨hypothesis⟩
(3) B ⟨(1, 2) + Eqn/Leib: A is $p[p := A]$ and B is $p[p := B]$⟩

(II) **Simulation of Leib by Eqn/Leib:** *Preparatory discussion*: We want to (meta)prove $A \equiv B \vdash C[\mathbf{p} := A] \equiv C[\mathbf{p} := B]$ using the axioms, but no rule other than Eqn/Leib. So we pick arbitrary A, B, C and $\mathbf{p}$. Let us also pick a $\mathbf{q}$ that is not $\mathbf{p}$ and does not occur in any of A, C. Clearly, the substitution $C[\mathbf{p} := A]$ has the same result as $C[\mathbf{p} := \mathbf{q}][\mathbf{q} := A]$.[58] That is, first change $\mathbf{p}$ into $\mathbf{q}$ everywhere in C and, *after that*, change $\mathbf{q}$ everywhere in the resulting formula into A. Similarly, the substitution $C[\mathbf{p} := B]$ has the same result as $C[\mathbf{p} := \mathbf{q}][\mathbf{q} := B]$.

(1) $A \equiv B$ ⟨hypothesis⟩
(2) $C[\mathbf{p} := A] \equiv C[\mathbf{p} := \mathbf{q}][\mathbf{q} := A]$ ⟨theorem (2.1.12); cf. discussion above⟩
(3) $C[\mathbf{p} := A] \equiv C[\mathbf{p} := \mathbf{q}][\mathbf{q} := B]$ ⟨(1, 2) + Eqn/Leib⟩

In view of the preparatory remarks, the formula in line (3) is $C[\mathbf{p} := A] \equiv C[\mathbf{p} := B]$ as needed, and line (2) is $\big(C[\mathbf{p} := A] \equiv C[\mathbf{p} := \mathbf{q}]\big)[\mathbf{q} := A]$ while line (3) is $\big(C[\mathbf{p} := A] \equiv C[\mathbf{p} := \mathbf{q}]\big)[\mathbf{q} := B]$ as required for the proper application of Eqn/Leib. By this I mean the requirement that "$[\mathbf{p} := A]$" and "$[\mathbf{p} := B]$" each apply to the *same entire formula*, not to some part thereof. Note that the assumption on $\mathbf{q}$ and A, C guarantees that $\mathbf{q}$ does not occur in $C[\mathbf{p} := A]$ and therefore $C[\mathbf{p} := A][\mathbf{q} := A]$ and $C[\mathbf{p} := A][\mathbf{q} := B]$ each give the same result as $C[\mathbf{p} := A]$.

Fatal Error! In the proof above, 2.1.12 was used to metaprove Leib (**Inf1**), yet Leib was used to prove $\vdash A \equiv A$ (cf. (d) on p. 47). So the above is an invalid (circular) proof!

Not really. It is *not* a circular proof because $\vdash A \equiv A$ is directly provable from Eqn/Leib and the given axioms! (I just wanted to tease you a bit.) Here's how:

(1) $A \lor A \equiv A$ ⟨axiom⟩
(2) $A \lor A \equiv A$ ⟨axiom⟩
(3) $A \equiv A$ ⟨Eqn/Leib + (1,2): The "C-part" is $\mathbf{p} \equiv A$—fresh $\mathbf{p}$⟩   □

2.1.18 Exercise. Prove by induction on the complexity of C that if $\mathbf{q}$ is not $\mathbf{p}$, nor does it occur in C, then for any formula A, $C[\mathbf{p} := A]$ and $C[\mathbf{p} := \mathbf{q}][\mathbf{q} := A]$ have the same result.   □

We conclude the section with some simple but important results.

2.1.19 Theorem. $\vdash A \equiv A \equiv B \equiv B$

Note. By earlier remarks that hinge on the associativity of "$\equiv$" (cf. axiom (1) and Theorem 2.1.9 as well as Remark 2.1.10), the above can be read in various ways: $\vdash (A \equiv A) \equiv (B \equiv B)$, $\vdash A \equiv (A \equiv (B \equiv B))$, $\vdash ((A \equiv A) \equiv B) \equiv B$, $\vdash A \equiv (A \equiv B) \equiv B$, etc. In the end, one will read it in the most convenient (goal-driven) way.
Proof. Brackets below are inserted for clarity so as to drive our argument:

$$(1) \quad (A \equiv B \equiv B) \equiv A \qquad\qquad\qquad\qquad \langle\text{axiom}\rangle$$

$$(2) \quad \Big((A \equiv B \equiv B) \equiv A\Big) \equiv \Big(A \equiv (A \equiv B \equiv B)\Big) \ \langle\text{the same axiom}\rangle$$

$$(3) \quad A \equiv (A \equiv B \equiv B) \qquad\qquad\qquad\qquad \langle (1) + \text{Eqn}\rangle \qquad □$$

2.1.20 Corollary. $\vdash \bot \equiv \bot \equiv B \equiv B$ *and* $\vdash A \equiv A \equiv \bot \equiv \bot$

Proof. Directly from the previous theorem schema, the first time making A specific (namely, $\bot$), the second time making B specific.   □

Of course, there is nothing in a name, and the corollary can be reformulated as "$\vdash \bot \equiv \bot \equiv A \equiv A$ and $\vdash A \equiv A \equiv \bot \equiv \bot$".

2.1.21 Corollary. (Redundant True) $\vdash \top \equiv A \equiv A$ *and* $\vdash A \equiv A \equiv \top$

Proof.

$$(1) \quad \top \equiv \bot \equiv \bot \qquad \langle\text{axiom}\rangle$$

$$(2) \quad \bot \equiv \bot \equiv A \equiv A \ \langle\text{abs. theorem}\rangle$$

$$(3) \quad \top \equiv A \equiv A \qquad \langle\text{Trans} + (1, 2)\rangle$$

As for the other one,

$$(1) \quad \top \equiv A \equiv A \qquad\qquad\qquad \langle\text{abs. theorem above}\rangle$$

$$(2) \quad \Big(\top \equiv A \equiv A\Big) \equiv \Big(A \equiv A \equiv \top\Big) \ \langle\text{axiom}\rangle$$

$$(3) \quad A \equiv A \equiv \top \qquad\qquad\qquad \langle\text{Eqn} + (1, 2)\rangle \qquad □$$

2.1.22 Remark. The import of "redundant true" is mostly felt in equational proofs that we will introduce in the next section. These proofs exploit the rule Leibniz in a process of "replacing equivalents by equivalents". Thus if we view our theorem schema "$A \equiv A \equiv \top$" as "$A \equiv (A \equiv \top)$" we see that—in terms of replacing equivalents by equivalents—A is as good as $A \equiv \top$. Thus in an expression such as "$A \equiv \top$" the part "$\equiv \top$" can be eliminated; it is "redundant" and its elimination simplifies the expression we are trying to prove. Conversely, it is often convenient to introduce "$\equiv \top$" (or "$\top \equiv$") replacing A by $A \equiv \top$ (or $\top \equiv A$). □

We state two easy (but again, important) metatheorems that flow directly from "redundant true" (2.1.21):

2.1.23 Metatheorem. *For any Γ and A, $\Gamma \vdash A$ iff $\Gamma \vdash A \equiv \top$.*

Proof. Only if: From $\vdash A \equiv A \equiv \top$ we get $\Gamma \vdash A \equiv A \equiv \top$ by hypothesis strengthening (2.1.1). Then, from $\Gamma \vdash A$, $\Gamma \vdash A \equiv A \equiv \top$, and $\{A, A \equiv A \equiv \top\} \vdash A \equiv \top$ (Eqn) we get $\Gamma \vdash A \equiv \top$ by transitivity of $\vdash$ (2.1.4).

If: From $\Gamma \vdash A \equiv \top$, $\Gamma \vdash A \equiv A \equiv \top$, and $\{A \equiv \top, A \equiv A \equiv \top\} \vdash A$ (Eqn) we get $\Gamma \vdash A$ by transitivity of $\vdash$ (2.1.4). □

2.1.24 Remark. The import of 2.1.23 lies within the special case $A \vdash A \equiv \top$:[59] Whenever we work with (special) assumptions, any such assumed formula A can be replaced via Leibniz by $\top$. We will see applications of this remark in the next section. □

2.1.25 Metatheorem. *For any Γ and A, B, if $\Gamma \vdash A$ and $\Gamma \vdash B$, then $\Gamma \vdash A \equiv B$.*

Proof. From $\Gamma \vdash A \equiv \top$, $\Gamma \vdash \top \equiv B$, and Trans (using 2.1.4). □

2.2 EQUATIONAL-STYLE PROOFS

In algebra and trigonometry we often prove identities[60] by calculating, systematically replacing equals for equals, thus—assuming the calculation started with a known identity—preserving equality at every step.

For this to work, we have an initial supply of identities (our "knowledge base"). The technique may be, depending on the problem, one of the following:

(1) Start with one side of "=" in "... = ···" and calculate (replacing equals for equals) until you reach the other side. These "equals for equals" are among our known supply of identities.

(2) Start with the entire "... = ···" and calculate until you reach a known identity. Then so is the original equality as it holds iff the one we reach does.

(3) Start with each side separately and calculate until you reach in both cases the same formula.

For example, to prove $1 + (\tan x)^2 = (\sec x)^2$ you work as follows, using as knowledge base the identities

$$\tan x = \frac{\sin x}{\cos x} \tag{i}$$

---

[59]Cf. 1.4.11.
[60]Recall that in algebra and trigonometry an "identity" is a formula of the type "... = ···" that is true for all values of the variables. So, $x^2 - y^2 = 0$ is *not* an identity, but $x^2 - y^2 = (x + y)(x - y)$ is.

$$\sec x = \frac{1}{\cos x} \qquad (ii)$$

$$(\sin x)^2 + (\cos x)^2 = 1 \text{ (Pythagorean Theorem)} \qquad (iii)$$

Here is our calculation. *Note the annotation!*

$$
\begin{aligned}
&1 + (\tan x)^2 \\
&= \langle \text{by } (i) \rangle \\
&\quad 1 + (\sin x / \cos x)^2 \\
&= \langle \text{arithmetic} \rangle \\
&\quad \frac{(\sin x)^2 + (\cos x)^2}{(\cos x)^2} \qquad\qquad (E) \\
&= \langle \text{by } (iii) \rangle \\
&\quad \frac{1}{(\cos x)^2} \\
&= \langle \text{by } (ii) \rangle \\
&\quad (\sec x)^2
\end{aligned}
$$

We can profitably mimic the above style of proof in logic. The presence of the Leibniz rule and the preponderance of axioms that involve "$\equiv$" make this possible, indeed easy.

In logic, the role of "$=$" is taken over by "$\equiv$". *I cannot emphasize enough that the two are entirely* different *symbols and we will* not *confuse them.*

So what is an *equational proof*? *It is a proof-layout methodology.*

(1) **What it does not do:** It does *not* supplement, amend, or replace the concept of *proof* or *theorem-calculation* of Definition 1.4.5. Nor does it do so for the concept of (Γ-) *theorem* (1.4.7). Both concepts remain the same.

Compare with algebra and trigonometry: The calculation (E) above does not define the concept of proof in these branches of mathematics, but it does provide a template for many nice proofs *within the normal framework of mathematical proof: certainly the proof (E) is acceptable as such.*

Compare also with programming. Some people using, say, the procedural language *Pascal* may choose to adopt the *structured programming* methodology and in particular never to use the **goto** instruction. Others may opt to use **goto** and follow program development by flowcharts, or indeed use a mixed approach, having their pie and eating it too: Do structured programming *with* **goto** (cf. [28]).

The programs written by the three groups, for the same problems, will look drastically different. However the existence of these groups and the two methodologies (along with the hybrid methodology) for writing programs does not detract from the fact that all use the very same programming language. Pascal programs have a formal, i.e., syntactic, definition of their structure that is independent of how people plan to use them.

The same goes for logic. Nevertheless, quite analogously, one can dogmatically stick to the Hilbert style of writing and annotating proofs (the "orthodoxy" according to Definition 1.4.5), or instead absolutely insist on writing every single proof under the sun in the equational-style.[61]

Then again, the smart user of logic will accept both styles. Such a person will judiciously choose the best tool for the task at hand each time, be it equational style or Hilbert style. The more tools we allow ourselves to use, the more effective "provers" we will be.

(2) **What it does:** In many cases it simplifies proofs by allowing a *goal-driven* approach toward the theorem.

(3) **What it is:** An equational-style proof is a sequence of (Γ-) *theorems* of the form

$$A_1 \equiv A_2, A_2 \equiv A_3, \ldots, A_{n-1} \equiv A_n, A_n \equiv A_{n+1} \tag{1}$$

Each of the individual theorems $A_i \equiv A_{i+1}$ must receive an independent, individual (Γ-) proof in order to be allowed to appear in sequence (1).

By an *independent, individual* proof I mean that this proof is *external* to the sequence, in general, and is not the result of things that we wrote to the left of $A_i \equiv A_{i+1}$ in the sequence. Sequence (1) is *not* a Hilbert-style proof!

Exactly how, and with what layout methodology, we obtained each individual proof for the various $A_i \equiv A_{i+1}$ is totally flexible: Some or all of these may have been proved by equational-style proofs. Some or all may have been proved by Hilbert-style proofs.

These individually proved results, $A_i \equiv A_{i+1}$, are from our (growing) database of theorems, which we may use in a proof like (1) just like the results (i)–(iii) were part of the database of independently obtained trigonometry facts that were used in our proof (E) on p. 61.

Pause. For one last time: A theorem is a theorem is a theorem, as per 1.4.7 *regardless of how, in Hilbert style or equationally, it was proved.*

Before we take a careful look at the layout of equational proofs, *which is extremely important*, let us take out of the way the "metatheory" part.

So, what does a sequence of equivalences like the above do for us?

The answer is provided by the following metatheorem. By the way, our relaxed terminology *theorem vs. metatheorem*, agreed to on p. 51, would have us label special cases like $A \equiv B, B \equiv C, C \equiv D \vdash A \equiv D$ "theorems". What makes the following definitely a "metatheorem" is its dependence on n.

2.2.1 Metatheorem.

$$A_1 \equiv A_2, A_2 \equiv A_3, \ldots, A_{n-1} \equiv A_n, A_n \equiv A_{n+1} \vdash A_1 \equiv A_{n+1} \tag{2}$$

---

[61]This can lead to some ridiculously long, mathematically ugly proofs of trivial results.

Proof. This is seen by repeating the (derived) rule "Trans". A rigorous proof is by induction on n:

Basis. For $n = 1$ we want $A_1 \equiv A_2 \vdash A_1 \equiv A_2$, which we got by 1.4.11.

Taking as I.H. the claim for n (it looks precisely as in (2) above) we establish the claim for $n + 1$. That is, we want:

$$A_1 \equiv A_2, A_2 \equiv A_3, \ldots, A_{n-1} \equiv A_n, A_n \equiv A_{n+1}, A_{n+1} \equiv A_{n+2} \vdash A_1 \equiv A_{n+2}$$
$$(3)$$

Here goes a proof of (3):

$$
\begin{array}{lll}
(1) & A_1 \equiv A_2 & \langle\text{hypothesis}\rangle \\
(2) & A_2 \equiv A_3 & \langle\text{hypothesis}\rangle \\
& \vdots & \\
(n) & A_n \equiv A_{n+1} & \langle\text{hypothesis}\rangle \\
(n+1) & A_1 \equiv A_{n+1} & \langle(1)\text{--}(n) + \text{I.H.}\rangle \\
(n+2) & A_{n+1} \equiv A_{n+2} & \langle\text{hypothesis}\rangle \\
(n+3) & A_1 \equiv A_{n+2} & \langle(n+1) + (n+2) + \text{Trans}\rangle \qquad \square
\end{array}
$$

2.2.2 Corollary. *In an equational proof (from assumptions Γ) such as (1) on p. 62 we have $\Gamma \vdash A_1 \equiv A_{n+1}$.*

Proof. By 2.2.1 and 2.1.4. $\qquad\square$

2.2.3 Corollary. *In an equational proof (from assumptions Γ) such as (1) on p. 62 we have $\Gamma \vdash A_1$ iff $\Gamma \vdash A_{n+1}$.*

Proof. By the previous corollary and Eqn. $\qquad\square$

In practice, just as in trigonometry, if we want to prove A_1 (from Γ), then an equational proof allows us to start with A_1 and be done as soon as we end up with some known Γ-theorem A_{n+1}, as 2.2.3 above makes clear. Of course, we do *not* have to start an equational proof of A with A, but we *may* do so if this is convenient, or makes things more intuitively clear.

Corollary 2.2.2 tells us that a chain like (1) (Γ-) proves the equivalence $A_1 \equiv A_{n+1}$. Equational proofs tend to be very natural when it comes to proving equivalences, but as 2.2.3 makes clear, they can also be used to prove formulae other than equivalences (A_1 and A_{n+1} could be anything at all).

2.3 EQUATIONAL PROOF LAYOUT

To emphasize the importance of layout, we devote a section to this topic. The layout is vertical, just like that of Hilbert-style proofs, but rather than writing (1) of p. 62 as

$$A_1 \equiv A_2$$

$$A_2 \equiv A_3$$
$$\vdots$$
$$A_{n-1} \equiv A_n$$
$$A_n \equiv A_{n+1}$$

(i)

we write it in the style of (E) on p. 61, that is, *in a first approximation*,

$$A_1$$
$$\equiv \langle \text{annotation} \rangle$$
$$A_2$$
$$\equiv \langle \text{annotation} \rangle$$
$$\vdots$$
$$A_{n-1}$$
$$\equiv \langle \text{annotation} \rangle$$
$$A_n$$
$$\equiv \langle \text{annotation} \rangle$$
$$A_{n+1}$$

(ii)

Several remarks are in order:

2.3.1 Remark. (1) Going from (i) to (ii) we have economized on writing, improving readability at the same time by not repeating the "joining formulae". By *joining formulae* I mean, for each i, the A_{i+1} that occurs to the right of "$\equiv$" in $A_i \equiv A_{i+1}$ and to the left of "$\equiv$" in the *immediately following equivalence*, $A_{i+1} \equiv A_{i+2}$.

Thus, (ii) implies a *conjunctional* use of the "$\equiv$" symbol that appears in the leftmost column; that is, it is meant to say *precisely* what (i) says, i.e., $A_1 \equiv A_2$ *and* $A_2 \equiv A_3$ *and* $A_3 \equiv A_4$, etc.

In other words, layout (ii) is a compact notation that depicts layout (i) and, therefore, e.g., we can then infer the theorem $A_1 \equiv A_{n+1}$ via 2.2.1.

This is totally consistent with normal use of symbols such as "$=$" and "$<$" in mathematics. In (E) (p. 61) we are using "$=$" conjunctionally. In an algebra course, when we write the short "$a < b < c$" we mean the long "$a < b$ *and* $b < c$".

But wait a minute! "$A \equiv B \equiv C$" does *not* mean "$A \equiv B$ and $B \equiv C$". The symbol "$\equiv$" is associative according to axiom schema (1) (cf. 1.4.4, p. 42), *not* conjunctional.

Pause. So it *is* associative. But this does not preclude it from *also* being conjunctional, does it? It does! We will come back to this question (cf. 3.1.6).

For now, we must take it simply on faith: "$\equiv$" is *not* conjunctional. We get around this obstacle by inventing a *new* symbol—*in the* meta*theory, of course!*—to denote conjunctional equivalence. This is an *informal* solution just as the practice in algebra where "$a < b < c$" means "$a < b$ *and* $b < c$" is informal.[62] That is, I will give neither definitions nor axioms for the new symbol, which we will denote by "$\Leftrightarrow$". This is our "conjunctional $\equiv$" and *will appear only in equational proofs and only on their leftmost column at that*.

Thus, "$A \Leftrightarrow B \Leftrightarrow C$" means *only* "$A \equiv B$ and $B \equiv C$".

Reference [17] uses "$=$" for the conjunctional "$\equiv$". As we will use "$=$" for formal equality of non-Boolean objects in the predicate calculus part of the volume, and we *are* already using it in the metatheory as the "ordinary" equals—for example, between strings—we prefer not to overload this symbol further with yet a third meaning.

So "$\Leftrightarrow$" it is, and (ii) becomes

$$A_1$$
$$\Leftrightarrow \langle\text{annotation}\rangle$$
$$A_2$$
$$\Leftrightarrow \langle\text{annotation}\rangle$$
$$A_3$$
$$\vdots$$

(Equational Proof Layout)

$$A_{n-1}$$
$$\Leftrightarrow \langle\text{annotation}\rangle$$
$$A_n$$
$$\Leftrightarrow \langle\text{annotation}\rangle$$
$$A_{n+1}$$

(2) *Informative annotation is mandatory!* The annotation, also called *hints* in some of the literature, is expected to clearly explain *at each step*—every $A_i \equiv A_{i+1}$ is *one step*[63]— precisely why $A_i \equiv A_{i+1}$ is a (Γ-) theorem. There may be any of the following reasons:

(a) Proved earlier (with either a Hilbert-style or an equational proof).

(b) It is a logical axiom.

(c) It is an assumption (from whatever "Γ" we have in mind).

(d) We just gave a proof of $A_i \equiv A_{i+1}$ on the spot, recorded in the annotation. Often such "on-the-spot" proofs are via the Leibniz rule. In this case, we must be clear (in the annotation) of what "$A \equiv B$"-part we used (the rule's "numerator") and why we are allowed to use it:

---

[62]There is a weird (old) programming language called "Programming Language One"—you can guess how old it is from its name—for short "PL/1", which is quite likely the only general purpose programming language where "$<$" is, *unlike its use in mathematics*, associative. In said language something like $2 < 4 < 2$ evaluates to true for these reasons: (1) Absence of brackets means, *in PL/1*, that we evaluate from *left to right*. (2) Thus $2 < 4$ evaluates to true. This is represented by the Boolean value "true" in PL/1, which a programmer writes down as "1B" (one nonzero bit). (3) PL/1 handles mixed-type expressions eagerly and has elaborate rules that convert from type to type so operations are possible as far as practicable. (4) Here "1B" is converted to the number "1" so that the comparison "$1B < 2$" can go ahead. But $1 < 2$ is true.

[63]Note how I wrote "$A_i \equiv A_{i+1}$". I use $\Leftrightarrow$ *only* in the leftmost column of an equational proof.

Pause. *Unless this* $A \equiv B$ *is an axiom or a* $(\Gamma\text{-})$ *theorem (this includes (c) above) we may* not *use it. The effect of using it outside these two cases is to introduce it as a new assumption. This is normally unacceptable as it changes (augments) the set of our original assumptions!*

When using Leibniz we must be also very clear as to what the "C-part" is (cf. 1.4.2, **Inf1**) and state any special requirements that we may have put on **p**, e.g., "freshness". Annotations will normally fit on one line. If not, we should put a note-mark in them and continue the explanation outside the body of the equational proof. For Leibniz, the suggested style of annotation is

$$\text{Leib} + \left\{ \begin{array}{cc} \text{axiom} & \text{if among 1.4.4} \\ \text{hypothesis} & \text{if in } \Gamma \\ \text{theorem, by its number in this volume}^{64} & \text{otherwise} \end{array} \right\} \; ; \text{``}C\text{-part''} \dots$$

$\square$

2.4 MORE PROOFS: ENRICHING OUR TOOLBOX

2.4.1 Theorem. $\vdash \neg(A \equiv B) \equiv \neg A \equiv B$

Proof. (Equational)

$\neg(A \equiv B)$

$\Leftrightarrow \langle \text{axiom} \rangle$

$A \equiv B \equiv \bot$

$\Leftrightarrow \langle \text{Leib} + \text{axiom: } B \equiv \bot \equiv \bot \equiv B; \text{``}C\text{-part'' is } A \equiv \mathbf{p}; \mathbf{p} \text{ fresh} \rangle$

$A \equiv \bot \equiv B$

$\Leftrightarrow \langle \text{Leib} + \text{axiom: } A \equiv \bot \equiv \neg A; \text{``}C\text{-part'' is } \mathbf{p} \equiv B; \mathbf{p} \text{ fresh} \rangle$

$\neg A \equiv B$ $\qquad\qquad\qquad\qquad\qquad\qquad\qquad\qquad\qquad\qquad\qquad\square$

2.4.2 Remark. (1) Cf. the comment following our presentation of the axioms on p. 43.

(2) Note that we use the minimum of brackets necessary in proofs. In particular, note how the brackets around $A \equiv B$ were dropped as soon as they became redundant (end of first step above). Indeed, associativity (axiom (1)) is almost always at work; unmentioned. Step two above views $A \equiv B \equiv \bot$ as $A \equiv (B \equiv \bot)$.

(3) Strictly speaking, "$A \equiv \bot \equiv \neg A$" in our last annotation above is not an axiom. It is a (trivial) theorem proved via axiom schema (2) like this (starting with what we want, and ending with an axiom):

$$A \equiv \bot \equiv \neg A$$

---

[64] In homework, this is fine. In tests/exams, one should either refer to the theorem by name—if it has one—or state it explicitly.

$$\Leftrightarrow \langle\text{axiom}\rangle$$
$$\neg A \equiv A \equiv \bot$$

Its trivial nature gave us "editorial license" to call "$A \equiv \bot \equiv \neg A$" an axiom. This kind of "abuse of nomenclature"—that expresses our unconcern for permutations of terms in a $\equiv$-chain—will persist and receive definitive and general justification in Remark 2.4.8.

(4) Just as with Hilbert-style proofs we need not specify which axiom we are using in steps annotated as "axiom". This should be obvious from the axiom's form.

(5) As is often the case, the condition "**p** fresh" is an expedient overkill. For example, in the last occurrence of the condition above, I could have given the (longer) condition "**p** does not occur in B", and that would still work—I just wanted the substitutions to leave B unchanged.

(6) *Increasingly I will be omitting the condition* **p** fresh *in obvious situations such as the above.* □

A trivial adaptation of the theorem's proof yields:

2.4.3 Corollary. $\vdash \neg(A \equiv B) \equiv A \equiv \neg B$

Proof. (Equational)

$$\neg(A \equiv B)$$
$$\Leftrightarrow \langle\text{axiom}\rangle$$
$$A \equiv B \equiv \bot$$
$$\Leftrightarrow \langle\text{Leib + axiom: } B \equiv \bot \equiv \neg B; \text{``}C\text{-part'' is } A \equiv \mathbf{p}; \mathbf{p} \text{ fresh}\rangle.$$
$$A \equiv \neg B$$

□

2.4.4 Theorem. (Double Negation) $\vdash \neg\neg A \equiv A$

Proof. (Equational)

$$\neg\neg A$$
$$\Leftrightarrow \langle\text{axiom}\rangle$$
$$\neg A \equiv \bot$$
$$\Leftrightarrow \langle\text{Leib + axiom: } \neg A \equiv A \equiv \bot; \text{``}C\text{-part'' is } \mathbf{p} \equiv \bot \rangle$$
$$A \equiv \bot \equiv \bot$$
$$\Leftrightarrow \langle\text{Leib + axiom: } \top \equiv \bot \equiv \bot; \text{``}C\text{-part'' is } A \equiv \mathbf{p}\rangle$$
$$A \equiv \top$$
$$\Leftrightarrow \langle\text{redundant true, i.e., } \vdash A \equiv A \equiv \top\rangle$$
$$A$$

□

(1) *General Hint*: Always plan to start from the more complex side of "$\equiv$" and use our database of results, and our rules, to get it simpler and simpler, until you reach the other side.

(2) A general technique—in the context of ≡-chains—worth imitating is to use axioms (1) and (2) without notice, to bracket/unbracket, to move brackets around, and to rearrange the order of subformulae separated by one "≡" (that occurs at either end of the chain).[65] The general annotation here, without details, would be "axioms (1, 2) + Leib".

(3) I have also made good on the earlier promise to be less pedantic and start omitting far-too-obvious **p** fresh conditions.

2.4.5 Theorem. $\vdash \top \equiv \neg\bot$

Proof. (Equational)

$$\top$$
$$\Leftrightarrow \langle \text{axiom} \rangle$$
$$\bot \equiv \bot$$
$$\Leftrightarrow \langle \text{axiom} \rangle$$
$$\neg\bot$$ □

2.4.6 Corollary. $\vdash \bot \equiv \neg\top$

Proof. (Equational)

$$\neg\top$$
$$\Leftrightarrow \langle \text{Leib} + 2.4.5; \text{``}C\text{-part'' is } \neg\mathbf{p} \rangle$$
$$\neg\neg\bot$$
$$\Leftrightarrow \langle 2.4.4 \rangle$$
$$\bot$$ □

2.4.7 Theorem. $\vdash A \lor \top$

Proof. (Equational)

$$A \lor \top$$
$$\Leftrightarrow \langle \text{Leib} + \text{axiom: } \top \equiv \bot \equiv \bot; \text{``}C\text{-part'' is } A \lor \mathbf{p}; \text{note inserted brackets!} \rangle$$
$$A \lor (\bot \equiv \bot)$$
$$\Leftrightarrow \langle \text{axiom} \rangle$$
$$A \lor \bot \equiv A \lor \bot$$ □

One normally does *not* draw attention to the obvious and leaves it unsaid in the proof: The last line is a theorem by 2.1.12.

---

[65]We soon show, in 2.4.8, that neither the hedging "separated by one" nor "(that is at either end of the chain)" are necessary.

2.4.8 Remark. Axioms (5) and (6) are the unsung heroes in the case of $\vee$-chains just like (1) and (2) are in the case of $\equiv$-chains. To begin with, axiom (5)—just as it was the case with $\equiv$-chains and axiom (1)—allows the insertion or omission of brackets in a $\vee$-chain in any manner we please. Of course, the proof of this fact, exactly like the one in 2.1.10, hinges also on theorem (†) below, which is the mirror image of axiom (5):

$$\vdash A \vee (B \vee C) \equiv (A \vee B) \vee C \tag{†}$$

One proves the above exactly analogously with 2.1.9.

Moreover, axiom (6) along with the Leibniz rule and the associativity of $\vee$ allows you to prove that in a chain of $\vee$-signs you can swap any two subformulae, i.e., $\vdash B \vee C \vee D \equiv D \vee C \vee B$ and, more generally, $\vdash A \vee B \vee C \vee D \vee E \equiv A \vee D \vee C \vee B \vee E$. Indeed,

$\quad B \vee C \vee D$

$\Leftrightarrow$ ⟨axiom (6) via (5), the latter allowing us to put brackets where we want them⟩

$\quad D \vee B \vee C \tag{*}$

$\Leftrightarrow$ ⟨Leib + axiom (6). "C-part" is $D \vee \mathbf{p}$⟩

$\quad D \vee C \vee B$

From this we can easily prove the general case:

$\quad\quad A \vee B \vee C \vee D \vee E$

$\quad\quad \Leftrightarrow$ ⟨Leib + special case just proved. "C-part" is $A \vee \mathbf{p} \vee E$⟩

$\quad\quad A \vee D \vee C \vee B \vee E$

Entirely similar comments hold for $\equiv$-chains (due to axioms (1) and (2)). This can be seen by repeating the above two proofs, replacing $\vee$ by $\equiv$ throughout, and replacing references to axioms (5) and (6) by references to (1) and (2). For example, $\vdash (B \equiv C \equiv D) \equiv (D \equiv C \equiv B)$ is proved by rephrasing (*) above:

$\quad B \equiv C \equiv D$

$\Leftrightarrow$ ⟨axiom (2) via (1), the latter allowing us to put brackets where we want them⟩

$\quad D \equiv B \equiv C$

$\Leftrightarrow$ ⟨Leib + axiom (2). "C-part" is $D \equiv \mathbf{p}$⟩

$\quad D \equiv C \equiv B \quad\quad\quad\quad\quad\quad\quad\quad\quad\quad\quad\quad\quad\quad\quad\quad\quad\quad \square$

The following occurs often. We might as well record it. The formulation is mindful of what we have just said in 2.4.8.

2.4.9 Proposition. $\vdash (A \equiv B) \vee (C \equiv D) \equiv A \vee C \equiv B \vee C \equiv A \vee D \equiv B \vee D$

Proof.

$\quad\quad (A \equiv B) \vee (C \equiv D)$

$\Leftrightarrow$ $\langle$axiom$\rangle$

$\quad (A \equiv B) \vee C \equiv (A \equiv B) \vee D$

$\Leftrightarrow$ $\langle$Leib + axiom; uses 2.4.8 implicitly!; "C-part" is $\mathbf{p} \equiv (A \equiv B) \vee D\rangle$

$\quad A \vee C \equiv B \vee C \equiv (A \equiv B) \vee D$

$\Leftrightarrow$ $\langle$Leib + axiom; uses 2.4.8 implicitly!; "C-part" is $A \vee C \equiv B \vee C \equiv \mathbf{p}\rangle$

$\quad A \vee C \equiv B \vee C \equiv A \vee D \equiv B \vee D$ $\qquad\qquad\qquad\qquad$ $\square$

2.4.10 Theorem. $\vdash A \vee \bot \equiv A$

Proof. (Equational) Here is a case where (after some thought) we find it convenient to deal with the entire formula.

$$A \vee \bot \equiv A$$
$\Leftrightarrow$ $\langle$Leib + axiom: $A \equiv A \vee A$; "C-part" is $A \vee \bot \equiv \mathbf{p}\rangle$
$$A \vee \bot \equiv A \vee A$$
$\Leftrightarrow$ $\langle$axiom$\rangle$
$$A \vee (\bot \equiv A)$$
$\Leftrightarrow$ $\langle$Leib + axiom: $\bot \equiv A \equiv \neg A$; "$C$-part" is $A \vee \mathbf{p}\rangle$
$$A \vee \neg A \qquad\qquad\qquad \square$$

Strictly speaking, $\bot \equiv A \equiv \neg A$ is a trivial theorem that axiom schema (4) yields via 2.4.8. Let us skip to some provable properties of $\rightarrow$:

2.4.11 Theorem. $\vdash A \rightarrow B \equiv \neg A \vee B$

Proof.

$$A \rightarrow B$$
$\Leftrightarrow$ $\langle$axiom$\rangle$
$$A \vee B \equiv B$$
$\Leftrightarrow$ $\langle$Leib + 2.4.10; "C-part" is $A \vee B \equiv \mathbf{p}\rangle$
$$A \vee B \equiv \bot \vee B$$
$\Leftrightarrow$ $\langle$axiom$\rangle$
$$(A \equiv \bot) \vee B$$
$\Leftrightarrow$ $\langle$Leib + axiom; "C-part" is $\mathbf{p} \vee B\rangle$
$$\neg A \vee B \qquad\qquad\qquad \square$$

2.4.12 Corollary. $\vdash \neg A \vee B \equiv A \vee B \equiv B$

Proof. Drop the first two lines (including the annotation) in the previous proof. $\quad\square$

2.4.13 Corollary. $\vdash A \to (B \equiv C) \equiv (A \to B \equiv A \to C)$[66]

Proof.

$$A \to (B \equiv C)$$
$$\Leftrightarrow \langle 2.4.11 \rangle$$
$$\neg A \vee (B \equiv C)$$
$$\Leftrightarrow \langle \text{axiom} \rangle$$
$$\neg A \vee B \equiv \neg A \vee C$$
$$\Leftrightarrow \langle \text{Leib} + 2.4.11\text{: ``}C\text{-part'' is } \mathbf{p} \equiv \neg A \vee C \rangle$$
$$A \to B \equiv \neg A \vee C$$
$$\Leftrightarrow \langle \text{Leib} + 2.4.11\text{: ``}C\text{-part'' is } A \to B \equiv \mathbf{p} \rangle$$
$$A \to B \equiv A \to C \qquad \qquad \square$$

2.4.14 Remark. (1) It is good form to apply one Leibniz at a time, hence the last *two* steps.

(2) [17] nickname 2.4.11 as "definition of $\to$"—even though they prove it. In the foundation of logic that we pursue in this volume 2.4.11 is, of course, *no* "definition" at all, but is a theorem that relates the connectives $\neg, \vee, \to$.

To be sure, there are *alternative foundations* of logic that employ only two *primitive* propositional connectives: $\vee$ and $\neg$. All the rest of the connectives, e.g., $\to$, are introduced by definitions such as

$$A \to B \stackrel{\text{def}}{=} \neg A \vee B \tag{1}$$

Such a definition says that the expression $A \to B$ is metatheoretical *argot*, an abbreviation of $\neg A \vee B$. This abbreviation introduces, as a side-effect, the metasymbol $\to$. But in our foundation the *formal* symbol $\to$ is a primitive of $\mathcal{V}$ and does not get (re)introduced! $\qquad \square$

Here is a real definition in our context. We introduce a new *informal* symbol—i.e., an abbreviation—named "$\not\equiv$" as follows:

2.4.15 Definition. $A \not\equiv B \stackrel{\text{def}}{=} \neg (A \equiv B)$ $\qquad \square$

How are abbreviations-by-definition used? Answer: *Expand and go!* That is, if I am asked to prove

$$\dots A \not\equiv B \dots$$

I prove instead

$$\dots \neg (A \equiv B) \dots$$

---

[66]The reader is reminded of the Boolean connectives' priorities; cf. 1.1.11.

Here's a rather surprising result:

2.4.16 Theorem. $\vdash \Big((A \not\equiv B) \not\equiv C \Big) \equiv \Big(A \not\equiv (B \not\equiv C) \Big)$

Proof. Expanding, we see that we really want to prove the formula schema

$$\neg\Big(\neg(A \equiv B) \equiv C \Big) \equiv \neg\Big(A \equiv \neg(B \equiv C) \Big)$$

$$\neg\Big(\neg(A \equiv B) \equiv C \Big)$$
$\Leftrightarrow \langle \text{axiom} \rangle$
$\quad \neg(A \equiv B) \equiv C \equiv \bot$
$\Leftrightarrow \langle \text{Leib + axiom; "C-part" is } \mathbf{p} \equiv C \equiv \bot \rangle$
$\quad A \equiv B \equiv \bot \equiv C \equiv \bot$
$\Leftrightarrow \langle \text{by Remark 2.4.8} \rangle$
$\quad A \equiv B \equiv C \equiv \bot \equiv \bot$
$\Leftrightarrow \langle \text{Leib + axiom; "C-part" is } A \equiv \mathbf{p} \equiv \bot \rangle$
$\quad A \equiv \neg(B \equiv C) \equiv \bot$
$\Leftrightarrow \langle \text{axiom} \rangle$
$\quad \neg\Big(A \equiv \neg(B \equiv C) \Big)$ $\square$

Here are some results for $\wedge$:

2.4.17 Theorem. (de Morgan 1) $\vdash A \wedge B \equiv \neg(\neg A \vee \neg B)$

Proof.   This is a lengthy, but totally straightforward calculation, nothing to write home about. Starting from the "complex" side, we have:

$\quad \neg(\neg A \vee \neg B)$
$\Leftrightarrow \langle \text{axiom} \rangle$
$\quad \neg A \vee \neg B \equiv \bot$
$\Leftrightarrow \langle \text{Leib + 2.4.12; "C-part" is } \mathbf{p} \equiv \bot \rangle$
$\quad A \vee \neg B \equiv \neg B \equiv \bot$
$\Leftrightarrow \langle \text{Leib + axiom; "C-part" is } A \vee \neg B \equiv \mathbf{p}\text{---2.4.8 used} \rangle$
$\quad A \vee \neg B \equiv B$
$\Leftrightarrow \langle \text{Leib + 2.4.12; "C-part" is } \mathbf{p} \equiv B \rangle$
$\quad A \vee B \equiv A \equiv B$
$\Leftrightarrow \langle \text{axiom---with the help of 2.4.8} \rangle$
$\quad A \wedge B$ $\square$

2.4.18 Corollary. (de Morgan 2) $\vdash A \lor B \equiv \neg(\neg A \land \neg B)$

Proof. This can be proved from scratch totally by imitating the above proof and swapping $\land$ and $\lor$. It is more instructive (and shorter) to see that it actually follows from 2.4.17.

$$\neg(\neg A \land \neg B)$$
$$\Leftrightarrow \langle \text{Leib} + 2.4.17; \text{``}C\text{-part'' is } \neg\mathbf{p}\rangle$$
$$\neg\neg(\neg\neg A \lor \neg\neg B)$$
$$\Leftrightarrow \langle 2.4.4\rangle$$
$$\neg\neg A \lor \neg\neg B$$
$$\Leftrightarrow \langle \text{Leib} + 2.4.4; \text{``}C\text{-part'' is } \mathbf{p} \lor \neg\neg B\rangle$$
$$A \lor \neg\neg B$$
$$\Leftrightarrow \langle \text{Leib} + 2.4.4; \text{``}C\text{-part'' is } A \lor \mathbf{p}\rangle$$
$$A \lor B \qquad\qquad\qquad \square$$

2.4.19 Theorem. $\vdash A \land A \equiv A$

Proof.

$$A \land A \equiv A$$
$$\Leftrightarrow \langle \text{axiom (and 2.4.8)}\rangle$$
$$A \lor A \equiv A \qquad\qquad\qquad \square$$

2.4.20 Theorem. $\vdash A \land \top \equiv A$

Proof.

$$A \land \top \equiv A$$
$$\Leftrightarrow \langle \text{axiom (and 2.4.8)}\rangle$$
$$A \lor \top \equiv \top$$
$$\Leftrightarrow \langle \text{redundant true (2.1.21)}\rangle$$
$$A \lor \top \qquad\qquad\qquad \square$$

2.4.21 Theorem. $\vdash A \land \bot \equiv \bot$

Proof.

$$A \land \bot \equiv \bot$$
$$\Leftrightarrow \langle \text{axiom (and 2.4.8)}\rangle$$
$$A \lor \bot \equiv A \qquad\qquad\qquad \square$$

2.4.22 Exercise. Prove

(1) $\vdash A \wedge (B \wedge C) \equiv (A \wedge B) \wedge C$

(2) $\vdash A \wedge B \equiv B \wedge A$

(3) State and prove for $\wedge$ the results corresponding to those proved in 2.4.8 for $\vee$ and $\equiv$. □

Distributivity of $\vee$ over $\wedge$ and of $\wedge$ over $\vee$ are major tools for calculations. Again there is nothing tricky about proving them; we just need to persevere because the calculations are long.

2.4.23 Theorem. (Distributivity: $\vee$ over $\wedge$ and $\wedge$ over $\vee$)

(i) $$\vdash A \vee B \wedge C \equiv (A \vee B) \wedge (A \vee C)$$

and

(ii) $$\vdash A \wedge (B \vee C) \equiv A \wedge B \vee A \wedge C$$

I could have written these with some redundant brackets to improve readability, but I thought it also a good opportunity to prompt the reader to review priorities (1.1.11, p. 15)

Proof. Just as is the case with the two de Morgan "laws" (2.4.17 and 2.4.18), we can prove (i) from (ii) (and conversely, we can prove (ii) from (i)). Alternatively, once one of the two is proved, a proof of the other can be extracted by systematically swapping $\vee$ and $\wedge$. Thus we only prove (i):

$$(A \vee B) \wedge (A \vee C)$$
$\Leftrightarrow$ $\langle$axiom; cf. 2.4.8 too!$\rangle$
$$A \vee B \vee A \vee C \equiv A \vee B \equiv A \vee C$$
$\Leftrightarrow$ $\langle$Leib + 2.4.8; "C-part" is $\mathbf{p} \equiv A \vee B \equiv A \vee C\rangle$
$$A \vee A \vee B \vee C \equiv A \vee B \equiv A \vee C$$
$\Leftrightarrow$ $\langle$Leib + axiom; "C-part" is $\mathbf{p} \vee B \vee C \equiv A \vee B \equiv A \vee C\rangle$
$$A \vee B \vee C \equiv A \vee B \equiv A \vee C$$

Let us now take a leaf from trig-proof methodology's book (expanding both sides and trying to show them equal) and expand $A \vee B \wedge C$:

$$A \vee B \wedge C$$
$\Leftrightarrow$ $\langle$Leib + axiom; "C-part" is $A \vee \mathbf{p}$; note inserted brackets!$\rangle$
$$A \vee (B \vee C \equiv B \equiv C)$$
$\Leftrightarrow$ $\langle$axiom; 2.4.8 remarks used to insert/remove brackets in any way we please [67]$\rangle$
$$A \vee B \vee C \equiv A \vee (B \equiv C)$$

---

[67]Here we viewed "$B \vee C \equiv B \equiv C$" as "$(B \vee C) \equiv (B \equiv C)$".

$\Leftrightarrow \langle$Leib + axiom; "C-part" is $A \vee B \vee C \equiv \mathbf{p}\rangle$
$\quad A \vee B \vee C \equiv A \vee B \equiv A \vee C$

We are done as both parts, left-hand side and right-hand side, are proved equivalent to the same formula. The reason that this constitutes a (single) equational proof is because we can write, say, the second subcalculation upside down and glue it to the end of the first subcalculation—not repeating $A \vee B \vee C \equiv A \vee B \equiv A \vee C$, of course. $\qquad \square$

2.4.24 Corollary. $\vdash A \vee B \to C \equiv (A \to C) \wedge (B \to C)$

Proof.

$$A \vee B \to C$$
$$\Leftrightarrow \langle 2.4.11 \rangle$$
$$\neg(A \vee B) \vee C$$
$$\Leftrightarrow \langle \text{Leib} + 2.4.18; \text{"}C\text{-part"}: \neg \mathbf{p} \vee C \rangle$$
$$\neg\neg(\neg A \wedge \neg B) \vee C$$
$$\Leftrightarrow \langle \text{Leib} + 2.4.4; \text{"}C\text{-part"}: \mathbf{p} \vee C \rangle$$
$$(\neg A \wedge \neg B) \vee C$$
$$\Leftrightarrow \langle 2.4.23 \rangle$$
$$(\neg A \vee C) \wedge (\neg B \vee C)$$
$$\Leftrightarrow \langle \text{obvious Leib, twice,} + 2.4.11 \rangle$$
$$(A \to C) \wedge (B \to C) \qquad \square$$

2.4.25 Corollary. $\vdash A \to B \wedge C \equiv (A \to B) \wedge (A \to C)$

Proof.

$$A \to B \wedge C$$
$$\Leftrightarrow \langle 2.4.11 \rangle$$
$$\neg A \vee B \wedge C$$
$$\Leftrightarrow \langle 2.4.23 \rangle$$
$$(\neg A \vee B) \wedge (\neg A \vee C)$$
$$\Leftrightarrow \langle \text{obvious Leib, twice,}[68] + 2.4.11 \rangle$$
$$(A \to B) \wedge (A \to C) \qquad \square$$

Here is a result connecting $\equiv$ with $\to$ and $\wedge$: Intuitively, one expects that "$A \equiv B$" is the same as "$(A \to B) \wedge (B \to A)$". Indeed, our toolbox allows us to prove that much.

---

[68]Okay: two simultaneous applications of Leibniz are allowed, if far too obvious.

2.4.26 Theorem. $\vdash A \equiv B \equiv (A \to B) \wedge (B \to A)$

Proof. Here is the routine calculation:

$$(A \to B) \wedge (B \to A)$$
$\Leftrightarrow \langle$Leib + axiom; "C-part" is $\mathbf{p} \wedge (B \to A)\rangle$
$$(A \vee B \equiv B) \wedge (B \to A)$$
$\Leftrightarrow \langle$Leib + axiom; "C-part" is $(A \vee B \equiv B) \wedge \mathbf{p}\rangle$
$$(A \vee B \equiv B) \wedge (B \vee A \equiv A)$$
$\Leftrightarrow \langle$Leib + axiom; "C-part" is $(A \vee B \equiv B) \wedge (\mathbf{p} \equiv A)\rangle$
$$(A \vee B \equiv B) \wedge (A \vee B \equiv A)$$
$\Leftrightarrow \langle$axiom (recall 2.4.8!)$\rangle$
$$(A \vee B \equiv B) \vee (A \vee B \equiv A) \equiv A \vee B \equiv A \vee B \equiv A \equiv B$$
$\Leftrightarrow \langle$Leib + (2.1.12, 2.1.23);
$$\text{"}C\text{-part": } (A \vee B \equiv B) \vee (A \vee B \equiv A) \equiv \mathbf{p} \equiv A \equiv B\rangle \qquad (1)$$
$$(A \vee B \equiv B) \vee (A \vee B \equiv A) \equiv \top \equiv A \equiv B$$
$\Leftrightarrow \langle$Leib + 2.1.21; "C-part" is $\mathbf{p} \equiv A \equiv B\rangle \qquad (2)$
$$(A \vee B \equiv B) \vee (A \vee B \equiv A) \equiv A \equiv B$$
$\Leftrightarrow \langle$Leib + 2.4.9, using 2.4.8 as needed!; "C-part" is $\mathbf{p} \equiv A \equiv B\rangle$
$$A \vee \dot{B} \vee A \vee B \equiv A \vee B \vee B \equiv A \vee A \vee B \equiv A \vee B \equiv A \equiv B$$
$\Leftrightarrow \langle$A lot of Leib and 2.4.8 + $X \vee X \equiv X$ all at once!$\rangle$
$$A \vee B \equiv A \vee B \equiv A \vee B \equiv A \vee B \equiv A \equiv B$$
$\Leftrightarrow \langle$Like step (1)$\rangle$
$$\top \equiv A \vee B \equiv A \vee B \equiv A \equiv B$$
$\Leftrightarrow \langle$Like step (2)$\rangle$
$$A \vee B \equiv A \vee B \equiv A \equiv B$$
$\Leftrightarrow \langle$Like step (1)$\rangle$
$$\top \equiv A \equiv B$$
$\Leftrightarrow \langle$Like step (2)$\rangle$
$$A \equiv B \qquad \qquad \square$$

2.5 USING SPECIAL AXIOMS IN EQUATIONAL PROOFS

At the heart of equational proofs where we have a nonempty Γ is Metatheorem 2.1.23. It states that "$\Gamma \vdash A$ iff $\Gamma \vdash A \equiv \top$", which, if $A \in \Gamma$, implies[69] $\Gamma \vdash A \equiv \top$.

---

[69]Cf. 1.4.11.

The key observation therefore is: *Under any assumptions that include the formula* A, "$A \equiv \top$" is a (*nonabsolute*) *theorem* and therefore can be used in an application of Leibniz (as the numerator of **Inf1**, 1.4.2) to replace occurrences of A by $\top$ anywhere that we find such a replacement to serve our goals. Conversely, any occurrence of $\top$ may be replaced by any hypothesis A (cf. 2.5.1 (4)).

We start with four trivial examples.

2.5.1 Example.

(1) $A, B \vdash A \wedge B$

(2) $A \vee A \vdash A$

(3) $A \vdash A \vee B$

(4) $A \wedge B \vdash A$

For (1) we calculate as follows:

$$A \wedge B$$
$$\Leftrightarrow \langle \text{Leib} + \text{assumption } B + 2.1.23; \text{``}C\text{-part''}: A \wedge \mathbf{p} \rangle$$
$$A \wedge \top$$
$$\Leftrightarrow \langle 2.4.20 \rangle$$
$$A$$

I will not normally say this, but here, for emphasis: "We are done, since A is a $\{A, B\}$-theorem" (of course, $\{A, B\}$ is our "Τ").

For (2) we calculate as follows:

$$A$$
$$\Leftrightarrow \langle \text{axiom} \rangle$$
$$A \vee A$$

For (3) we calculate as follows:

$$A \vee B$$
$$\Leftrightarrow \langle \text{Leib} + \text{assumption } A + 2.1.23; \text{``}C\text{-part''}: \mathbf{p} \vee B \rangle$$
$$\top \vee B \qquad\qquad\qquad \langle \text{cf. } 2.4.7 \rangle$$

Example (4) has the trickiest proof, but is still very short. We calculate as follows:

$$A$$
$$\Leftrightarrow \langle 2.4.20 \rangle$$
$$A \wedge \top$$

⇔ ⟨Leib + assumption $A \wedge B$ + 2.1.23; "C-part": $A \wedge$ **p**—2.4.22 is used below⟩

$A \wedge A \wedge B$

⇔ ⟨Leib + 2.4.19; "C-part": **p** $\wedge B$⟩

$A \wedge B$ □

Results (1) and (4) above lead to the extremely important (for Hilbert-style and especially resolution-style proofs) metatheorem:

2.5.2 Metatheorem. (Splitting/Merging Hypotheses) *For any formulae* A, B, C *and set* Γ, *we have* $\Gamma \cup \{A, B\} \vdash C$ *iff* $\Gamma \cup \{A \wedge B\} \vdash C$.

Proof. (Hilbert-style)

(I) Assume $\Gamma \cup \{A, B\} \vdash C$ and prove $\Gamma \cup \{A \wedge B\} \vdash C$.

(1) $\boxed{\Gamma}$ ⟨a *finite subset* of Γ^{70} *used to establish* $\Gamma \cup \{A, B\} \vdash C$⟩

(2) $A \wedge B$ ⟨hypothesis⟩

(3) A ⟨(2), and (4) from 2.5.1⟩

(4) B ⟨(2), and (4) from 2.5.1⟩

(5) C ⟨hypothesis I, using (1), (3) and (4) ⟩

‣ (II) Assume $\Gamma \cup \{A \wedge B\} \vdash C$ and prove $\Gamma \cup \{A, B\} \vdash C$.

(1) $\boxed{\Gamma}$ ⟨a *finite subset* of Γ *used to show* $\Gamma \cup \{A \wedge B\} \vdash C$⟩

(2) A ⟨hypothesis⟩

(3) B ⟨hypothesis⟩

(4) $A \wedge B$ ⟨(2, 3), and (1) from 2.5.1⟩

(5) C ⟨hypothesis II, using (1), and (4) ⟩ □

The above allows us (by a trivial induction on n) to merge any n separate hypotheses into a single one and still derive the same theorems, and conversely to split any assumption that is a conjunction of n conjuncts into n separate assumptions (the n conjuncts). That is, it generalizes to

$$\Gamma \cup \{A_1, A_2, \dots, A_n\} \vdash B \text{ iff } \Gamma \cup \{A_1 \wedge A_2 \wedge \dots \wedge A_n\} \vdash B$$

The following is an extremely important derived rule for use in Hilbert-style proofs. We will prove it equationally. It has a name: *Modus Ponens* (MP). You will note that it is a stronger version of *Eqn*, in the sense that the hypothesis is weaker—$A \rightarrow B$ rather than $A \equiv B$—therefore MP works harder, or is "smarter" than Eqn, as it concludes the same with weaker hypotheses.

[70] In Hilbert-style proofs a line must hold exactly one formula. Here, line (1) holds finitely many from Γ, but we gave them one line number, collectively. The box around Γ is in recognition that we deviated from the correct notation.

2.5.3 Theorem. (Modus Ponens) $A, A \rightarrow B \vdash B$

Proof.

$$A \rightarrow B$$
$$\Leftrightarrow \langle 2.4.11 \rangle$$
$$\neg A \vee B$$
$$\Leftrightarrow \langle \text{Leib + assumption } A + 2.1.23; \text{``}C\text{-part''}: \neg \mathbf{p} \vee B \rangle$$
$$\neg \top \vee B$$
$$\Leftrightarrow \langle \text{Leib + 2.4.6}; \text{``}C\text{-part''}: \mathbf{p} \vee B \rangle$$
$$\bot \vee B$$
$$\Leftrightarrow \langle 2.4.10 \rangle$$
$$B \qquad\qquad \square$$

A generalization of MP is also extremely important in Hilbert-style proofs, especially those that we do by the so-called *resolution* technique. It is

$$A \vee B, \neg A \vee C \vdash B \vee C$$

and originated in Gentzen's *natural deduction*-style proofs. It is called the *cut rule* since it "cuts out" a subformula that occurs both "positively" (A) and "negatively" ($\neg A$) in the two hypotheses, and then it glues what remains together, using a "$\vee$" as glue.

2.5.4 Theorem. (Cut Rule) $A \vee B, \neg A \vee C \vdash B \vee C$

Proof. We start with a subcalculation (a lemma) analyzing the most complicated hypothesis, $\neg A \vee C$:

$$\neg A \vee C$$
$$\Leftrightarrow \langle 2.4.12 \rangle$$
$$A \vee C \equiv C$$

Since $\neg A \vee C$ is a theorem *under our assumptions*, so is $A \vee C \equiv C$ and it can be used in the following proof of $B \vee C$ from our two assumptions:

$$B \vee C$$
$$\Leftrightarrow \langle \text{Leib + lemma}; \text{``}C\text{-part''}: B \vee \mathbf{p} \rangle$$
$$B \vee (A \vee C)$$
$$\Leftrightarrow \langle 2.4.8 \rangle$$
$$(A \vee B) \vee C$$
$$\Leftrightarrow \langle \text{Leib + assumption } A \vee B + 2.1.23; \text{``}C\text{-part''}: \mathbf{p} \vee C \rangle$$
$$\top \vee C \qquad\qquad \square$$

2.5.5 Corollary. $A \lor B, \neg A \lor B \vdash B$

Proof. By 2.5.4, we have $A \lor B, \neg A \lor B \vdash B \lor B$. By 2.5.1(2), we have $B \lor B \vdash B$. We are done by 2.1.4. □

2.5.6 Corollary. $A \lor B, \neg A \vdash B$

Proof. By 2.5.1(3), we have $\neg A \vdash \neg A \lor B$. We are done by 2.5.5. □

2.5.7 Corollary. $A, \neg A \vdash \bot$

Proof. By 2.5.5, where we take B to be the specific formula $\bot$, using 2.4.10. □

2.5.8 Exercise. The proofs of the three previous corollaries deviated from our usual pedantic Hilbert style. They look more like everyday proofs that a mathematician might write.

Give Hilbert-style proofs for each. □

2.5.9 Corollary. (Transitivity of $\rightarrow$) $A \rightarrow B, B \rightarrow C \vdash A \rightarrow C$

Proof. (Hilbert-style)

$$
\begin{array}{lll}
(1) & A \rightarrow B & \langle \text{assumption} \rangle \\
(2) & B \rightarrow C & \langle \text{assumption} \rangle \\
(3) & A \rightarrow B \equiv \neg A \lor B & \langle 2.4.11 \rangle \\
(4) & B \rightarrow C \equiv \neg B \lor C & \langle 2.4.11 \rangle \\
(5) & \neg A \lor B & \langle (1, 3) + \text{Eqn} \rangle \\
(6) & \neg B \lor C & \langle (2, 4) + \text{Eqn} \rangle \\
(7) & \neg A \lor C & \langle (5, 6) + 2.5.4 \ (\text{using } 2.4.8) \rangle \quad \square
\end{array}
$$

2.5.10 Theorem. $A \rightarrow C, B \rightarrow D \vdash A \lor B \rightarrow C \lor D$

Proof. As in 2.5.4, analysis of the two hypotheses yields the theorems (theorems *from* the hypotheses, that is!)

$$A \lor C \equiv C \tag{1}$$

and

$$B \lor D \equiv D \tag{2}$$

Informed by the above we calculate:

$$
\begin{array}{l}
A \lor B \rightarrow C \lor D \\
\Leftrightarrow \langle \text{axiom} + 2.4.8 \rangle \\
A \lor C \lor B \lor D \equiv C \lor D \\
\Leftrightarrow \langle \text{Leib} + (1); \text{``}C\text{-part''}: \mathbf{p} \lor B \lor D \equiv C \lor D \rangle
\end{array}
$$

$$C \vee B \vee D \equiv C \vee D$$
$$\Leftrightarrow \langle \text{Leib} + (2); \text{``}C\text{-part'': } C \vee \mathbf{p} \equiv C \vee D \rangle$$
$$C \vee D \equiv C \vee D \qquad \qquad \Box$$

2.5.11 Corollary. (Proof by Cases) $A \to C, B \to C \vdash A \vee B \to C$

Proof. By 2.5.10, $A \to C, B \to C \vdash A \vee B \to C \vee C$. Since $C \vee C \equiv C$ is an axiom, we are done via an obvious application of ($Leib$). $\Box$

2.5.12 Remark. (1) The name of 2.5.11 derives from what it says: To establish that a formula C is implied by a disjunction $A \vee B$ it is sufficient to establish separately *both cases*: $A \to C$ and $B \to C$.

(2) Actually there is more to it: By 2.5.2, the result in 2.5.11 is equivalent to

$$(A \to C) \wedge (B \to C) \vdash A \vee B \to C \qquad \qquad (*)$$

However ($*$) is a direct result of 2.4.24 via one application of equanimity. The same process also proves "the other direction":

$$A \vee B \to C \vdash (A \to C) \wedge (B \to C)$$

(3) The following special case follows trivially, since $\neg A \vee A$ is an axiom, and therefore $\vdash \neg A \vee A \equiv \top$ by 2.1.21. It follows as easily from 2.5.4. We leave the details to the reader. $\Box$

2.5.13 Corollary. $A \to C, \neg A \to C \vdash C$

2.6   THE DEDUCTION THEOREM

The main result in this section is, strictly speaking, a *Meta*theorem. But the above title is its established nickname. It states:

2.6.1 Metatheorem. (Deduction Theorem) *If* $\Gamma \cup \{A\} \vdash B$, *then also* $\Gamma \vdash A \to B$.

Note. Due to the notational convention given on p. 53, the deduction theorem can be also stated as "If $\Gamma + A \vdash B$, then also $\Gamma \vdash A \to B$". Much of the proof makes use of the following result. So let us prove it separately, to avoid obscuring the proof of the deduction theorem.

2.6.2 Lemma. $A \to (B \equiv C) \vdash A \to (D[\mathbf{p} := B] \equiv D[\mathbf{p} := C])$

Proof. This is a theorem schema, of course, and we have handled several such already. However, here we employ a new technique: We prove it not directly, but instead by induction on the complexity of D, in essence *meta*proving it, first for the least complex D, and then pushing the proof forward from less to more complex D.[71]

---

[71]I will be interested to know if you can come up with a direct proof.

That this constitutes a metaproof is clear: Our logic does not "know" induction!

Basis. D has complexity 0: So it is one of:

(1) **p**: Then we must show $A \to (B \equiv C) \vdash A \to (B \equiv C)$ and we are done by 1.4.11.

(2) **q** (other than **p**): Then we must show $A \to (B \equiv C) \vdash A \to (\mathbf{q} \equiv \mathbf{q})$. Well, start with $\vdash \mathbf{q} \equiv \mathbf{q}$ by 2.1.12. By (3) in 2.5.1, and 2.1.4, $\vdash A \to (\mathbf{q} \equiv \mathbf{q})$. We are done by 2.1.1.

(3) $\top$ or $\bot$: Same argument as in (2) above.

We now take the complex case, where D has complexity $n + 1$, on the I.H. that the claim is true for all D (or whatever else you want to call them) with complexity n or less. *Throughout the rest of the argument we will not forget that $A \to (B \equiv C)$ is an assumption.*

We have several cases of how the formula, D, of complexity $n + 1$, was built:

(i) D is $\neg E$. We calculate as follows (cf. 1.3.15):

$$A \to (\neg E[\mathbf{p} := B] \equiv \neg E[\mathbf{p} := C])$$
$$\Leftrightarrow \langle \text{Leibniz, twice along with 2.4.1} \rangle$$
$$A \to \neg\neg(E[\mathbf{p} := B] \equiv E[\mathbf{p} := C])$$
$$\Leftrightarrow \langle \text{Leibniz} + 2.4.4; \text{``}C\text{-part''}: A \to \mathbf{q} \rangle$$
$$A \to (E[\mathbf{p} := B] \equiv E[\mathbf{p} := C])$$

The last formula is an $A \to (B \equiv C)$-theorem by the I.H., so this is true for the top formula, too.

(ii) D is $E \vee G$. In view of 2.4.11, we have the (absolute) theorem

$$\Big(A \to (D[\mathbf{p} := B] \equiv D[\mathbf{p} := C])\Big) \equiv \Big(\neg A \vee (D[\mathbf{p} := B] \equiv D[\mathbf{p} := C])\Big)$$

thus we will prove $\neg A \vee (D[\mathbf{p} := B] \equiv D[\mathbf{p} := C])$ via the "$\vee$ over $\equiv$" distributivity axiom, i.e., we will prove instead

$$\neg A \vee D[\mathbf{p} := B] \equiv \neg A \vee D[\mathbf{p} := C]$$

To this end, we calculate as follows (cf. 1.3.15):

$$\neg A \vee E[\mathbf{p} := B] \vee G[\mathbf{p} := B]$$
$$\Leftrightarrow \langle \text{Leibniz} + \text{I.H.}; \text{``}C\text{-part''}: \mathbf{q} \vee G[\mathbf{p} := B]; 2.4.8 \text{ used as well} \rangle$$
$$\neg A \vee G[\mathbf{p} := B] \vee E[\mathbf{p} := C]$$
$$\Leftrightarrow \langle \text{Leibniz} + \text{I.H.}; \text{``}C\text{-part''}: \mathbf{q} \vee E[\mathbf{p} := C]; 2.4.8 \text{ used as well} \rangle$$
$$\neg A \vee E[\mathbf{p} := C] \vee G[\mathbf{p} := C]$$

(iii) D is $E \wedge G$.   We calculate as follows (cf. 1.3.15):

$$A \to E[\mathbf{p} := B] \wedge G[\mathbf{p} := B]$$
$$\Leftrightarrow \langle 2.4.25 \rangle$$
$$(A \to E[\mathbf{p} := B]) \wedge (A \to G[\mathbf{p} := B])$$
$$\Leftrightarrow \langle \text{obvious Leib twice, using the I.H.} \rangle$$
$$(A \to E[\mathbf{p} := C]) \wedge (A \to G[\mathbf{p} := C])$$
$$\Leftrightarrow \langle 2.4.25 \rangle$$
$$A \to E[\mathbf{p} := C] \wedge G[\mathbf{p} := C]$$

(iv) D is $E \to G$.   We calculate similarly to the "D is $E \vee G$" case (cf. also 1.3.15):

$$\neg A \vee \neg E[\mathbf{p} := B] \vee G[\mathbf{p} := B]$$
$$\Leftrightarrow \langle \text{Leib + I.H. on } E + \text{case (i); "}C\text{-part": } \mathbf{q} \vee G[\mathbf{p} := B]; 2.4.8 \text{ applied implicitly} \rangle$$
$$\neg A \vee G[\mathbf{p} := B] \vee \neg E[\mathbf{p} := C]$$
$$\Leftrightarrow \langle \text{Leib + I.H. on } G; \text{"}C\text{-part": } \mathbf{q} \vee \neg E[\mathbf{p} := C]; 2.4.8 \text{ applied implicitly} \rangle$$
$$\neg A \vee \neg E[\mathbf{p} := C] \vee G[\mathbf{p} := C]$$

Finally,

(v) D is $E \equiv G$.   We calculate as follows, mindful of 2.4.13 (cf. also 1.3.15):

$$A \to (E[\mathbf{p} := B] \equiv G[\mathbf{p} := B])$$
$$\Leftrightarrow \langle 2.4.13 \rangle$$
$$A \to E[\mathbf{p} := B] \equiv A \to G[\mathbf{p} := B]$$
$$\Leftrightarrow \langle \text{obvious Leib + I.H. twice} \rangle$$
$$A \to E[\mathbf{p} := C] \equiv A \to G[\mathbf{p} := C]$$
$$\Leftrightarrow \langle 2.4.13 \rangle$$
$$A \to (E[\mathbf{p} := C] \equiv G[\mathbf{p} := C]) \qquad\qquad \square$$

Proof. (**Of the Deduction Theorem**) This is a metatheorem about proofs (equivalently, about theorems). It says, essentially, that a proof of B from Γ *and* A must be somehow *transformable* into a proof of $A \to B$ *from* Γ *alone*.

So how does one (meta)prove[72] a (meta)theorem like this, that a *proof* from $\Gamma + A$ is so transformable?

By constructing the transformation by *induction on the length of proof where B occurs (1.4.5)*, of course!

---

[72]Even a casual observer will see that this is a proof *outside* Boolean logic, using such extraneous tools as induction.

Important! *A (meta)proof by induction on the length of formal proofs of our given logic is to use the formal definition of proof as in 1.4.5!* Specifically, such a metaproof will deal, in the induction step, *only* with the two primary rules of inference (1.4.2) since each use of a derived rule in a formal proof *can be eliminated* by "unwinding" (replacing the invocation of) the rule itself into a formal proof, rigorously and completely written according to 1.4.5.

Basis. For $\Gamma + A$-proofs of length $n = 1$: Such a proof consists only of B, right? But then, B must be one of the following (see 1.4.5!):

(i) A. Since $\vdash A \to A$ (this is the same as stating $\vdash \neg A \lor A$, by 2.4.11 and Eqn), we have that $\Gamma \vdash A \to B$ in this case by 2.1.1, remembering that A *is* B.

(ii) In $\Gamma \cup \Lambda$. Then (1.4.7), $\Gamma \vdash B$. Since $B \vdash \neg A \lor B$ by 2.5.1(3), we are done—i.e., once more $\Gamma \vdash A \to B$—by 2.1.4.

We take for I.H. that the claim of the metatheorem is true for all $\Gamma + A$-proofs of lengths n or less.

We now look at the induction step, the case where we have a $\Gamma + A$-proof of B, one that has length $n + 1$. Our aim is to prove that under the I.H. $\Gamma \vdash A \to B$.

So in the very last step of this proof, we either wrote down B, or we did not:

Case where we did not: Then B—which we assumed at the outset that this proof proves—must have appeared earlier. Since we can truncate a proof by dropping any length of tail (cf. 1.4.8) and still have a proof, it follows that (dropping, say, the last formula of the proof) we have a proof of B of length n or less. By the I.H. we conclude $\Gamma \vdash A \to B$ in this case.

Case where we wrote B at the very last step: Now, there are two reasons why we may have legitimately written B (1.4.5):

Subcase: B is one of: A, or is in $\Gamma \cup \Lambda$. If so, then we have already argued in the Basis that we will then have $\Gamma \vdash A \to B$.

Subcase: B was written down because we applied Eqn in the last step, so $C \equiv B$ and C have appeared in the proof *earlier*. By I.H. we have

$$\Gamma \vdash A \to C \tag{1}$$

and

$$\Gamma \vdash A \to (C \equiv B)$$

By 2.4.13 and Eqn, the latter yields

$$\Gamma \vdash A \to C \equiv A \to B \tag{2}$$

(1) and (2) yield $\Gamma \vdash A \to B$ by Eqn.

Subcase: B was written down because we applied Leibniz in the last step, so B is $D[\mathbf{p} := C] \equiv D[\mathbf{p} := E]$ and $C \equiv E$ has appeared in the proof *earlier*. By I.H.

$\Gamma \vdash A \rightarrow (C \equiv E)$. Now Lemma 2.6.2, via 2.1.4, yields $\Gamma \vdash A \rightarrow (D[\mathbf{p} := C] \equiv D[\mathbf{p} := E])$, i.e., $\Gamma \vdash A \rightarrow B$. ☐

2.6.3 Corollary. $\Gamma + A \vdash B$ *iff* $\Gamma \vdash A \rightarrow B$

Proof. The left-to-right direction ("only if") is 2.6.1. The "if" is by MP (2.5.3). ☐

2.6.4 Remark. What is the deduction theorem good for?

(1) The metatheoretician, who studies rather than uses logic, will proclaim: It shows that all the theorems that we ever need to study are absolute, for whenever we need to show $A \vdash B$ we can simply show $\vdash A \rightarrow B$, since the two *are equivalent (metatheoretical) statements* by the above corollary.

As *users*, of logic we dismiss this clinical point of view. Indeed, even in metatheoretical work the deduction theorem comes in handy—e.g., see the proof of Post's theorem (3.2.1) in the next chapter.

(2) For the user, it is exactly the opposite point of view that is important:

To prove "$\vdash A \rightarrow B$" is in general harder to do than to prove "$A \vdash B$" because the latter asks us to prove a *less complex formula* than $A \rightarrow B$—just B—and it throws in as a bonus an extra assumption—A—that is *not available* if we are proving $A \rightarrow B$. An extra assumption almost always makes life easier; for it allows us to know more before we start the proof.

The user will (almost) always prefer to tackle $A \vdash B$, over $\vdash A \rightarrow B$. Indeed, the deduction theorem is used extremely frequently by the practicing mathematician, computer scientist, logician, and any other person who reasons logically. ☐

2.6.5 Exercise. Assume the deduction theorem, and prove Lemma 2.6.2 from it. ☐

2.6.6 Metatheorem. *The following are equivalent:*

(1) $\Gamma \vdash \perp$.
(2) For all A, we have $\Gamma \vdash A$.
(3) For some B, we have $\Gamma \vdash B \wedge \neg B$.

 A Γ such as the one in the metatheorem is called *inconsistent* or *contradictory*. A Γ that does *not* have the property—e.g., it fails to prove at least one formula, which is the negation of (2)—is called *consistent*.

A formula of the form $B \wedge \neg B$ is called a *contradiction*.

Proof. We show that from (1) follows (2); from (2) follows (3); from (3) follows (1).

(1) to (2): Let A be arbitrary. From 2.5.1(3), we have $\perp \vdash \perp \vee A$. By the assumption and 2.1.4, we get $\Gamma \vdash \perp \vee A$. We are done by 2.4.10 via Eqn.

(2) to (3): Since $\Gamma \vdash A$ is valid for any A, then for any[73] B—since $B \wedge \neg B$ is a formula—we have $\Gamma \vdash B \wedge \neg B$.

---

[73] We have proved more than we were asked to: for all B, rather than for some B.

(3) to (1): Let B be chosen to satisfy (3). We argue that

$$B \wedge \neg B \vdash \bot \qquad\qquad (*)$$

With $(*)$ out of the way, we are done by 2.1.4. As for $(*)$, it follows from 2.5.2 and 2.5.7. $\qquad\square$

2.6.7 Corollary. (Proof by Contradiction) $\Gamma \vdash A$ *iff* $\Gamma + \neg A \vdash \bot$

Proof. *If*-part (right to left): By 2.6.1, the assumption yields

$$\Gamma \vdash \neg A \to \bot$$

But

$$\neg A \to \bot$$
$$\Leftrightarrow \langle 2.4.11 \rangle$$
$$\neg\neg A \vee \bot$$
$$\Leftrightarrow \langle 2.4.10 \rangle$$
$$\neg\neg A$$
$$\Leftrightarrow \langle 2.4.4 \rangle$$
$$A$$

Only if-part (left to right):

(1) $\boxed{\Gamma}$ $\langle$a finite subset of Γ, collectively numbered (1), and such that $\Gamma \vdash A\rangle$

(2) $\neg A$ $\langle$assumption$\rangle$

(3) A $\langle$from (1), since we assume $\Gamma \vdash A\rangle$

(4) $\bot$ $\langle$(2,3) and 2.5.7$\rangle$ $\qquad\qquad\square$

2.7 ADDITIONAL EXERCISES

1. Give a proof of $\vdash A \vee B \equiv A \vee \neg B \equiv A$.

2. Give a proof of $\vdash A \wedge (A \vee B) \equiv A$.

3. Give a proof of $\vdash A \vee A \wedge B \equiv A$.

4. Give a proof of $\vdash A \wedge B \vee A \wedge \neg B \equiv A$.

5. Give a proof of $\vdash A \equiv B \equiv (A \wedge B) \vee (\neg A \wedge \neg B)$.

6. Give a proof of $\vdash A \to (B \to C). \equiv (A \to B) \to (A \to C)$.

7. Give a proof of $A, B \vdash A \equiv B$.

8. Give a direct proof of $A, \neg A \vdash \bot$, not one via the cut rule.

9. Prove that $A \to B \vdash C \lor A \to C \lor B$.

10. Prove that $A \to (B \to C),\ B \vdash A \to C$.

11. Prove that $\vdash A \lor (B \to A) \equiv B \to A$.

12. Prove that $A \lor A \lor A \vdash B \to A$.

13. Prove that if two logics have the same absolute theorems, then they have the same relative theorems as well; that is, for every Γ and A, Γ proves A in one logic iff it does so in the other.

14. Prove (ii) of 2.4.23 as a consequence of (i)—i.e., using (i) as a hypothesis.

15. Prove (i) of 2.4.23 as a consequence of (ii).

16. Prove the theorem schema $X \to Y \equiv \neg Y \to \neg X$.

17. Prove the theorem schema $A \to \neg B \to \neg(A \to B)$.

18. Prove that $\vdash (A \to B) \to (\neg A \to B) \to B$.

19. Prove that $\vdash ((A \to B) \to A) \to A$.

20. Prove that $\vdash (A \to C) \to (B \to C) \to (A \lor B \to C)$.

CHAPTER 3

THE INTERPLAY BETWEEN SYNTAX AND SEMANTICS

We have promised that syntactic proofs provide us with an alternative (nondeterministic, that is, based on [educated] guessing) tool that we apply toward discovering tautological implications and, in particular, tautologies (cf. discussion at the onset of Section 1.4). We make good on this promise in this chapter, proving two metatheorems, *soundness* and *completeness* (Post's theorem) for Boolean logic. The first states that our calculus is truthful, or *sound*, as people say technically. That is, whenever ⊢ A, then also $\models_{\text{taut}} A$, or more generally, if $\Gamma \vdash A$, then also $\Gamma \models_{\text{taut}} A$ (3.1.3 below).

The second is a deeper result due to Post and provides the converse: If $\models_{\text{taut}} A$, then also ⊢ A, or more generally, if $\Gamma \models_{\text{taut}} A$, then also $\Gamma \vdash A$.

In short, there is nothing that we can establish via truth tables that we cannot do syntactically (via proofs), and vice versa.

Another way to say this is that the axioms (and rules) are *well chosen* (cf. discussion on p. 38, in particular footnote 42):

On one hand the axioms are "true" (technically, tautologies) and the rules preserve truth. This yields soundness.

On the other, the chosen axioms (and rules) are "just the right ones" to ensure that syntactic proofs are able to generate *all* tautologies (completeness).

To put it colloquially, our logic not only tells the truth (soundness), but it tells the whole truth (completeness).

3.1 SOUNDNESS

Clearly, to show that we have soundness in Boolean logic, we need to show that our rules propagate truth, and that the logical axioms are true (tautologies).

3.1.1 Lemma. *The two primary rules of inference preserve truth. That is,*

$$A, A \equiv B \models_{taut} B \tag{1}$$

and

$$A \equiv B \models_{taut} C[\mathbf{p} := A] \equiv C[\mathbf{p} := B] \tag{2}$$

Proof. The reader will want to review Definition 1.3.11.

(1) Let s be a state such that $s(A) = \mathbf{t}$ and $s(A) = s(B)$.[74] But then $s(B) = \mathbf{t}$.

(2) Let s be a state such that $s(A \equiv B) = \mathbf{t}$, i.e., $s(A) = s(B)$. We want to show that $s(C[\mathbf{p} := A]) = s(C[\mathbf{p} := B])$.

Let us view the value $s(C)$ as the result of substitution of values from the set $\{\mathbf{f}, \mathbf{t}\}$ into the variables $\mathbf{q}_i$ of a Boolean-type "function" $f(\mathbf{q}_1, \mathbf{q}_2, \ldots, \mathbf{q}_n)$, where the variables $\mathbf{q}_i$, $i = 1, \ldots, n$, are precisely those that occur in C. Without loss of generality, say $\mathbf{q}_1$ is $\mathbf{p}$. Thus, $s(C) = f(s(\mathbf{p}), s(\mathbf{q}_2), s(\mathbf{q}_3), \ldots, s(\mathbf{q}_n))$.

Therefore,

$$s(C[\mathbf{p} := A]) = f(s(A), s(\mathbf{q}_2), s(\mathbf{q}_3), \ldots, s(\mathbf{q}_n))$$

and

$$s(C[\mathbf{p} := B]) = f(s(B), s(\mathbf{q}_2), s(\mathbf{q}_3), \ldots, s(\mathbf{q}_n))$$

Using the hypothesis we get $s(C[\mathbf{p} := A]) = s(C[\mathbf{p} := B])$.   □

For the demanding reader who found the argument in (2) not rigorous enough here is a totally rigorous one, by induction on the complexity of C: We are given that $s(A \equiv B) = \mathbf{t}$, i.e., $s(A) = s(B)$.

Basis. If C is any of $\top$, $\bot$ or $\mathbf{q}$ (other than $\mathbf{p}$), then $C[\mathbf{p} := A] \equiv C[\mathbf{p} := B]$ is $C \equiv C$; hence $s(C \equiv C) = \mathbf{t}$ since $s(C) = s(C)$. If on the other hand C is $\mathbf{p}$, then $C[\mathbf{p} := A] \equiv C[\mathbf{p} := B]$ is $A \equiv B$; hence $s(A \equiv B) = \mathbf{t}$ again.

For the induction step, we pick an arbitrary nonatomic C and prove

$$s\Big(C[\mathbf{p} := A] \equiv C[\mathbf{p} := B]\Big) = \mathbf{t}$$

---

[74]I will remind the reader that, in Part I, "=" is informal equality, *outside* our logic, and it is *not* to be confused with "$\equiv$", nor with "$\Leftrightarrow$". By the way, you will recall that $s(A \equiv B) = F_{\equiv}(s(A), s(B))$, thus—cf. the truth table on p. 29—$s(A \equiv B) = \mathbf{t}$ iff $s(A) = s(B)$.

that is

$$s(C[\mathbf{p} := A]) = s(C[\mathbf{p} := B]) \tag{1}$$

on the I.H. that the claim $s(E[\mathbf{p} := A]) = s(E[\mathbf{p} := B])$ is true for all formulae less complex than C.

We have cases according to what are the i.p. of C (cf. 1.1.10).

Case1: C is $\neg D$.

> Now, $C[\mathbf{p} := A] \equiv C[\mathbf{p} := B]$ is $\neg(D[\mathbf{p} := A]) \equiv \neg(D[\mathbf{p} := B])$ (cf. 1.3.15), and therefore, by I.H.,
>
> $$s(\neg D[\mathbf{p} := A]) = s(\neg D[\mathbf{p} := B]) \text{ iff } s(D[\mathbf{p} := A]) = s(D[\mathbf{p} := B])$$

Case2: Let next C be $D \vee E$. The I.H. applies on D and E (i.p. of C).

> Now, $C[\mathbf{p} := A] \equiv C[\mathbf{p} := B]$ is (cf. 1.3.15) $D[\mathbf{p} := A] \vee E[\mathbf{p} := A] \equiv D[\mathbf{p} := B] \vee E[\mathbf{p} := B]$; hence we get (1) as follows:
>
> $$\begin{aligned} s(D[\mathbf{p} := A] \vee E[\mathbf{p} := A]) &= F_\vee\Big(s(D[\mathbf{p} := A]), s(E[\mathbf{p} := A])\Big) \\ &= F_\vee\Big(s(D[\mathbf{p} := B]), s(E[\mathbf{p} := B])\Big) \text{ (by I.H.)} \\ &= s(D[\mathbf{p} := B] \vee E[\mathbf{p} := B]) \end{aligned}$$

The cases where C is any of $D \equiv E$, $D \wedge E$ or $D \rightarrow E$ are entirely similar to the above and are omitted.

3.1.2 Exercise. Using truth tables, verify that *all* the logical axioms (1.4.4) are tautologies.   □

3.1.3 Metatheorem. (Soundness of Propositional Calculus) $\Gamma \vdash A$ *implies that* $\Gamma \models_{taut} A$.

The reader will observe that the proof below—the very first sentence in the Basis case—is oblivious to exactly what Γ we have in mind. Thus it is valid for any Γ, from empty to infinite.

Proof. We do induction on the length of Γ-proofs where A occurs.

Basis. Length $n = 1$. If A is in Γ, then certainly $\Gamma \models_{taut} A$ (cf. 1.3.11; any state that satisfies Γ will do so for A in particular). If A is in Λ, then $\models_{taut} A$ as you have verified in the exercise above. But then $\Gamma \models_{taut} A$, since again any state s that satisfies Γ will still make $s(A) = \mathbf{t}$ (*any state whatsoever will make* $s(A) = \mathbf{t}$).

We now assume the claim for lengths n or less (I.H.)

Consider now the case where A occurs in a proof of length $n + 1$. There are subcases:

(1) A is *not* the last formula in the proof. So the proof ending with A—obtained by deleting the formulae following A (cf. 1.4.8)—has length n or less. By I.H. $\Gamma \models_{\text{taut}} A$ in this case.

(2) A is the last formula.

(2.1) Subcase where $A \in \Gamma \cup \Lambda$ is handled exactly as in the Basis.

(2.2) Subcase where A was written as a result of an application of Eqn. That is, the proof contains some B and also $B \equiv A$, to the left of A. By I.H., $\Gamma \models_{\text{taut}} B$ and $\Gamma \models_{\text{taut}} B \equiv A$. Let now s be any state such that $s(X) = \mathbf{t}$ for all X in Γ. Thus, $s(B) = \mathbf{t}$ and $s(B) = s(A)$. Hence, $s(A) = \mathbf{t}$.

(2.3) Subcase where A was written as a result of an application of Leib. So $B \equiv C$ occurs to the left of A, and A is $D[\mathbf{p} := B] \equiv D[\mathbf{p} := C]$. By I.H., $\Gamma \models_{\text{taut}} B \equiv C$. Let now s be any state such that $s(X) \doteq \mathbf{t}$ for all X in Γ. Thus, $s(B) = s(C)$. Hence, $s\big(D[\mathbf{p} := B] \equiv D[\mathbf{p} := C]\big) = \mathbf{t}$ by Lemma 3.1.1. $\qquad\square$

3.1.4 Corollary. *If* $\vdash A$, *then* $\models_{taut} A$.

Proof. Take $\Gamma = \emptyset$ in the above. $\qquad\square$

3.1.5 Remark. (Counterexample Constructions in Boolean Logic) In what ways is soundness useful?

(1) It tells us that we are on track with our "program" to have syntactic tools to verify tautologies: Whatever these tools obtain (as absolute theorems) *are* tautologies.

(2) It allows us to disprove fallacious "results" of the form "such-and-such formula is formally provable—in Boolean logic—from such-and-such assumptions". For example, "$p \vee q \vdash p$" is a false statement in the metatheory. It says that from the assumption $p \vee q$ we can write a proof that contains (or ends with) p: Impossible! Why? Well, if the claim were true, then we would also have $p \vee q \models_{\text{taut}} p$. This is readily seen to be a false claim: Take any state s where $s(p) = \mathbf{f}$ and $s(q) = \mathbf{t}$.

Similarly, $\vdash \perp$ is false (because $\models_{\text{taut}} \perp$ is). By the way, we indicate that "$\Gamma \vdash A$ is false" (false metatheoretical statement) by writing $\Gamma \not\vdash A$. $\qquad\square$

3.1.6 Example. We already stated on p. 64 that $\equiv$ is not conjunctional—which is why we invented its conjunctional cousin $\Leftrightarrow$. Here is why:

The statement means that (1) below is *not a theorem schema*; i.e., for some specific choices of A, B, C we end up with a nontheorem.

$$A \equiv B \equiv C \equiv (A \equiv B) \wedge (B \equiv C) \tag{1}$$

Our job is to find *specific* formulae A, B, C that verify (2) below:

$$\not\vdash A \equiv B \equiv C \equiv (A \equiv B) \wedge (B \equiv C) \tag{2}$$

Let us try $\top$, $\perp$, and $\perp$ respectively and verify (2) by showing

$$\not\models_{\text{taut}} \top \equiv \perp \equiv \perp \underset{\uparrow}{\equiv} (\top \equiv \perp) \wedge (\perp \equiv \perp) \tag{3}$$

The subformula $\top \equiv \bot \equiv \bot$ to the left of the "$\equiv$" connective marked with "$\uparrow$" evaluates as **t** in any state. Yet, the subformula $(\top \equiv \bot) \wedge (\bot \equiv \bot)$ to the right evaluates as **f**. This verifies (3).   □

3.2   POST'S THEOREM

3.2.1 Metatheorem. (Post's Tautology Theorem) *If* $\Gamma \models_{taut} A$, *then* $\Gamma \vdash A$.

Proof. The proof of the metatheorem is, of course, informal and uses any tools we may be pleased to employ from the metatheory.

It is most convenient to prove the *contrapositive*, namely,

$$\text{If } \Gamma \not\vdash A, \text{ then } \Gamma \not\models_{taut} A \qquad (1)$$

Digression. The term *contrapositive* refers to an implication. The contrapositive of the *formal* implication "$X \rightarrow Y$" is "$\neg Y \rightarrow \neg X$". It is trivial to show (exercise!) both

$$\vdash X \rightarrow Y \equiv \neg Y \rightarrow \neg X$$

and

$$\models_{taut} X \rightarrow Y \equiv \neg Y \rightarrow \neg X$$

Thus, *in formal logic, guided by 1.4.2, 1.4.4, 1.4.5 and 1.4.7, proving* $\vdash X \rightarrow Y$ is as good as proving $\vdash \neg Y \rightarrow \neg X$ (by Eqn).

The term *contrapositive* also applies to all sorts of implications (including $\vdash$ and $\models_{taut}$) in informal mathematics (metatheory). Thus, the contrapositive of "if $[\cdots]$, then $(\ldots)$" is "if *not* $(\ldots)$, then *not* $[\cdots]$". (Meta)proving one is as good as proving the other.

Thinking commonsensically: Suppose I can prove the last of these two metastatements.

$$\text{Suppose I assume now that "}[\cdots]\text{" is true} \qquad (*)$$

Then it also must be that "$(\ldots)$" is true, for if not, then I must have the opposite: "*not* $(\ldots)$" is true. As this implies *not* $[\cdots]$ it cannot be, for it contradicts my assumption $(*)$.

Returning from our digression, which introduced a commonly used term of logic, we embark on our proof. This will consist of a few constructions along with a few claims—and their (meta)proofs—about the properties of the objects we construct.

First, let us argue that

Claim One. *There is an enumeration*

$$G_0, G_1, G_2, \ldots \qquad (2)$$

of a all formulae of Boolean logic. That is, every formula appears in the infinite array (2), and no string that is not a formula appears there.

Proof. [of **Claim One**] We may make a retroactive adjustment to the alphabet $\mathcal{V}$ that makes it finite. This will change nothing that was said so far in this volume, except a minute remark on p. 9: "Most variable symbols are formed through the use of 'subsymbols'—such as $0, 1, 2, '$— that are not members of the alphabet $\mathcal{V}$ themselves", I said there. Well, let me backtrack over this comment, now including $0, 1, 2, 3, 4, 5, 6, 7, 8, 9, '$ in an *amended* $\mathcal{V}$. But I am going to remove all the Boolean variables, except the three p, q, and r, because I am going to *build* all the rest!

This idea is hardly revolutionary, and is entirely analogous to that of building the infinite set of natural numbers by using just two symbols, 0 and 1 (binary notation), or the infinitely many variables of a programming language such as Algol from a finite set of symbols. In the latter case we start with the letters $a, b, \ldots, z, A, B, \ldots, Z$ and the digits $0, 1, \ldots, 9$ and use the algorithm in quotes to build any variable: "A variable is a string that starts with a letter and continues (to the right) as far as we wish, using any letter or digit."

So, for the purposes of this section, we take $\mathcal{V}$ to be (commas not included)

$$p, q, r, \prime, 0, 1, 2, 3, 4, 5, 6, 7, 8, 9, \top, \bot, (,), \neg, \wedge, \vee, \rightarrow, \equiv \tag{3}$$

The formation (syntax) of Boolean variables is now defined faithfully to what we said in **A1** (p. 9), but here we are more precise:

In **A1** we implied that all variables are given at once—"donated" as it were— and gave a couple of examples. Here instead we give a variable-construction rule similar to the one for Algol's variables, which will generate all Boolean variables, in essence giving us a new variable any time we need one:

A Boolean variable over the alphabet $\mathcal{V}$ given by (3) above is a string that starts with one of the letters p, q, r and continues with a block of zero or more primes ($\prime$) and then—optionally—with a string over the subalphabet $\{0, 1, 2, 3, 4, 5, 6, 7, 8, 9\}$ that does not begin with 0.

In writing a variable we write the block of primes as a superscript and the block of digits as a subscript. Thus, rather than $p_{\prime\prime\prime}123$ we write, as on p. 9, p'''_{123}.

This view of variables facilitates the proof of **Claim One**. We note that we have not changed our set of variables, which corroborates the earlier claim that we need change nothing that we have said and proved so far—except for our concept of where variables come from. We might as well consider this "new" definition as the "firming up" of the one on p. 9 rather than a revision of it.

The reader who has seen *regular sets* (e.g., in courses about UNIX, or theory of computation) would agree that the word-definition above can be captured by the following notation:[75]

$$\{p, q, r\}\{\prime\}^* \Big(\{\epsilon\} \cup \{1, 2, 3, 4, 5, 6, 7, 8, 9\}\{0, 1, 2, 3, 4, 5, 6, 7, 8, 9\}^* \Big) \tag{4}$$

---

[75]Here are two or three quick words about "regular sets". Let us denote, for the benefit of this footnote, *sets of strings* by script capital Latin letters, $\mathscr{A}$, $\mathscr{B}$ etc. Then, (1) $\mathscr{A}\mathscr{B}$ names the set of all strings we can get by concatenating a string $x \in \mathscr{A}$ to the left of a string $y \in \mathscr{B}$; (2) $\mathscr{A}^*$ (the so-called "Kleene star") names the set of all strings of any length (even zero) that we can build using as building blocks any finite number of strings from $\mathscr{A}$; (3) $\mathscr{A}\mathscr{B}\mathscr{C}$ means $(\mathscr{A}\mathscr{B})\mathscr{C}$.

We can now build (2) simultaneously with the sequence of *all* strings over the alphabet (3). The latter sequence we build alphabetically (*lexicographically*) by listing strings *by groups of increasing string-length*, and within each length group sorting them alphabetically. For the latter task we take the order of the symbols in (3) as going from smaller to larger ("p" is smallest and "$\equiv$" is largest).

So how is (2) built? According to the following procedure, which runs forever:
repeat forever:
      build the next string of the all-strings (over the alphabet (3)) sequence
     if it is a formula **then** write it as the next formula in sequence (2)
[End of proof of **Claim One**].[76]   □

Thus the sequence of all strings over $\mathcal{V}$ (as in (3)!) looks like (commas not included)

$$p, q, r, \prime, 0, 1, 2, 3, 4, 5, 6, 7, 8, 9, \top, \bot, (,), \neg, \wedge, \vee, \rightarrow, \equiv, pp, pq, pr, p\prime, p0, p1, \ldots$$

and the first few entries of sequence (2) are

$$p, q, r, \top, \bot, p\prime, p_1, p_2, \ldots$$

These first eight are the $G_0, G_1, G_2, G_3, G_4, G_5, G_6, G_7$ of (2), in this order.

We now turn to the proof of (1) proper and assume the hypothesis side,

$$\Gamma \nvdash A \tag{5}$$

We next construct a set of formulae, Δ, which is *as large as possible* with the properties that it includes Γ, but also

$$\Delta \nvdash A \tag{6}$$

We build Δ by stages, $\Delta_0, \Delta_1, \Delta_2, \ldots$ by an inductive definition, adding no more than one formula at each step and aiming to satisfy **Claim Six** below.

Pause. The reader must have seen inductive definitions, at least of number-sequences: For example, for a number $x \neq 0$, the sequence of the nonnegative powers of x—$x^0, x^1, x^2, \ldots$—is given by $x^0 = 1$ and, for $n \geq 0$, $x^{n+1} = x \cdot x^n$. Another example is the famous Fibonacci sequence defined by $F_0 = 0$, $F_1 = 1$ and, for $n \geq 1$, $F_{n+1} = F_n + F_{n-1}$.

The Δ_n sequence:

$$\Delta_0 = \Gamma$$

---

[76]If we were to assume some knowledge of set theory, then all preceding acrobatics for the proof of **Claim One** would become redundant: A string of length $n \geq 1$ over $\mathcal{V}$—whether $\mathcal{V}$ is finite or not—is a member of the Cartesian power $\mathcal{V}^n$, which by a known theorem of set theory admits an enumeration (is an "enumerable" set) because $\mathcal{V}$ does. But then, the set of all nonempty strings, i.e., $\bigcup_{n \geq 1} \mathcal{V}^n$, is enumerable by a known theorem of set theory, and thus so is its infinite subset **WFF** by yet another theorem. However, our original elementary proof is better in that it gives more information: Clearly, the effected enumeration is algorithmic.

For $n \geq 0$

$$\Delta_{n+1} = \begin{cases} \text{if } \Delta_n \cup \{G_n\} \not\vdash A & \text{then } \Delta_n \cup \{G_n\} \\ \text{else if } \Delta_n \cup \{\neg G_n\} \not\vdash A & \text{then } \Delta_n \cup \{\neg G_n\} \\ \text{else} & \Delta_n \end{cases}$$

Thus, at each stage we add to the set that we are constructing at most one formula, which is a member or a negation of a member of the sequence (2).

We define Δ by $\Delta = \bigcup_{n \geq 0} \Delta_n$, meaning "$\Delta = \Delta_0 \cup \Delta_1 \cup \Delta_2 \cup \cdots$", that is, forming Δ as the set of *all* the members found in *all* the Δ_i.

We state and prove a few claims about the Δ_n sequence and about Δ that contains *precisely all* formulae found in all the Δ_n.

Claim Two. $\Gamma \subseteq \Delta$. This follows at once from $\Delta_0 = \Gamma$.

Claim Three. *For $n \geq 0$, $\Delta_n \not\vdash A$.* This follows by a quick induction on n: For $n = 0$ (Basis) the claim is true by (5). On the I.H. that the claim holds for n we see that it so does for $n + 1$ by construction of Δ_{n+1} (note that only the last "else" uses the I.H.)

Claim Four. *The last "else" case in the definition of Δ_{n+1} is never applicable.* Indeed, the condition for that case is "$\Delta_n \cup \{G_n\} \vdash A$ and $\Delta_n \cup \{\neg G_n\} \vdash A$". By the deduction theorem (2.6.1) these two lead to $\Delta_n \vdash G_n \rightarrow A$ and $\Delta_n \vdash \neg G_n \rightarrow A$. These, by 2.5.13, give $\Delta_n \vdash A$, which by **Claim Three** cannot happen.

So why bother *having* the last case? Because it is proper mathematical manners, when we give a definition by cases, to have all possible cases present—including the "otherwise" (last "else")—to ensure that what we are defining is defined under all possible circumstances. It is best to check whether the definition can be simplified (e.g., by dropping redundant cases) *only after* the definition is given rather than analyzing it to death *a priori*.

Claim Five. $\Delta \not\vdash A$. Indeed, if we think otherwise, then, since proofs have finite length and trivially $\Delta_n \subseteq \Delta_{n+1}$ for all n, there is an n—large enough—so that all the Δ formulae used in the proof of $\Delta \vdash A$ lie in Δ_n. So, $\Delta_n \vdash A$ as well, contrary to **Claim Three**.

Claim Six. *For every formula B, either B is in Δ, or $\neg B$ is in Δ, but not both.* Indeed, every B is some G_m in the sequence (2). By the construction of the Δ_n sequence—and since the last "else" never applies (cf. **Claim Four**)—we note that at least one of B or $\neg B$ will be added to Δ_m to form Δ_{m+1}. How about both B and $\neg B$ being in Δ?[77] Then (2.5.7) $\Delta \vdash \bot$, and hence $\Delta \vdash A$ by 2.6.6, which cannot be by **Claim Five**.

Claim Seven. Δ *is deductively closed, that is, if $\Delta \vdash B$, then $B \in \Delta$.* Indeed, if we thought for a minute that for some B it is possible to have $\Delta \vdash B$, and yet also

---

[77] As some G_m and some G_k, $m \neq k$. Naturally, they cannot be inserted at the same step, since a step adds *one* formula to Δ.

have $B \notin \Delta$, then (by **Claim Six**) $\neg B$ will be in Δ. The latter implies (cf. 1.4.7) that $\Delta \vdash \neg B$, which along with $\Delta \vdash B$ and 2.5.7 yield $\Delta \vdash \bot$ and thus $\Delta \vdash A$ (2.6.6), contradicting **Claim Five**.

The previous claim may be understood as saying that Δ is so big a set of assumptions that anything you can prove from them, with any proof, can also be proved by a proof of length one.

We are ready to define a state v that verifies the conclusion of (1).

Define a state v by setting, for each variable $\mathbf{p}$, $v(\mathbf{p}) = \mathbf{t}$ iff $\mathbf{p} \in \Delta$. (7)

Main Claim. *For all formulae B, $v(B) = \mathbf{t}$ iff $B \in \Delta$.*

Note: The claim also reads, by looking at the contrapositive (informally), "For all formulae B, $v(B) = \mathbf{f}$ iff $B \notin \Delta$."

The proof is by induction on the complexity of B. We have the following cases:

(i) B is a variable. The claim is (7).

(ii) B is the formula $\top$. Since $v(\top) = \mathbf{t}$, we want $\top \in \Delta$. By **Claim Seven**, it suffices to have $\Delta \vdash \top$. We have this by 2.1.15 and 2.1.1.

(iii) B is the formula $\bot$. Since $v(\bot) = \mathbf{f}$, we want $\bot \notin \Delta$. Well, in the opposite case, we would have $\Delta \vdash \bot$, from which, via 2.6.6, we would also have $\Delta \vdash A$, contradicting **Claim Five**.

(iv) B is $\neg C$. Say $v(\neg C) = \mathbf{t}$. Then $v(C) = \mathbf{f}$ and the I.H. yields (cf. Note following the main claim) $C \notin \Delta$; hence $\neg C$ is in Δ by **Claim Six**.

Conversely, if $\neg C$ is in Δ, then $C \notin \Delta$ by **Claim Six**. By the I.H. we have $v(C) = \mathbf{f}$; hence $v(\neg C) = \mathbf{t}$.

(v) B is $C \vee D$. Say $v(C \vee D) = \mathbf{t}$. There are two cases, but we deal with one, the other being similar: $v(C) = \mathbf{t}$. By the I.H. $C \in \Delta$. Hence (1.4.7) $\Delta \vdash C$. It follows that $\Delta \vdash C \vee D$ by 2.5.1 and 2.1.4. Hence $C \vee D \in \Delta$ by **Claim Seven**.

Conversely, let $C \vee D \in \Delta$. It must be that at least one of C or D is in Δ, for if not, then $\neg C$ and $\neg D$ are in Δ (**Claim Six**). Why is this impossible? Because, by 2.5.6, $\Delta \vdash D$; hence $\Delta \vdash \bot$ by 2.5.7. We have seen already that this cannot be. Say then $C \in \Delta$. By I.H. $v(C) = \mathbf{t}$; hence $v(C \vee D) = \mathbf{t}$.

(vi) B is $C \wedge D$. Say $v(C \wedge D) = \mathbf{t}$. Then $v(C) = \mathbf{t}$ and $v(D) = \mathbf{t}$ so that C and D are in Δ by I.H. Then $\Delta \vdash C \wedge D$ by 2.5.1; thus $C \wedge D \in \Delta$ by **Claim Seven**.

Conversely, let $C \wedge D \in \Delta$. Hence $\Delta \vdash C$ and $\Delta \vdash D$ by 2.5.1. Thus $C \in \Delta$ and $D \in \Delta$; hence (I.H.) $v(C) = \mathbf{t} = v(D)$.

(vii) B is $C \to D$. Say $v(C \to D) = \mathbf{t}$. There are two similar cases, $v(C) = \mathbf{f}$ or $v(D) = \mathbf{t}$. We just consider the first: By I.H., $C \notin \Delta$; thus $\neg C$ is in Δ.

By 2.5.1 $\Delta \vdash \neg C \vee D$; hence $\Delta \vdash C \rightarrow D$ by 2.4.11. Thus $C \rightarrow D$ is in Δ (**Claim Seven**).

Conversely, let $C \rightarrow D$ be in Δ. Thus, $\Delta \vdash \neg C \vee D$, by 2.4.11. By case (v), we have $\neg C$ or D (possibly both) are in Δ. If the former, then $C \notin \Delta$ by **Claim Six**; hence $v(C) = \mathbf{f}$ by I.H. It follows that $v(C \rightarrow D) = \mathbf{t}$. The other case is as simple.

(viii) B is $C \equiv D$. Say $v(C \equiv D) = \mathbf{t}$. There are two similar cases.

Case where $v(C) = v(D) = \mathbf{t}$. By I.H. C and D are in Δ. Thus $\Delta \vdash C \equiv D$ as per calculation

$$C \equiv D$$
$$\Leftrightarrow \;\langle \text{Leib twice, using the assumptions } C, D \text{ and redundant true (2.1.23)}\rangle$$
$$\top \equiv \top$$

Case where $v(C) = v(D) = \mathbf{f}$. By I.H. neither of C and D are in Δ. Thus both $\neg C$ and $\neg D$ are in, and, as before, $\Delta \vdash \neg C \equiv \neg D$. Using 2.4.3 twice and 2.4.4 we conclude $\Delta \vdash C \equiv D$ and are reduced to the previous case. Via **Claim Seven**, both yield that $C \equiv D$ is in Δ.

Conversely, let $C \equiv D$ be in Δ. We argue that it is impossible to have *exactly one* of C and D in Δ. Indeed, say that C is in and D is not. Thus $\neg D$ is in. As above, this entails $\Delta \vdash C \equiv \neg D$ and—by 2.4.3—$\Delta \vdash \neg(C \equiv D)$. Along with the assumption this yields (2.5.7) $\Delta \vdash \bot$, which we know is impossible.

Thus, either both C and D are in, where the I.H. furnishes $v(C) = \mathbf{t} = v(D)$, or neither is in, where the I.H. furnishes $v(C) = \mathbf{f} = v(D)$. Both alternatives yield $v(C \equiv D) = \mathbf{t}$.

At the end of all this the reader is entitled to a coffee (no sugar, no milk) break.

After that, he can easily conclude the proof as follows: By the **Main Claim**, every formula B in Δ—and hence every formula B in Γ since $\Gamma \subseteq \Delta$—satisfies $v(B) = \mathbf{t}$. On the other hand, as $\Delta \not\vdash A$ it must be $A \notin \Delta$; thus, again via the **Main Claim**, $v(A) = \mathbf{f}$. Therefore $\Gamma \not\models_{\text{taut}} A$. This establishes (1). $\qquad\qquad\square$

The proof of the **Main Claim** had too many inductive cases (corresponding to the various cases of i.p.) because having the best interests of the user in mind (rather than those of the metatheoritician), we adopted too many Boolean connectives as *primitive* (just as [17] did, presumably for the same reason). Books and articles in logic that write mostly about the metatheory often employ just $\neg$ and $\vee$ as primitive connectives, which reduces the induction steps above to only two rather than five.

Post's theorem is often called the *completeness theorem* of propositional calculus. It shows that the syntactic manipulation apparatus completely captures the notion of "truth" (tautologyhood) and "preservation of truth" (tautological implication) in the Boolean case.

3.2.2 Corollary. *If* $\models_{taut} B$, *then* $\vdash B$.

Proof. Case of $\Gamma = \emptyset$.   □

3.2.3 Exercise. Prove that the Δ constructed in the proof of Post's theorem is infinite even if Γ is finite.

Hint. Prove that $\Delta_n \neq \Delta_{n+1}$ for all n.   □

3.2.4 Exercise. Prove that if $\Gamma \models_{taut} A$, then, for some finite $\Sigma \subseteq \Gamma$, we also have $\Sigma \models_{taut} A$.   □

3.2.5 Exercise. (Compactness of Sentential Logic) Prove that if *every finite* subset of a set of formulae Γ is satisfiable (cf. 1.3.11), then so is Γ.   □

3.2.6 Exercise. Prove that a set of formulae Γ is satisfiable (cf. 1.3.11) iff it is consistent (cf. p. 85).   □

3.2.7 Exercise. Fully prove or disprove in the metatheory:

"For any set of formulae Γ and any formulae A and B, if $\Gamma \vdash A \vee B$, then it must be $\Gamma \vdash A$ or $\Gamma \vdash B$."   □

3.3   FULL CIRCLE

Post's theorem is very convenient. It says that any (correct) schema $A_1, \dots, A_n \models_{taut} B$ leads to a *derived rule of inference*, $A_1, \dots, A_n \vdash B$. In particular, combining with 2.1.4, we get

3.3.1 Corollary. *If* $\Gamma \vdash A_i$, *for* $i = 1, \dots, n$, *and if* $A_1, \dots, A_n \models_{taut} B$, *then* $\Gamma \vdash B$.

This is a very important result. It frees the user of logic to use *any* tautological implication *schema* as a *derived rule of inference* in the progress of a proof. That is, while the rules **Inf1** and **Inf2** of 1.4.2 *suffice* to construct all "truths" starting from the "eleven original truths" (1.4.4)—and we already augmented them with all sorts of derived rules such as *cut*, *MP*, etc.—nevertheless, if convenience dictates, we can employ as an *additional* derived rule of inference *any* tautological implication schema that we happen to know, or happen to invent easily on the spur of the moment. Unless, of course, for the higher purpose of *learning through hardship*(!) we are constrained otherwise in an assigned question of a problem set or of a test/exam!

In sum—*unless otherwise requested!*—we can, *and will from now on*, rigorously mix syntactic with semantic justifications of our Boolean proof steps.

We have come full circle. We have started semantically, indicating that what matters in Boolean logic is to identify the "true" (tautologies) and "relatively true" (tautological implications of given premises) formulae. We indicated that in the present (and foreseeable) state of the art this is an extremely laborious process in general.

To compensate, logicians have long ago discovered a systematic *syntactic* way to exploit *educated guessing* in such verifications, and therefore often make such verifications *shorter*.[78] Such methodology of educated guessing is what we have called *proofs*.

The question remained whether such methodology—proofs—fully captures the *ability* of truth tables. With the settling of Post's theorem (and soundness, 3.1.3) we saw that it does.

Thus the *semantic* approach—using the truth values **t** and **f** and the "operations" $F_\vee, F_\equiv$, etc., on $\{\mathbf{f}, \mathbf{t}\}$—and the *syntactic* one (using proofs) are totally *equivalent* and *interchangeable*.

This interchangeability is subject only to the caveat mentioned in the ⚠-note above.

3.3.2 Example. Here is an example of a tautology, hence a theorem by 3.2.2, which we can easily verify semantically ("easily" means without going into truth tables or proofs).

$$\vdash ((A \to B) \to A) \to A \tag{1}$$

So let us verify

$$\models_{\text{taut}} ((A \to B) \to A) \to A \tag{2}$$

How easily? I show that there is no state v that makes the formula in (2) **f**. Well, if some state v does make it **f**, then A must be **f**,[79] but $(A \to B) \to A$ must be **t**. Thus $A \to B$ must be **f**. With A being **f**, this is impossible.

It is instructive for the reader to attempt a proof of (1) without using semantic notions at all.   □

3.4   SINGLE-FORMULA LEIBNIZ

3.4.1 Example. The following is readily verifiable:[80]

$$\models_{\text{taut}} (A \equiv B) \to (E[\mathbf{p} := A] \equiv E[\mathbf{p} := B])$$

Thus, by 3.2.1,

$$\vdash (A \equiv B) \to (E[\mathbf{p} := A] \equiv E[\mathbf{p} := B]) \tag{SFL}$$

In [17] (SFL)—where it is surprisingly presented as the *Leibniz Axiom* despite the fact that it is provable—plays an active role in a number of applications.   □

---

[78]The hedging "often" is appropriate at the present state of knowledge, as we have already remarked: We do not know whether there is a "fast" (polynomial time) *nondeterministic* algorithm that recognizes tautologies.

[79]"A is **f**" is colloquial for $v(A) = \mathbf{f}$.

[80]Let v be a state where $v(A \equiv B) = \mathbf{t}$. Thus, $v(A) = v(B) = \mathbf{t}$, or $v(A) = v(B) = \mathbf{f}$. Now apply the proof of Lemma 3.1.1 to see that $v(E[\mathbf{p} := A]) = v(E[\mathbf{p} := B])$. The demanding reader may glimpse at the detailed proof that supplements that of the lemma.

Two other ways—*without reliance on Post's Theorem*—to prove SFL in the Boolean logic as it is founded in this volume, that is, *on the axioms 1.4.4 and rules 1.4.2*, are

(1) Applying the deduction theorem (2.6.1) to

$$A \equiv B \vdash (E[\mathbf{p} := A] \equiv E[\mathbf{p} := B]) \qquad (Leib)$$

(2) Using 2.6.2, which says

$$D \to (A \equiv B) \vdash D \to (E[\mathbf{p} := A] \equiv E[\mathbf{p} := B])$$

Indeed, taking "D" to be $A \equiv B$ in the above and noting that $\vdash (A \equiv B) \to (A \equiv B)$, we are done.

Can one prove SFL in the system presented in [17]? *Yes, indeed.*

However, in [17] the deduction theorem is badly compromised by the presence of the *substitution rule* as a *primary* rule:

$$\frac{A}{A[\mathbf{p} := B]} \quad \text{(We call $\mathbf{p}$ the \emph{eigenvariable})} \qquad (Sub)$$

Thus, in order to obtain a proof of SFL within the logic of [17], we would rather *not* use the deduction theorem.[81]

Yet—and this is a fact not proved in [17], nor here, but a fact nevertheless for Boolean logics that *do allow* the substitution rule—Post's theorem holds in the logic of [17] in the form of Corollary 3.2.2.

In fact, *any correctly founded Boolean logic will have as (absolute) theorems precisely all tautologies.* Therefore the reason we gave in the preceding example for the theoremhood of SFL is good, not only in our logic, but also in that of [17].

Let us derive a few interesting results from SFL.

3.4.2 Example. By SFL (which is an absolute theorem schema) we have

$$\vdash (\mathbf{p} \equiv \top) \to (C \equiv C[\mathbf{p} := \top])$$

and

$$\vdash (\mathbf{p} \equiv \bot) \to (C \equiv C[\mathbf{p} := \bot])$$

where we note that $C[\mathbf{p} := \mathbf{p}]$ is just C. Using redundant true and the axiom $\neg A \equiv A \equiv \bot$ (and Leib, of course), the above yield the equivalent formulations

$$\vdash \mathbf{p} \to (C \equiv C[\mathbf{p} := \top]) \qquad (1)$$

and

$$\vdash \neg\mathbf{p} \to (C \equiv C[\mathbf{p} := \bot]) \qquad (2)$$

---

[81] For the record, the correct formulation of the deduction theorem in the system of [17] is: "If $\Gamma + A \vdash B$ with a proof that never used an eigenvariable that occurs anywhere in A or the formulae of Γ, then $\Gamma \vdash A \to B$."

(1) and (2) yield

$$\vdash (C \equiv C[\mathbf{p} := \top]) \vee (C \equiv C[\mathbf{p} := \bot]) \tag{3}$$

via the cut rule (2.5.4). ▫

3.4.3 Example. Using SFL we can readily prove $\vdash (A \equiv B) \rightarrow C[\mathbf{p} := A] \equiv (A \equiv B) \rightarrow C[\mathbf{p} := B]$. Indeed, SFL and 2.4.13 imply the above via Eqn. ▫

3.4.4 Example. We next verify $\vdash (A \equiv B) \wedge C[\mathbf{p} := A] \equiv (A \equiv B) \wedge C[\mathbf{p} := B]$.

$$(A \equiv B) \wedge C[\mathbf{p} := A] \equiv (A \equiv B) \wedge C[\mathbf{p} := B]$$
$$\Leftrightarrow \langle \text{deMorgan and two applications of Leib (at once)} \rangle$$
$$\neg(\neg(A \equiv B) \vee \neg C[\mathbf{p} := A]) \equiv \neg(\neg(A \equiv B) \vee \neg C[\mathbf{p} := B])$$
$$\Leftrightarrow \langle \text{2.4.1 and two applications of Leib (at once)} \rangle$$
$$\neg(A \equiv B) \vee \neg C[\mathbf{p} := A] \equiv \neg(A \equiv B) \vee \neg C[\mathbf{p} := B]$$
$$\Leftrightarrow \langle \text{2.4.11 and two applications of Leib (at once)} \rangle$$
$$(A \equiv B) \rightarrow \neg C[\mathbf{p} := A] \equiv (A \equiv B) \rightarrow \neg C[\mathbf{p} := B]$$

The last line is the result of the previous example using $\neg C$ rather than C. ▫

3.4.5 Example. It is instructive to offer a Hilbert-style proof of the above—without invoking SFL—as it will introduce a general technique that some people call a *Ping-Pong argument*. Ping-Pong arguments will be especially useful in Part II. This technique of proving equivalences is extremely widespread outside the equational methodology and is based on the theorem schema below (cf. 2.4.26):

$$\vdash (A \equiv B) \equiv (A \rightarrow B) \wedge (B \rightarrow A)$$

By equanimity,

$$\Gamma \vdash A \equiv B \text{ iff } \Gamma \vdash (A \rightarrow B) \wedge (B \rightarrow A) \tag{1}$$

By 2.5.1, (1) is equivalent to

$$\Gamma \vdash A \equiv B \text{ iff we have } both \, \Gamma \vdash A \rightarrow B \text{ and } \Gamma \vdash B \rightarrow A \tag{2}$$

Thus, to prove $\Gamma \vdash A \equiv B$, one *equivalently* proves the *two directions*, "$(\rightarrow)$" (nickname of $\Gamma \vdash A \rightarrow B$) and "$(\leftarrow)$" (nickname of $\Gamma \vdash B \rightarrow A$)

The technique is almost always used in conjunction with the deduction theorem. That is, rather than showing $\Gamma \vdash A \rightarrow B$, one proves instead $\Gamma + A \vdash B$.

Let us now re-prove $\vdash (A \equiv B) \wedge C[\mathbf{p} := A] \equiv (A \equiv B) \wedge C[\mathbf{p} := B]$.

$(\rightarrow)$

$$(1) \quad (A \equiv B) \wedge C[\mathbf{p} := A] \quad \langle \text{assumption} \rangle$$

$$(2) \quad A \equiv B \qquad\qquad \langle (1) + 2.5.1 \rangle$$
$$(3) \quad C[\mathbf{p} := A] \qquad\qquad \langle (1) + 2.5.1 \rangle$$
$$(4) \quad C[\mathbf{p} := B] \qquad\qquad \langle (2, 3) + \text{Eqn/Leib} (2.1.16) \rangle$$
$$(5) \quad (A \equiv B) \wedge C[\mathbf{p} := B] \; \langle (2, 4) + 2.5.1 \rangle$$

$(\leftarrow)$

Entirely similar to $(\rightarrow)$, therefore omitted. □

3.4.6 Exercise. Give a Ping-Pong argument (Hilbert-style) proof of 3.4.3, without invoking SFL. □

3.4.7 Example. Using redundant true (use the special case $\top$ for B) on the above two examples (3.4.3 and 3.4.4), we get:

$$\vdash A \rightarrow C[\mathbf{p} := A] \equiv A \rightarrow C[\mathbf{p} := \top]$$

and

$$\vdash A \wedge C[\mathbf{p} := A] \equiv A \wedge C[\mathbf{p} := \top]$$

If, moreover, we specialize A to $\mathbf{p}$ and note that $C[\mathbf{p} := \mathbf{p}]$ is just C, then we get

$$\vdash \mathbf{p} \rightarrow C \equiv \mathbf{p} \rightarrow C[\mathbf{p} := \top]$$

and

$$\vdash \mathbf{p} \wedge C \equiv \mathbf{p} \wedge C[\mathbf{p} := \top] \qquad\qquad (S1)$$

□

3.4.8 Example. (Shannon) Using 3.4.4 with $\bot$ for B, we get —via $\vdash \neg X \equiv X \equiv \bot$ and Leib

$$\vdash \neg A \wedge C[\mathbf{p} := A] \equiv \neg A \wedge C[\mathbf{p} := \bot]$$

In particular, if A is $\mathbf{p}$,

$$\vdash \neg \mathbf{p} \wedge C \equiv \neg \mathbf{p} \wedge C[\mathbf{p} := \bot] \qquad\qquad (S2)$$

The $(S1)$ of 3.4.7 and $(S2)$ lead to the following simple calculation:

$\mathbf{p} \wedge C[\mathbf{p} := \top] \vee \neg \mathbf{p} \wedge C[\mathbf{p} := \bot]$
$\Leftrightarrow \langle$ two obvious applications of Leib using $(S1)$ of 3.4.7 and $(S2) \rangle$
$\mathbf{p} \wedge C \vee \neg \mathbf{p} \wedge C$
$\Leftrightarrow \langle$ distributivity $\rangle$
$(\mathbf{p} \vee \neg \mathbf{p}) \wedge C$
$\Leftrightarrow \langle$ Leib + excl. middle via 2.1.23; "C-part" is $\mathbf{q} \wedge C \rangle$
$\top \wedge C$
$\Leftrightarrow \langle 2.4.20 \rangle$

$$C$$

Thus we have obtained $\vdash \mathbf{p} \wedge C[\mathbf{p} := \top] \vee \neg\mathbf{p} \wedge C[\mathbf{p} := \bot] \equiv C$ (Shannon).   □

In the presence of Post's theorem and of the deduction theorem, SFL is highly redundant as a tool. We will not use it beyond these examples.

Here is another highly redundant tool that we have already discussed—cf. pp. 42 and 101—and promised never to use: The substitution *rule*:

$$\frac{A}{A[\mathbf{p} := B]}$$—applicable when $\mathbf{p}$ does not appear in the special axioms   (*Sub*)

It turns out that it *is* a derived rule in our logic.

3.4.9 Exercise. Prove that if $\vdash A$, then also $\vdash A[\mathbf{p} := B]$ for any Boolean variable $\mathbf{p}$ and formula B.

Hint. Either by induction on length of proofs, or using Post's theorem.   □

3.4.10 Exercise. Prove that if $\Gamma \vdash A$, then also $\Gamma \vdash A[\mathbf{p} := B]$ for any Boolean variable $\mathbf{p}$ and formula B—as long as there is a proof of A from Γ where $\mathbf{p}$ occurs in none of the formulae used from Γ.   □

3.5   APPENDIX: RESOLUTION IN BOOLEAN LOGIC

Resolution is a simple way to establish the validity of a particular configuration, or configuration-schema, of the type $\Gamma \vdash A$ by essentially using just one rule, the *cut rule* (cf. 2.5.4). It is a proof technique introduced by Robinson ([40]) and is based on the metatheorem "$\Gamma \vdash A$ iff $\Gamma + \neg A \vdash \bot$" (cf. 2.6.7). It has been popular with automatic theorem provers, that is, computer programs that prove theorems.

Of course, by the nature of the cut rule, in order to apply it easily on the premises $\Gamma + \neg A$ (which, in general, are schemata due to the presence of syntactic variables) these must be, or must be *brought into*, the form

$$\{C_1, C_2, \ldots, C_n\}$$

where each C_i—called a *clause*—is a *disjunction* of *simple formulae*, specifically of type: atomic, negation of atomic, formula-variable,[82] negation of formula-variable. We call the formulae in these four categories *literals*.

If the premises are not of that form, one will apply a combination of simple semantic or syntactic tools to convert them *on an as-needed basis*. For example, a premise such as "$A \to B$" would be replaced by the equivalent (cf. 2.4.11) "$\neg A \vee B$", which is a disjunction.

One additional feature of such proofs is that they are normally written in a two-dimensional manner—as opposed to linear—and since, essentially, there is only

[82]I.e., syntactic variable, such as $A, B, \ldots$

one rule in resolution, and the circumstances of its applicability are evident, one dispenses with detailed annotation and uses as such two lines connecting the two premises, $A \vee B$ and $\neg A \vee C$ with the conclusion $B \vee C$—that is, like this:

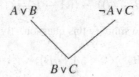

The technique is best illustrated via examples. Recall that the cut rule shown above has special cases, namely 2.5.5, 2.5.6, and 2.5.7. In the context of resolution, they are all instances of the cut rule.

3.5.1 Example. Using resolution we prove the most general rule of proof by cases (2.5.10), namely:

$$A \to B, C \to D \vdash A \vee C \to B \vee D$$

Using the deduction theorem, we need to show

$$A \to B, C \to D, A \vee C \vdash B \vee D$$

that is, prove $\perp$ from $\neg A \vee B, \neg C \vee D, A \vee C; \neg(B \vee D)$, or, in other words (cf. 2.6.6 and the remark following it), prove that the set of formulae $\{\neg A \vee B, \neg C \vee D, A \vee C, \neg(B \vee D)\}$ is *inconsistent*. Here it goes:

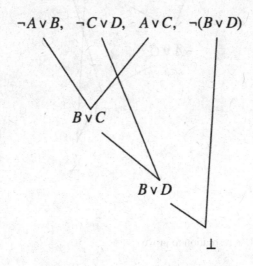

3.5.2 Example. We next show

$$\vdash (A \to (B \to C)) \to ((A \to B) \to (A \to C))$$

We do some preprocessing to simplify this question. By deduction theorem, prove instead

$$A \to (B \to C) \vdash (A \to B) \to (A \to C)$$

By two more applications of the deduction theorem, prove instead

$$A \to (B \to C), A \to B, A \vdash C$$

Therefore we need to show that the set

$$\{\neg A \vee \neg B \vee C, \neg A \vee B, A, \neg C\}$$

is inconsistent.

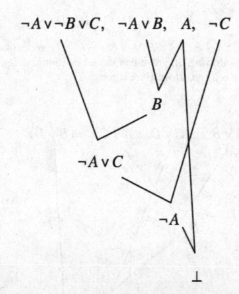

3.5.3 Example. Use resolution to prove

$$\vdash (A \wedge \neg B) \to \neg(A \to B)$$

By the deduction theorem prove instead $A \wedge \neg B \vdash \neg(A \rightarrow B)$. Split hypotheses (2.5.2) and move the negation (cf. 2.4.4 and 2.4.11) of the sought conclusion to the hypotheses side. We get the hypotheses:

$$A, \neg B, \neg A \vee B$$

Cut 1st and 3rd to get B. Cut this with $\neg B$ to get $\bot$. This (simple) case did not require us to draw any lines. □

3.6 ADDITIONAL EXERCISES

1. We say that a Boolean formula A is in *disjunctive normal form*, or DNF, iff it is of the form

$$D_1 \vee D_2 \vee \ldots \vee D_n \qquad (DNF)$$

where each *disjunct* D_i *is a conjunction of the form* $C_1 \wedge C_2 \wedge \ldots \wedge C_k$, and *each* C_j *is a variable or a negated variable, and moreover all the variables of A appear in each disjunct, and do so once.*

Correspondingly, we say that a Boolean formula A is in *conjunctive normal form*, or CNF, iff it is of the form

$$C_1 \wedge C_2 \wedge \ldots \wedge C_n \qquad (CNF)$$

where each *conjunct* C_i *is a disjunction of the form* $D_1 \vee D_2 \vee \ldots \vee D_k$, and *each* D_j *is a variable or a negated variable, and moreover all the variables of A appear in each conjunct, and do so once.*

Prove

- Every formula B is provably equivalent either to $\bot$ or to a formula A in DNF that has the same variables as B.

  Hint. Do induction on the number of variables in B. The induction step can be helped by 3.4.8.

- Every formula B is provably equivalent either to $\top$ (hence is a theorem) or to a formula A in CNF that has the same variables as B.

2. (a) Let A be a formula in which the variables p, q, r occur, but no others, and whose truth table has a result $\mathbf{t}$ *only* in the rows $\mathbf{f}, \mathbf{t}, \mathbf{f}$ (state for p, q, r in that order) and $\mathbf{t}, \mathbf{f}, \mathbf{f}$. Show that A is provably equivalent to the formula $\neg p \wedge q \wedge \neg r \vee p \wedge \neg q \wedge \neg r$.

    Hint. Prove that $\models_{\text{taut}} A \equiv \neg p \wedge q \wedge \neg r \vee p \wedge \neg q \wedge \neg r$ instead.

   (b) Generalize the above to give an alternative proof for the first bullet in Exercise 1.

3. (a) Let B be a formula in which the variables p, q, r occur, but no others, and whose truth table has a result $\mathbf{f}$ *only* in the rows $\mathbf{f}, \mathbf{t}, \mathbf{f}$ (state for p, q, r in

that order) and $\mathbf{t}, \mathbf{f}, \mathbf{f}$. Show that B is provably equivalent to the formula $(p \vee \neg q \vee r) \wedge (\neg p \vee q \vee r)$.

Hint. Prove that $\models_{\text{taut}} B \equiv (p \vee \neg q \vee r) \wedge (\neg p \vee q \vee r)$ instead.

(b) Generalize the above to give an alternative proof for the second bullet in Exercise 1.

4. What is the DNF of $p \rightarrow p \vee q \vee r$?

5. What is the DNF of $\neg p \vee p \vee q$?

6. What is the CNF of $\neg p \wedge p$?

7. What is the CNF of $p \wedge q \wedge r \wedge p_1 \rightarrow \bot$?

8. Use resolution (in combination with the deduction theorem)—but *not* Post's theorem—to prove $\vdash A \vee (B \wedge C) \rightarrow A \vee B$.

9. Use resolution (in combination with the deduction theorem)—but *not* Post's theorem—to prove $\vdash (A \rightarrow B) \rightarrow (A \rightarrow C) \rightarrow (A \rightarrow B \wedge C)$.

10. Use resolution (in combination with the deduction theorem)—but *not* Post's theorem—to prove $\vdash (p \vee q \vee r) \wedge (p \rightarrow p') \wedge (q \rightarrow p') \wedge (r \rightarrow p') \rightarrow p'$.

11. Use resolution (in combination with the deduction theorem)—but *not* Post's theorem—to prove $\vdash (p \rightarrow (q \rightarrow r)) \rightarrow (q \rightarrow (p \rightarrow r))$.

12. Suppose that Γ is a set of assumptions, and A, B are two formulae.

    We know that if $\Gamma \vdash A \wedge B$, then $\Gamma \vdash A$ and $\Gamma \vdash B$.

    Is it also true that if $\Gamma \vdash A \vee B$, then $\Gamma \vdash A$ or $\Gamma \vdash B$?

    If yes, then give a (meta)proof for any Γ, A, B.

    If no, then use soundness (3.1.3) to give a definitive counterexample for appropriately chosen Γ, A, B.

13. Give a proof of $\vdash A \rightarrow (B \rightarrow C) \equiv (A \rightarrow B) \rightarrow (A \rightarrow C)$ by a Ping-Pong argument.

14. Prove the following absolute theorem schemata. The use of Post's theorem is *not* allowed in this exercise.

    - $A \vee A \rightarrow A$
    - $A \rightarrow A \vee B$
    - $A \vee B \rightarrow B \vee A$
    - $(A \rightarrow B) \rightarrow (C \vee A \rightarrow C \vee B)$

15. For all A, B show that $\vdash A \rightarrow B \vee A$. The use of Post's theorem is *not* allowed in this exercise.

16. Show that for all A and B, we have $\vdash A \to B \to A$. The use of Post's theorem is *not* allowed in this exercise.

17. Use the deduction theorem and resolution (but *not* Post's theorem!) to prove

$$\vdash (p \to (q \to r)) \to ((p \to q) \to (p \to r))$$

18. Use the deduction theorem and resolution (but *not* Post's theorem!) to prove

$$\vdash (p \wedge \neg q) \to \neg(p \to q)$$

19. Use the deduction theorem and resolution (but *not* Post's theorem!) to prove

$$p \to q \to r \to p_1, A \to B \to r \vdash A \to p \to B \to q \to p_1$$

20. Use the deduction theorem and resolution (but *not* Post's theorem!) to prove

$$(\neg B \to \neg A) \to (\neg B \to A) \to B$$

21. Use the deduction theorem and resolution (but *not* Post's theorem!) to prove

$$\vdash (A \vee B \vee \neg C) \wedge (A \to B) \to (C \to B)$$

22. Use the deduction theorem and resolution (but *not* Post's theorem!) to prove

$$\vdash ((A \to B) \to A) \to A$$

PREDICATE LOGIC

CHAPTER 4

EXTENDING BOOLEAN LOGIC

By now we must possess (assuming we did a lot of exercises!) a pretty solid technique for proving theorems of Boolean logic. Is this skill (toolbox) sufficient toward reasoning in mathematics and computer science?

I regret to say that it is totally insufficient. You see, computer science and mathematics are talking—and contain reasoning and theorems—about *objects* such as sets, strings, numbers, matrices, trees, graphs, programs, models of computation (such as "Turing machines"), and many others.

On the other hand, Boolean logic talks only about the Boolean connectives, and how using them we can formulate the truth of *extremely general* statements, which do not express any *specific* statement that involves any of the previously mentioned *objects*. For example, we cannot formulate the statement, and much less reason about its truth: "Every natural number greater than 1 has a prime factor". Propositional logic does *not* know what is a "number"—even less so which numbers are "natural"—what is the meaning of "greater", what is "1", what is a "prime", and what is a "factor". The statements that we *can* write down in Boole's logic (and then derive conclusions about them using logic's proving tools) are *abstractions* of mathematical statements, that is, statements where all the details about what *mathematical objects* we are talking about—and exactly what we are saying—have been deleted.

Mathematical Logic. By George Tourlakis
Copyright © 2008 John Wiley & Sons, Inc.

In particular, we may think of an arbitrary Boolean variable as an abstract mathematical statement—one whose details we do not know, or *ignore and hide*, doing so in the interest of simplifying reasoning (cf. p. 6). Boolean logic is unable to further analyze these *atomic* statements.

4.0.1 Example. The two statements of mathematics

$$\text{Any } two \text{ sets } y \text{ and } z \text{ are equal if they have exactly the same elements} \tag{1}$$

and

$$\text{An object } x \text{ is equal to itself} \tag{2}$$

have the mathematical formulations[83]

$$(\forall y)(\forall z)\Big((\forall x)(x \in y \equiv x \in z) \rightarrow y = z\Big) \tag{1'}$$

and

$$x = x \tag{2'}$$

respectively, in the language of everyday mathematics.

Statement $(1')$ is about sets, and it happens to be true in set theory. Statement $(2')$ is a philosophical principle true everywhere, in all of mathematics, not just set theory.

Yet, if we attempted to formulate $(1')$ and $(2')$ in the language of Boolean logic, we could just manage to say that both are captured(!) by the same Boolean "statement": p (where, of course, any other Boolean variable will do). Thus, in the Boolean formulation (which is a high level of abstraction), we would totally lose the intrinsic—and very different—meanings of these two mathematical statements!

What is the reason? Logic exploits the *logical structure* of statements (i.e., formulae) and through the use of axioms and syntactic proof-writing rules aims to verify those statements that are true.[84]

Now *propositional* logic can only see and reason about the *propositional structure* of statements, i.e., how they are put together via Boolean connectives.

This logic sees no connectives in $(2')$ because there are none. It sees no connectives in $(1')$ either, because all such are hidden inside the so-called *scope* of "$(\forall y)(\forall z)$", that is, the area between the two big brackets. Boolean logic cannot get into this scope since it can neither see nor manipulate its "gate-keepers", the so-called *quantifiers* $(\forall y)(\forall z)$; moreover it cannot see the mathematical objects y and z that these quantifiers refer to.

What is the result? Boolean logic, in its inability to see *any* logical structure *of the type it understands* in either of $(1')$ or $(2')$, "believes" that it just sees atomic formulae in both cases! □

---

[83]"$(\forall x)$" is pronounced "for all x", thus "Any two (sets) y and z" is mathematized by "for all y, for all z", that is, "$(\forall y)(\forall z)$".

[84]These aims are not fully realized because not all true statements of set theory or Peano number theory can be so verified, as Gödel showed in [16].

Clearly, in order to do mathematics, we need to expand the *language* of (Boolean) logic so that we can write down statements about *objects* such as trees, numbers, sets, and the like.

So we need—at a minimum[85]—symbols for specific objects (constants) and unspecified objects (object variables). We will also need—at a minimum—a way to say that two objects are the same.[86]

We should be happy to know that all that we have learned about Boolean logic will be useful and readily applicable in predicate logic. There is nothing that we need to discard or unlearn.

4.1 THE FIRST-ORDER LANGUAGE OF PREDICATE LOGIC

As already remarked, we need to extend the *language* of Boolean logic, if we are to use logic to reason in mathematics and computer science.

First, we will need an infinite supply of *object variables*, that is, variables that are anything but Boolean. For such we will use the letters x, y, z, u, v, w with or without subscripts or primes. For example, x, u, w_{301}''' are all acceptable (names of) object variables.

 As on p. 94 where we generated the infinite list of Boolean variables using just a finite set of subsymbols—$p, q, r, \prime, 0, 1, 2, 3, 4, 5, 6, 7, 8, 9$—we may do so here for object variables, using $x, y, z, u, v, w, \prime, 0, 1, 2, 3, 4, 5, 6, 7, 8, 9$.[87] But we need not worry about this in the main body of Part II (see, however, the Appendix, Section A.4). This "do-not-read-me" comment is solely for the benefit of the picky reader.

By the way, we will drop the qualifier *object* from now on, but we will continue using the qualifier *Boolean* for Boolean variables.

When we use logic to reason in computer science or mathematics, we need, besides variables, additional objects. For example, when we do number theory—that is, the theory of the natural numbers $\mathbb{N} = \{0, 1, 2, 3, \ldots\}$—we also need symbols for *constants* (e.g., 0), *functions* (e.g., $+, \times$), and *predicates*—that is, *names* of relations—(e.g., $<$).

Thus, the first-order language[88] of predicate logic will build on that of Boolean logic by adding to the Boolean alphabet (cf. p. 9) the following:

(1) Symbols for object variables $(x, y, z_{12}'', \ldots)$.

(2) A symbol for equality between *non-Boolean* objects. We use "$=$".

(3) A symbol for "for all"—the quantifier "$\forall$".

---

[85]"At a minimum" because we need also axioms and rules so we can verify by *syntactic means* the "truth" of what we write down.
[86]For objects of *Boolean* type, that is, formulae, "$\equiv$" does that for us.
[87]In this approach, the object variables will be the members of the regular set $\{x, y, z, u, v, w\}\{\prime\}^*(\{\epsilon\} \cup \{1, 2, 3, 4, 5, 6, 7, 8, 9\}\{0, 1, 2, 3, 4, 5, 6, 7, 8, 9\}^*)$.
[88]Why "first-order"? I will explain soon.

(4) Symbols for constants (*non-Boolean!*).

(5) Symbols for functions.

(6) Symbols for predicates.

The qualifier "symbol(s) for" will be henceforth understood and therefore omitted.

Items (1)–(3) are *mandatory* in the alphabet regardless of where we plan to apply our predicate logic. Thus, since they are independent of application, we call them *logical symbols*, just as we so call all the symbols we inherited from the Boolean alphabet.

However, what precise symbols of types (4)–(6) we employ in a first-order language depends on the branch of mathematics, or computer science, where we want to apply predicate logic (as a proving tool), and for that reason these are called *nonlogical symbols*.

4.1.1 Example. (Some Examples of Nonlogical Symbols) To do number theory we employ one constant, 0, three functions, $+$, $\times$ and S—where S is the name of the "$+1$ function" (the so-called *successor*)—and one predicate, $<$.

Pause. But what about other constants (1, 2, 11, 100056) and other functions (e.g., exponentiation) and other predicates (e.g., $\leq$)?

All these can be introduced in terms of the given primitive symbols *using definitions*. It is not in our scope to say how this is done—some such definitions are quite tricky, e.g., the one for exponentiation—but here are some easy ones: "1" is defined as $S0$, "2" as $SS0$, "5" as $SSSSS0$, where we wrote "Sx" for "$S(x)$". Recalling that S is *intended to signify*—hence *intended to behave* like—the successor function, it is clear that these definitions are sensible. After all, $(0+1)+1 = 2$, etc.

Also, $\leq$ is easily introduced by the definition "$x \leq y$ abbreviates $x = y \vee x < y$".

To do set theory one need employ no constants and no functions, but just one predicate, $\in$. All other familiar (and all the unfamiliar) symbols of set theory are built *via definitions*, using just $\in$. For example, constants such as $\emptyset$, predicates such as $\subseteq$, and functions such as $\cup$ are all built in terms of $\in$. □

The *intended behavior* of nonlogical symbols is enforced in each *application* of predicate calculus—such as number theory—by *special axioms*. These behavior-enforcing special axioms are naturally called *special axioms*, but also *nonlogical axioms*. By their nature they are *not* universally applicable to all of predicate logic; rather they are specific to *one* application.

I should mention that what I call an *application* here—or *applied first-order logic* if you will—we nowadays normally call a *theory*. That is, a theory is a toolbox using which we can, in principle, generate all the theorems that together describe the behavior and properties of selected mathematical objects (for example, the sets and the relation $\in$). This toolbox consists of:

(i) A first-order language that has a hand-picked, specific, set of nonlogical symbols that is appropriate for the intended application (e.g., for set theory we just include the predicate $\in$; nothing else).

(ii) Special axioms that give the basic properties of the nonlogical symbols. Intuitively, these state selected fundamental "relative truths" that characterize the theory.

(iii) The logical axioms that are common to all theories. Intuitively, these state "absolute truths" that are valid in all theories.

(iv) Rules of inference (see the next section)

With tools (i)–(iv) we can generate theorems, as will be discussed in the next section.

The purpose of Part II of this volume is to thoroughly acquaint the reader with what *predicate calculus* is, and to equip the reader with a solid technique in using this calculus in *any application*. For this reason, when we teach the calculus, and train the reader in its use, we are obliged to formulate it in terms of unspecified nonlogical symbols, so that we can talk about all possible first-order languages in a unified manner.

Thus, we will denote—in the general description of the alphabet below—the constants by generic names a, b, c, the functions by generic names f, g, h, and the predicates by generic names ϕ, ψ.

These notational conventions will not stop us from giving examples from time to time, where symbols from a specific first-order language are used, e.g., from the language of number theory.

In summary, we have:

4.1.2 Definition. (The Alphabet of the General First-Order Language)   A first-order alphabet consists of a logical part (logical symbols) and a nonlogical part (nonlogical symbols).

The *fixed part* for all first-order alphabets is the *set of logical symbols*, which is an extension of the Boolean alphabet (p. 9). All such alphabets include precisely **L1–L7**:

L1. *Boolean variables.*[89]   These are p, q, r, with or without primes or subscripts— e.g., p', q_{13}, r_{51}'''.

We also have a supply of Boolean metavariables, $\mathbf{p}, \mathbf{q}, \mathbf{r}, \mathbf{q}_{76}''', \ldots$ precisely as on p. 9 and for the reasons already explained there.

L2. *Object variables*—or just "variables". These are x, y, z, u, v, w, with or without primes or subscripts—e.g., $x, u, x', w_{13}, v_{51}'''$.

As in the case for Boolean variables, we will often need to write down expressions such as "if $\vdash A$, then $\vdash (\forall x) A$, for any x" (cf. 6.1.3 later on).

This immediately creates a difficulty: What do we mean by "any x"? There is only one specific x. Thus we employ object metavariables to name or *point to* arbitrary object variables. The symbols for those are chosen analogously

---

[89]Recall that we promised not to say "symbol for". Of course, everything in the alphabet *is* just a symbol.

with those for the Boolean case: $\mathbf{x}, \mathbf{y}, \mathbf{z}, \mathbf{u}, \mathbf{v}, \mathbf{w}$, with or without primes or subscripts—e.g., $\mathbf{x}, \mathbf{u}, \mathbf{x}', \mathbf{w}_{13}, \mathbf{v}_{51}'''$.

Now it is all right to say "if $\vdash A$, then $\vdash (\forall \mathbf{x})A$, for any $\mathbf{x}$" or indeed just "if $\vdash A$, then $\vdash (\forall \mathbf{x})A$" since $\mathbf{x}$ points to an arbitrary object variable, thus rendering the part "for any $\mathbf{x}$" redundant. Remember: "for any $\mathbf{x}$" refers to the variety of what $\mathbf{x}$ *names*, *not* to a variety of metavariables!

L3. Two *symbols* for Boolean *constants*, namely $\top$ and $\bot$.

L4. Brackets, namely, (and).

L5. Boolean connectives, namely, the symbols listed below, separated by commas[90]

$$\neg, \wedge, \vee, \rightarrow, \equiv \qquad (i)$$

L6. The *equality* of objects symbol, "$=$".

L7. The *universal quantifier* symbol, "$\forall$".

The *variable part* for all first-order alphabets is the *set of nonlogical symbols*, a set that is application-specific. We have a different alphabet for each different application.

We can talk about any first-order language without being pinned down to any specific application by using generic symbols. Thus a first-order alphabet must also contain:

NL1. Zero or more *object constants*,[91] which in an application-independent fashion are denoted generically by a, b, c, i.e., lowercase Latin letters from the beginning of the alphabet, with or without primes or subscripts; e.g., a''', b_{19}, c_{519}''.

NL2. Zero or more *functions*, denoted generically by f, g, h—just these lower case Latin letters!— with or without primes or subscripts; e.g., $f', g_{1009}, h_{1000519}''$. Each function has an *arity*, and this strange word means "the number of arguments that the function can take". This is significant; see definition of terms below.

NL3. Zero or more *predicates*, denoted generically by the letters ϕ, ψ—just these two lower case Greek letters!—with or without primes or subscripts; e.g., $\phi', \phi_{123409}''', \psi_{1509}''$. Each predicate too has an arity, and this also is significant; see definition of atomic formulae below. □

4.1.3 Remark. (1) The term *arity* was made up by mathematicians and logicians. It is derived from words such as "bin*ary*", "un*ary*", "tern*ary*". A ternary function has three arguments: Its arity is 3.

(2) How does one *know* the arity of, say, f_{15}'''? Well, this is a silly question; "f_{15}'''" is one of (infinitely) many generic symbols that we use to denote functions *when we*

---

[90] The commas are not part of the alphabet.
[91] Symbols for constants.

do not want to work in a specific theory and its language. Just as in algebra we say "Let f be a function of ten variables ... " and thus introduce into the discussion the symbol f, accordingly, when we are training in predicate logic we can say similar things: "Let f_{15}''' be a function of arity 9 ... " Thus it is not an issue of *knowing*—f_{15}''' is what we say it is; it has no fixed status. In a different *application-independent* discussion we may find ourselves saying "Let f_{15}''' be a function of arity 99 ... ", and this is perfectly fine.

On the other hand, if one speaks a *specific language*, say, that of set theory, then the symbol $\in$ is the same throughout set theory. We cannot say today "Let $\in$ have arity 3" and say tomorrow "Let $\in$ have arity 1". In fact, its arity is fixed once and for all at the time the alphabet for the language of set theory is given (in this case it is 2).

(3) We were tempted to use the generic names P, Q, R for predicates, but that clashes with the naming of formulae (see 1.1.2). Hence we suggested ϕ and ψ.

You might ask: So, what is wrong with using P, Q, R and living with the clash? Is not a predicate a formula, after all?

No, not really. For example, "$<$" is a predicate, and it is clear (on intuitive grounds, even before we give Definition 4.1.13 below) that it is not a formula. On the other hand, "$2 < 3$" *is* a formula, but it is not a predicate.

A configuration consisting of a predicate *acting on arguments is* a formula (a very simple one at that). □

We have talked about alphabets and languages. We have defined *alphabet*. So, what is a *language*? Well, the language consists of all those "important" *strings* that we can build using the symbols of our alphabet. In Boolean logic, *the* set of important strings—the language—is the set of all formulae (**WFF**). In predicate logic, we have a richer language. Not only can we use the alphabet to write down *statements* (formulae), but we can also use it to write down *objects* (*terms*).

We will need to define the syntax of terms first, as it will become obvious.

Intuitively, the simplest objects are the variables and the constants. We can build more complicated objects by applying functions to objects that we already have. For example, in the language of number theory, x and y are simple objects; $x + y$ is a bit more complicated.

This is the idea behind the definition below, which is formulated as a "calculation" in the style of 1.1.3.

4.1.4 Definition. (Term-Calculation or Term-Parse)

A *term-calculation* (or term-parse) is any finite (ordered) sequence of strings that we may write respecting the following two requirements:

(1) At any step we may write any symbol from **L2** or **NL1** of the alphabet (4.1.2).

(2) At any step, if f is a function of arity n and we *already have written down* the strings (without the quotes, of course) "t_1", "t_2", ..., "t_n", then we may write the string "$ft_1t_2 \ldots t_n$". □

Imitating 1.1.5, we next define terms:

4.1.5 Definition. (Terms) A string t over the alphabet of 4.1.2 will be called a *term* iff it is a string written at some step of some term-calculation.

The set of terms we will denote by **Term**.   □

4.1.6 Remark. (1) We will use the generic symbols t and s, with or without subscripts or primes, to denote terms. Thus, t'''_{99} denotes some term. Therefore, these names are syntactic variables (metavariables) for terms.

(2) In practice, we use a more friendly notation than "$ft_1t_2 \ldots t_n$". We write instead "$f(t_1, t_2, \ldots, t_n)$". Note that the comma "," is not in our alphabet, but in the metalanguage we have infinite leeway when it comes to serving user-friendliness.

Similar conventions apply to specific languages. For example, in the language of number theory the correct notation is "$+xy$". However, one sacrifices absolute syntactic correctness in the interest of user-friendliness and writes "$x + y$" instead. These conventions are in the same spirit as the conventions regarding the elimination of "redundant brackets" and are made in the metatheory.

(3) Analogous remarks to those in 1.1.6 apply here; thus, on one hand, one can do induction on the *complexity of terms* to metaprove properties of terms. We define the complexity of a term to be *the number of function symbols*—counting each repetition as a new occurrence—appearing in it. This is a natural measure as the complexity increases with every step such as (2) of Definition 4.1.4. Thus, x has complexity 0, fx (assuming the arity of f is 1) has complexity 1, $fffx$ has complexity 3, and $gyfffx$[92]—where g has arity 2—has complexity 5.

On the other hand, an inductive (recursive) definition of terms is possible, as follows.   □

4.1.7 Definition. (Alternative (Recursive) Definition of Terms) The set of terms is the *smallest* set of strings, **Term**, that satisfies

(1) *All* variables and constants are in **Term**.

(2) If f is an n-ary (of arity n) function and $t_1, t_2, \ldots, t_n$ are in **Term**, then so is $ft_1t_2 \ldots t_n$.   □

We can now define the simplest possible formulae, the ones we can write down without using any Boolean connective.

4.1.8 Definition. (Atomic Formulae) The following are the *atomic formulae* of predicate calculus:

(a) Any Boolean variable and any Boolean constant.

(b) The string $t = s$ for any terms t and s (possibly, t and s name the same term).[93]

(c) For any predicate ϕ of arity n, and any n terms $t_1, t_2, \ldots, t_n$ the string $\phi t_1 t_2 \ldots t_n$.

---

[92] $gyfffx$ in friendly (!) notation is written as $g(y, f(f(f(f(x)))))$.

[93] Let us recall that t and s are meta*variables* for terms.

We denote the set of all atomic formulae by **AF**. □

4.1.9 Remark. (1) As in the case of terms, we opt for the friendly *informal* notation *with* brackets and commas: Rather than the correct "$\phi t_1 t_2 \ldots t_n$" we abuse syntax and write "$\phi(t_1, t_2, \ldots, t_n)$".

(2) "=" is a *very special logical* (the *only* logical one!) 2-ary (or binary) predicate, that of equality. Our syntactic rule uses a so-called *infix* notation for the associated formula; "$t = s$" rather than "$= ts$".

Some texts use "$\approx$" (e.g., [13]) instead of "=" in order to avoid confusion with the informal (metamathematical) "=". We will not do that; rather we will allow the context to fend for itself. Note that [17] uses "=" for at least three different roles: informal equality, formal equality, and as a conjunctional alias of "$\equiv$" in equational proofs.

We overload the "=" symbol a bit less by letting "$\Leftrightarrow$" perform the last role. *We will never let "=" stand for "$\equiv$".* □

We can finally define *all* formulae of predicate calculus exactly as we did with the Boolean case, starting with the concept of *formula-calculation*.

4.1.10 Definition. (Formula-Calculation or Formula-Parse; First-Order Case)
A *formula-calculation* (or formula-parse) is any finite (ordered) sequence of strings that we may write respecting the following *four* requirements:

(1) At any step we may write any atomic formula (member of **AF** defined in 4.1.8).

(2) At any step we may write the string $(\neg A)$, *provided* we have already written the string A.

(3) At any step we may write any of the strings $(A \wedge B)$, $(A \vee B)$, $(A \rightarrow B)$, $(A \equiv B)$, *provided* we have already written the strings A and B.

(4) At any step—and for *any choice* of variable **x**—we may write the string $((\forall \mathbf{x})A)$, *provided* we have already written the string A.

"$\forall$" is called the universal quantifier and is pronounced "for all". The string "$(\forall \mathbf{x})$" we pronounce "for all **x**". We say that the subformula A in $((\forall \mathbf{x})A)$ is the *scope* "$(\forall \mathbf{x})$". □

4.1.11 Remark. (a) Case (4) in 4.1.10 is *new*. It did not occur in Definition 1.1.3. We say that **x** in $((\forall \mathbf{x})A)$ is a quantified variable (we also call it *bound*, but more on this shortly). In a *first-order language*, we are allowed to quantify only first-order variables, as we call the object variables. We are *not* allowed to quantify second- (or higher-) order variables such as names of predicates or functions. For example, we have no way, in a first-order language, to write down a formula that says "for all functions $f \ldots$".

(b) Since cases (2)–(3) in the above definition are the same as in 1.1.3 and since, moreover, **AF** includes all Boolean variables and constants, it follows that *every formula-calculation in the sense of 1.1.3 is also valid according to 4.1.10.* □

The picky reader will probably ask: "You said '[F]or example, we have no way, in a first-order language, to write down a formula that says "for all functions f ...".' But surely in set theory, which has been repeatedly mentioned as an 'applied case' of first-order logic, we *must* be able to say things like 'for all functions f ...'?" Well, yes, we can. What I meant is that a first-order language does not have the ability to say "$(\forall f)$", where f is a *function symbol*.

However, in set theory, *functions* are not symbols of the alphabet, but are *defined*, as we say *extensionally*,[94] i.e., as *sets* of ordered pairs, pairs themselves being defined (implemented) as certain sets.[95]

Now, axiomatic set theory has normally just one type of variable, "set". We can certainly say "$(\forall x)$" if x is an object (set) variable. Since a function, extensionally, is a set, I can say "(for all functions)(...)" by saying instead "$(\forall x)(x$ is a function $\rightarrow$ ...)".

4.1.12 Example. In the *first* step of any formula-calculation, only requirement (1) of Definition 4.1.10 is applicable, since the other three require the existence of prior steps. Thus in the first step we may write *only* an atomic formula. In all other steps, *all* the requirements (1)–(4) are applicable.

Here is a calculation (the comma is not part of the calculation; it just separates strings written in various steps):

$$p, \top, (\neg\top), q$$

Verify that the above obeys Definition 4.1.10.

Here is a more interesting one:

$$p, q, (p \vee q), (p \wedge q), ((p \vee q) \equiv q), ((p \wedge q) \equiv p), \Big(((p \vee q) \equiv q) \equiv ((p \wedge q) \equiv p)\Big)$$

Both previous calculations are also calculations in the sense of 1.1.3 and were written down in an earlier example. Here are a few that are not, but which are valid calculations as far as 4.1.10 is concerned:

$$p, ((\forall z)p)$$

$$x = a, ((\forall x)x = a), p, (((\forall x)x = a) \wedge p), u = v, (u = v \rightarrow (((\forall x)x = a) \wedge p))$$

Recall that "a" is a constant.

$$x = y, (\neg x = y), ((\forall z)(\neg x = y))$$

$$x = y, ((\forall x)x = y), ((\forall x)((\forall x)x = y))$$

---

[94]That is, by what they include as members—not behaviorally or "intentionally".

[95]Normally via Kuratowski's definition, by which the ordered pair "(x, y)" is an abbreviation of the set $\{\{x\}, \{x, y\}\}$.

Particularly important are the first—$p, ((\forall z)p)$—and the last two calculations: The first two of these exemplify that *we do not care* whether a variable z occurs in "A" before we form "$(\forall z)A$". The "A" here were p and $(\neg x = y)$ respectively.

The last one illustrates that it *is* allowed to place one $(\forall x)$ immediately in front of another.

Of course, why these actions are legal is implicit in (4) of 4.1.10: Absolutely *no restrictions* are placed on either **x** or A (other than A must be already written).   □

4.1.13 Definition. (First-Order Formulae) A string A over the alphabet 4.1.2 will be called a *first-order formula* or a *well-formed-formula* iff it is a string written at some step of some formula-calculation conducted as per 4.1.10.

The set of first-order formulae we will denote by **WFF**. A member of **WFF** is often called a "wff" (or just a formula).   □

My apologies to the reader for using "**WFF**" and "wff" both for the *first-order formulae* of Part II and for the *Boolean expressions* of Part I.

In my defense, "**WFF**" from now on will be in the former (Part II) sense exclusively, and, in any case, any Boolean expression *is* also a wff in the new (4.1.13) sense as we remarked above (cf. 4.1.11(b)).

As in the cases of Boolean formulae, and members of **Term**, a string is a wff (member of the set **WFF** defined 4.1.13) iff this can be certified by showing that the string is put together using certain strings that we *already know* are in **WFF**. And again, as was done twice before, this leads to a recursive (inductive) definition of **WFF**:

4.1.14 Definition. (Recursive Definition of WFF) **WFF** is defined as the smallest set of strings that contains all the members of **AF** and moreover satisfies:
(1) If A is in **WFF**, then so are $(\neg A)$ and $((\forall \mathbf{x})A)$.
(2) If A and B are in **WFF**, then so are $(A \wedge B)$, $(A \vee B)$, $(A \rightarrow B)$, and $(A \equiv B)$.   □

Clearly, the complexity of wff increases every time we perform a step (2), (3), or (4), in a formula-calculation (4.1.10). Thus we define

4.1.15 Definition. (Complexity of Members of WFF) The complexity of a wff is the total number of occurrences of $\forall, \neg, \wedge, \vee, \rightarrow, \equiv$ in the formula, counting each repetition as a new occurrence.   □

4.1.16 Example. $x = y$ and p have complexity 0 each. $((\forall x)((\forall y)(\neg x = y)))$ has complexity 3. $((\forall y)((\neg x = y) \wedge p))$ and $(((\forall y)(\neg x = y)) \wedge p)$ each have complexity 3.

But as an aside, note that in the first of the last two examples p is in the scope of $(\forall y)$, whereas in the second it is not.   □

4.1.17 Definition. (The Existential Quantifier) We introduce a new symbol in the *metatheory*—an abbreviation, that is, not a formal symbol—$\exists$, called the *existential quantifier*, pronounced "there exists" or "for some":

For any formula A, the string $((\exists x)A)$—a string of the metatheory this is; not a string of **WFF**—abbreviates the formal string (member of **WFF**) $(\neg(\forall x)(\neg A))$. $\square$

4.1.18 Remark. (For the reader who will consult [17]: Notation Translation)
Where we write $((\forall x)A)$ and $((\exists x)A)$, [17] uses

$$(\forall x \mid : A) \text{ and } (\exists x \mid : A) \text{ respectively} \tag{1}$$

If you are wondering, "What on earth do both "|" and ":" do, one next to the other?" the reason is that (1) is a special case of

$$(\forall x \mid B : A) \text{ and } (\exists x \mid B : A)$$

which in the standard notation of the computer science and mathematical literature—which we follow—are written as

$$((\forall x)(B \rightarrow A)) \text{ and } ((\exists x)(B \wedge A))$$

respectively.

An expression like $((\forall x)A)$ we pronounce "for all x, A holds". An expression like $((\exists x)A)$ we pronounce "for some x, A holds", but also "there exists an x, such that A holds".

In an expression such as $((\forall x)A)$, x stands for a *specific* variable among x, y, z_{899}''', etc., where we either do not know which, or do not care. Thus, *intuitively*, it says "for all *values* of x ...", *not "for all* variables $x, x'', y_{67} ...$". Similarly, by "for some x, A holds" we understand "for some value of x, A holds".

This pronunciation, as well as the clarification regarding "value", are consistent with the *intended meaning* of quantifiers. This intended meaning not only tells us what is the right way to pronounce $((\forall x)A)$ but will also guide us to choose appropriate logical axioms (next section) that *capture this intended meaning purely syntactically*.

I emphasize that our main task in predicate calculus will be to *calculate theorems*, that is, to write proofs. Semantic ideas—with the exception of those borrowed from Boolean logic—will have absolutely no role in the *writing* of our proofs.

However, keeping our intuition informed and active, in particular understanding what an expression like $((\forall x)A)$ is meant to say, will often assist our imagination toward *discovering* proofs. Anything goes in the discovery stage, even consulting the Oracle at Delphi.[96] The actual writing stage is, however, restricted to be syntactic (except as noted in the previous paragraph). $\square$

---

[96]*How* we guess the next step is our business. However, we write and document a proof according to *rules*.

Before we go on, and in the interest of making notation friendlier (and sloppier), we augment Remark 1.1.11 here to take care of the additional symbols of first-order languages:

4.1.19 Remark. (Priorities and Bracket Reduction) Our previous agreement (in Remark 1.1.11) on how to be sloppy and get away with it remains essentially the same, now augmented to take care of $\forall$ as well:

Certain brackets are *redundant*—and hence can be removed—from a formula written according to Definitions 4.1.13 and 4.1.14, *still allowing it to say the same thing as before*:

(1) Outermost brackets are redundant.

   As in 1.1.11, for the next two cases it is easiest to think of the process in reverse: how to reinsert correctly (as per Definition 4.1.13) any omitted brackets.

(2) Any other pair of brackets is redundant, if its presence (as dictated by 4.1.13) can be understood from the *priority*, or *precedence*, of the connectives. Higher-priority connectives *bind before* lower-priority ones. That is, if we have a situation where a subformula A of a formula has already been reconstructed as per 4.1.13, and is claimed by two distinct connectives $\circ$ and $\diamond$, among those in $(*)$ below, as in "$\ldots \circ A \diamond \ldots$", then the higher-priority connective "glues" first. This means that the implied brackets are (reinserted as) "$\ldots \circ A) \diamond \ldots$" or "$\ldots \circ (A \diamond \ldots$" according as $\circ$ or $\diamond$ has the higher priority respectively.

   The order of priorities (decreasing from left to right, but $(\forall x)$ and $\neg$ having equal priority) is *agreed to be*:

   $$\left\{ \begin{matrix} (\forall x) \\ \neg \end{matrix} \right\}, \wedge, \vee, \rightarrow, \equiv \tag{$*$}$$

(3) In a situation like "$\ldots \diamond A \diamond \ldots$"—where A has already been reconstructed as per 4.1.13, and $\diamond$ is any connective listed in $(*)$ above, other than $\neg$ or $(\forall x)$—the right $\diamond$ acts before the left. Thus the implied bracketing is "$\ldots \diamond (A \diamond \ldots$".

   Similarly, $\neg\neg A$ is short for $\neg(\neg A)$, $\neg(\forall x)A$ is short for $\neg((\forall x)A)$, $(\forall x)\neg A$ is short for $(\forall x)(\neg A)$, and $(\forall x)(\forall y)A$ is short for $(\forall x)((\forall y)A)$.

   We say that *all connectives are right associative*. This applies to $(\exists x)$ as well when this abbreviation is used; after all, the convention of this remark applies to metatheoretical notation.

   It is important to emphasize:

(a) This "agreement" results in a shorthand notation. Most of the strings depicted by this notation are *not* correctly written formulae, but this is fine: Our agreement allows us to decipher the shorthand and *uniquely recover* the correctly written formula we had in mind.

(b) The agreement on removing brackets is a *syntactic* agreement.

In particular, right associativity says simply that, e.g., $p \vee q \vee r$ *is shorthand for* $(p \vee (q \vee r))$ *rather than* $((p \vee q) \vee r)$. □

4.1.20 Example. Here are some examples of simplified notation:

(1) Instead of $(u = v \rightarrow (((\forall x)x = a) \wedge p))$ we may write simply

$$u = v \rightarrow (\forall x)x = a \wedge p$$

In the simplified notation the inexperienced reader may have some trouble readily seeing that p is *not* in the scope of $(\forall x)$. The rule of thumb is

Whenever in doubt, use extra brackets!

(2) Instead of $((\forall z)(\neg x = y))$ we write simply

$$(\forall z)\neg x = y$$

(3) Instead of $((\forall x)((\forall x)x = y))$ we may write

$$(\forall x)(\forall x)x = y$$

Note that we do not want to eliminate the brackets around a quantifier, treating "$(\forall \mathbf{x})$"—*and the defined* "$(\exists \mathbf{x})$"—*as compound symbols.* □

If "$(\forall \mathbf{x})A$" is meant to say that "for all values of $\mathbf{x}$, A holds", then this is analogous to

$$\sum_{i=1}^{4} i^2 \tag{1}$$

which says, "For all integer values of i—from 1 to 4 inclusive—compute the result i^2 and then add all these four results." That is, compute

$$1^2 + 2^2 + 3^2 + 4^2$$

Note that in expression (1) we are not allowed to substitute values into i. That is, something like

$$\sum_{3=1}^{4} 3^2 \tag{2}$$

is totally meaningless as it would say, "For all integer values of 3—from 1 to 4 inclusive—compute the result 3^2 and then add all these four results." Nonsense!— "3" cannot obtain any values other than 3.

Thus, i in (1) is *unavailable for substitution*. We say that it is a *bound variable*. On the other hand, the expression

$$\sum_{i=1}^{4} (i + x)^2 \tag{3}$$

means

$$(1 + x)^2 + (2 + x)^2 + (3 + x)^2 + (4 + x)^2$$

Thus we *may*, if we wish, substitute specific values into x, say, 9, and compute

$$(1+9)^2 + (2+9)^2 + (3+9)^2 + (4+9)^2$$

for short

$$\sum_{i=1}^{4}(i+9)^2$$

Thus x in (3) *is* available for substitution, or as we say, *free*.

Analogously, x is bound in the expression below, while z is free:

$$(\forall x)x = z$$

We capture this by a definition.

4.1.21 Definition. (Bound and Free Occurrences) An occurrence of a variable $\mathbf{x}$ in a formula A is characterized by the position—from left to right: first, second, etc.—where $\mathbf{x}$ occurs as a substring in A (cf. 1.3.6). For example, the 3rd occurrence of x in the following formula is shown boxed:

$$x = y \rightarrow x = y \vee (\forall \boxed{x})x = z$$

We say that an occurrence of $\mathbf{x}$ in A is *bound* iff the occurrence is *the* $\mathbf{x}$ that occurs in a substring $(\forall \mathbf{x})$ of A, or if it occurs in the *scope* of some $(\forall \mathbf{x})$ that occurs in A.

An occurrence of $\mathbf{x}$ in A that is not bound is called *free*. Below I show all bound occurrences of x in the foregoing example (singly boxed):

$$\boxed{\boxed{x}} = y \rightarrow \boxed{\boxed{x}} = y \vee (\forall\boxed{x})\boxed{x} = z$$

We usually count *bound occurrences*—and *free occurrences*—from left to right, *separately*, thus, we have above two bound occurrences, shown boxed, and two free occurrences, shown doubly boxed. We also have two free occurrences of y and one of z. □

4.1.22 Remark. We sometimes say, "The nth bound occurrence of $\mathbf{x}$ *belongs to*—or is *bound to*—the mth occurrence of $(\forall \mathbf{x})$ in the formula A." What do we mean by that?

We mean one of the following:

(1) The nth bound occurrence of $\mathbf{x}$ is *the* $\mathbf{x}$ in the mth occurrence of $(\forall \mathbf{x})$, or

(2) The scope of the mth occurrence of $(\forall \mathbf{x})$ is the shortest among the scopes, of *any* $(\forall \mathbf{x})$, which contain the nth bound occurrence of $\mathbf{x}$.

Thus, in

$$(\forall\boxed{x})\left(\boxed{x} = y \vee (\forall\boxed{x})\boxed{\boxed{x}} = z \wedge \boxed{x} = y\right)$$

the three boxed bound x belong to the first (leftmost) $(\forall x)$ while the doubly boxed ones belong to the second one.

Similarly, in

$$(\forall x)(\forall x)\boxed{x} = y \tag{1}$$

the boxed x belongs to the second (rightmost) $(\forall x)$.

This process of finding where a bound variable **x** belongs—to which $(\forall \mathbf{x})$—hinges on formula-calculations: *An* **x** *belongs to that* $(\forall \mathbf{x})$ *that took away its freedom in the course of a formula-calculation.*

For example, a formula-calculation for (1)—with reduced bracket notation—is

$$x = y, (\forall x)x = y, (\forall x)(\forall x)x = y \qquad \square$$

 This process of finding to which "$(\forall \mathbf{x})$" a bound variable **x** belongs is analogous to that of finding the "block-head" where a local variable **x** is declared in a language like Algol or Pascal: One goes back (leftward in the program string) until one reaches the first occurrence of a declaration for **x**.

4.1.23 Definition. (Subformulae) We define the concept "B is a subformula of A" inductively:

(1) A is atomic. Means that the strings A and B are identical.

(2) A is $\neg C$. Means that either B is the same string as A, or it is a subformula of C.

(3) A is $(\forall \mathbf{x})C$. Means that either B is the same string as A, or it is a subformula of C.

(4) A is $C \circ D$, where $\circ \in \{\wedge, \vee, \rightarrow, \equiv\}$. Means that either B is the same string as A, or it is a subformula of C, or of D (or both). $\qquad \square$

4.1.24 Exercise. Show by induction on A, that if we replace any (possibly all) occurrences of a subformula of A by some Boolean variable, then the resulting string is a formula. $\qquad \square$

 4.1.25 Remark. (Abstractions) We saw that every Boolean expression as defined in 1.1.5 is also a first-order wff, as defined in 4.1.13 (cf. 4.1.11). The former is a special case of the latter.

Much more useful to us, it turns out, is the fact that Boolean expressions are *abstractions* of first-order formulae. We already said so in the preamble of the current chapter (p. 113), but we will now make the case; the *why* and *how*.

First the *how*: To "abstract" means to discard information that you do not need so that you can focus on what really matters unhindered by unnecessary details. We can view (first-order) formulae as Boolean formulae if we become oblivious to the presence of all *non-Boolean elements*—those that speak of *objects*—and concentrate instead on the Boolean structure, that is, how the Boolean connectives $\neg, \wedge, \vee, \rightarrow, \equiv$ connect things up. The non-Boolean elements are, of course:

The (object) variables, constants, predicates, functions, "$=$", and "$\forall$"   (1)

How do we become oblivious to such elements of syntax—as enumerated in (1)—that occur in a first-order formula A?

We implement our unconcern by covering those elements up! That is, *by first identifying the* shortest *possible subformulae of A that contain such symbols* and then *replacing* said subformulae by *new*[97] Boolean variables.

Important! *In the actual implementation of this procedure we* do not *really replace these subformulae by new symbols. They themselves, as written, are (names of) these* new Boolean variables, *a fact that automatically satisfies the remark in the preceding footnote.*

Just like the existing Boolean variables of the alphabet, such as $p_{777999}^{''''}$, these too are compound symbols—*such as "$x = y$" or "$(\forall x)x = z$"—the* elementary *subsymbols of which, e.g., "$($, $\forall$, x", are invisible to Boolean logic as separate syntactic entities, just like the individual subsymbols "$7, 9, {}'$" of* $p_{777999}^{''''}$ *are invisible as symbols of any structural significance vis à vis 1.1.3.*

Exercise 4.1.24 guarantees that these substitutions of subformulae by Boolean variables generate formulae. Of course, viewing a first-order subformula as a new Boolean variable, *in effect, substitutes the subformula by the said variable.*

In essence we are saying:

I do not care what these subformulae of A say about objects. I am only interested in how these subformulae interconnect via $\neg, \wedge, \vee, \rightarrow$ *and* $\equiv$, *that is, in the Boolean structure of A.*

For example (using simplified notation), if A is

$$p \rightarrow x = y \vee (\forall x)\phi x \wedge q \text{ (Note that } q \text{ is } not \text{ in the scope of } (\forall x))$$

then the abstraction is

$$p \rightarrow \mathbf{p}' \vee \mathbf{p}'' \wedge q$$

where I used metavariables, in order to emphasize the form of the "abstraction", $\mathbf{p}'$ for "$x = y$" and $\mathbf{p}''$ for "$(\forall x)\phi x$".

Needless to say, I view $\mathbf{p}'$ and $\mathbf{p}''$ as distinct, and different from any Boolean variables of alphabet 4.1.2—the former because, by inspection, the actual names "$x = y$" and "$(\forall x)\phi x$" these metavariables stand for are distinct strings; the latter by the "Important" italicized remark above. *This obvious comment does not deserve repetition.*

If A is

$$x = y \rightarrow x = y \vee z = v \tag{1}$$

then the abstraction is

$$\mathbf{p} \rightarrow \mathbf{p} \vee \mathbf{q}$$

again using the metavariables only for notational emphasis that indicates our indifference to the exact first-order structure of $x = y$ and $z = v$.

---

[97]"New" means that they do not already occur in A as members of first-order alphabet 4.1.2.

Note that the abstraction of (1) is a tautology.

A more interesting example is the following, which shows that some of the shortest subformulae that contain non-Boolean elements will be totally suppressed in the final abstraction:

$$(\forall x)(x = y \rightarrow (\forall z)z = a \vee q)$$

is just a (new) Boolean variable, **p**, since the shortest subformulae that contain non-Boolean elements have been identified (via enclosing boxes) as follows:

$$(\forall x)(\boxed{x = y} \rightarrow \boxed{(\forall z)\boxed{z = a}} \vee q) \tag{2}$$

so that the abstraction process, in slow motion, and working from inside out is: First obtain

$$(\forall x)(\mathbf{q}' \rightarrow \boxed{(\forall z)\mathbf{q}''} \vee q) \tag{3}$$

then

$$\boxed{(\forall x)(\mathbf{q}' \rightarrow \mathbf{q}''' \vee q)} \tag{4}$$

and finally take care of this last box and call it, say, "**p**".

Similarly we handle an A that is

$$p \rightarrow x = y \vee (\forall x)(\phi x \wedge q)\text{—}q \text{ in the scope of } (\forall x) \tag{5}$$

Then the abstraction is marked as

$$p \rightarrow \boxed{x = y} \vee (\forall x)(\boxed{\phi x} \wedge q)$$

and in the final stage yields

$$p \rightarrow \mathbf{p}' \vee \mathbf{p}''$$

It should be clear (cf. 4.1.26) that the subformulae of the following two types, (a) atomic—but not Boolean or (b) of the form $((\forall \mathbf{x})A)$, are precisely the ones that get abstracted into (i.e., name) *new Boolean variables. Not all such identified subformulae of some formula A need appear in the final result of the abstraction of A (cf. (3) and (4) above; $\mathbf{q}', \mathbf{q}'', \mathbf{q}''$ are lost). Subformulae of the types (a) and (b) are called* prime *(e.g., [45, 53]).*

But *why do we want to abstract*? Abstractions are extremely useful in predicate logic. On one hand, we have the obvious benefit, *simplification*. On the other, by allowing us to *view first-order formulae as Boolean formulae*, abstractions enable us to use, *in predicate logic, all the semantic (truth table) and syntactic (proof) techniques that we have learned in Part I—including 3.2.1.*

To be sure, additional techniques that allow us to handle "$\forall$" and "$=$"— which abstraction totally hides from view—will be necessary, so we must not *expect to reduce predicate calculus totally to propositional calculus!*

Thus, in the context of first-order logic, we have a new view of Boolean variables:

A Boolean variable denotes a statement about objects, but we either do not know what it says, or we do not care what it says about these objects.

With only a modicum of practice, we will find that we do not have to rewrite a formula, introducing new Boolean metavariables, in order to view it abstractly. For example, we should be able to see at once the (final) boxed stage that identifies the Boolean structure of (5):

$$p \rightarrow \boxed{x = y} \lor \boxed{(\forall x)(\phi x \land q)}$$

By the way, boxing the *shortest* subformulae in our process of abstraction is important as it maximizes the number of Boolean connectives that remain "uncovered". We suppress details about *objects* but keep *all* the detail of the *Boolean structure*.   □

4.1.26 Exercise. Verify the statement I made earlier, in 4.1.25, namely, "It should be clear that the subformulae of the following two types, (a) atomic—but not Boolean or (b) of the form $((\forall \mathbf{x})A)$, are precisely the ones that get abstracted into (*i.e., name*) new Boolean variables."

Namely, show that any non-Boolean symbol of a formula belongs to a shortest subformula of the type (a) or (b).   □

We may now define:

4.1.27 Definition. (Tautologies and Tautological Implications) We say that a first-order formula A is a tautology, and write $\models_{\text{taut}} A$, iff the *abstraction* of A is a tautology. In first-order logic, we write $\Gamma \models_{\text{taut}} A$ iff the abstractions of the formulae in Γ tautologically imply the abstraction of A.   □

Before we leave the section on language, we need a few more definitions, notably concepts of substitution.

As in algebra, we want to be able to substitute an object for a variable that accepts such substitutions (i.e., it is a free variable). But for the purpose of applying Leibniz we also want to be able to substitute a formula A into a Boolean variable **p** (cf. 1.3.15).

For any terms s and t and variable $\mathbf{x}$, the notation (in the metatheory) "$s[\mathbf{x} := t]$" will denote the result of replacing all original occurrences of $\mathbf{x}$ in s by t. Similarly, for any formula A, variable $\mathbf{x}$ and term t, the (informal) name $A[\mathbf{x} := t]$ will mean the result of replacing all original free occurrences of $\mathbf{x}$ in A by t.

In the case of a formula A, we must be careful when to allow such a substitution to take place. For example, $(\exists x)\neg x = y$ says, *intuitively*, that *no matter what the (value of) y, there is a (value of) x that is different*.

We expect that the meaning that we just expressed in English should be independent of what name we use for the free variable y.

Yet, if we allowed substitutions of terms into y recklessly we would in particular allow the substitution

$$\Big((\exists x)\neg x = y\Big)[y := x]$$

which results into $(\exists x)\neg x = x$. But this has a totally different meaning from the original: It says that there is a (value) x that is not equal to itself, a clearly absurd suggestion!

This motivates us to disallow the *completion* of the operation of substitution if it results in a free variable $\mathbf{x}$ getting in the scope of $(\forall \mathbf{x})$—getting *captured*, as we say.

This is taken care of in the following (recursive!) definition, which has two parts: one for $s[\mathbf{x} := t]$ and one for $A[\mathbf{x} := t]$.

In order to read the following definitions on substitution correctly—"operations" such as $[\mathbf{x} := t], [\mathbf{p} := B]$ and $[\mathbf{p} \setminus B]$—we emphasize that the operation takes place in the metatheory and has the highest priority against all other "formal or informal operations" such as $\forall, \exists, =, \neg, \wedge, \vee, \rightarrow, \equiv$. For example, $(\exists \mathbf{x})A[\mathbf{p} := B]$ means $(\exists \mathbf{x})\big\{A[\mathbf{p} := B]\big\}$ and $t = s[\mathbf{x} := s']$ means $t = \big\{s[\mathbf{x} := s']\big\}$, where the symbols "$\{, \}$" are here meta-brackets inserted to indicate order of application of "operations". By abuse of notation—"$(t = s)$" being illegal—we write $(t = s)[\mathbf{x} := s']$ for $\big\{t = s\big\}[\mathbf{x} := s']$. Once more, these operations are *left*-associative, e.g., $A[\mathbf{p} := B][\mathbf{q} := C]$ means $\big\{A[\mathbf{p} := B]\big\}[\mathbf{q} := C]$.

4.1.28 Definition. (Substitution of Terms into Variables) *In what follows, we allow "$=$" to appear only formally, thus* unlike *Definition 1.3.15, instead of also using "$=$" metatheoretically, we will use the verb* is *for equality between strings.*

In the interest of generality, the definitions are given in terms of the metavariables $\mathbf{x}$ and $\mathbf{y}$ that name arbitrary variables. Since $\mathbf{x}$ and $\mathbf{y}$ name arbitrary variables, it is conceivable that they name the same variable. If we want to claim the opposite, we must explicitly say so: "$\mathbf{x}$ and $\mathbf{y}$ are different" or "$\mathbf{x}$ (not $\mathbf{y}$)".

(1) The meaning of "$s[\mathbf{x} := t]$"—i.e., its expansion—is given by induction (recursion) on the complexity of the term s:

$$s[\mathbf{x} := t] \text{ is } \begin{cases} s & \text{if } s \text{ is a constant or a variable (not } \mathbf{x}) \\ t & \text{if } s \text{ is } \mathbf{x} \\ f(s_1[\mathbf{x} := t], \ldots, s_n[\mathbf{x} := t]) & \text{if } s \text{ is } f(s_1, \ldots, s_n) \end{cases}$$

where we used the notation with brackets and commas (cf. 4.1.6) in the general case above.

(2) The definition says that we are to replace every *free* occurrence of $\mathbf{x}$ in A by t, *but* if for some $\mathbf{y}$ that occurs in t—free, of course—there is a subformula $(\forall \mathbf{y})B$ of A where $\mathbf{x}$ occurs free, then we abort the operation and declare $A[\mathbf{x} := t]$ *undefined*.

The meaning (expansion) of "$A[\mathbf{x} := t]$" is given by induction (recursion) on the complexity of the formula A. We use reduced-brackets notation, as usual:

$$
A[\mathbf{x} := t] \text{ is }
\begin{cases}
\phi(s_1[\mathbf{x} := t], \ldots, s_n[\mathbf{x} := t]) & \text{if } A \text{ is } \phi(s_1, \ldots, s_n) \\
s_1[\mathbf{x} := t] = s_2[\mathbf{x} := t] & \text{if } A \text{ is } s_1 = s_2 \\
\neg C[\mathbf{x} := t] & \text{if } A \text{ is } \neg C \\
C[\mathbf{x} := t] \circ D[\mathbf{x} := t] & \text{if } A \text{ is } C \circ D \\
A & \text{if } A \text{ is one of } p, \top, \bot, (\forall \mathbf{x})B \\
(\forall \mathbf{y})B[\mathbf{x} := t] & \text{if } A \text{ is } (\forall \mathbf{y})B, \\
& \text{where } \mathbf{y}(\text{not } \mathbf{x}) \text{ does } not \text{ occur in } t \\
& or \ \mathbf{x} \text{ is not free in } B \\
\text{undefined} & \text{if } A \text{ is } (\forall \mathbf{y})B, \\
& \text{where } \mathbf{y}(\text{not } \mathbf{x}) \ does \text{ occur in } t \\
& and \ \mathbf{x} \text{ is free in } B
\end{cases}
$$

where $\circ$ above is one of $\wedge, \vee, \rightarrow, \equiv$.

In each case above, the left-hand side, $A[\mathbf{x} := t]$, is defined iff all the needed right-hand side substitutions are defined—e.g., $C[\mathbf{x} := t]$ and $D[\mathbf{x} := t]$ in the case of $\circ$. □

The definition is pretty natural:

In (1) the middle case (where s is $\mathbf{x}$) is obvious. The case where s a constant is too: You cannot change the constant! As for when s is a $\mathbf{y}$—that is, other than $\mathbf{x}$—then there is no change in s either, for we are asking to change $\mathbf{x}$ but s neither contains, nor is, $\mathbf{x}$.

The last (inductive) case is also pretty understandable: How would one go about plugging a 3 into x in, say, $\cos(\sin(x))$—i.e., what are the steps to do $\cos(\sin(x))[x := 3]$? Well, we work from inside out: We first plug the 3 into the x of $\sin(x)$—to obtain $\sin(3)$— and then we apply cos to $\sin(3)$ to get $\cos(\sin(3))$.

The last case of (1) simply generalizes this example: Think of cos as "f", take $n = 1$, and then think of $\sin(x)$ as "s_1" and 3 as "t".

In (2) the points to emphasize are:

(a) If A is $(\forall \mathbf{x})B$ then, *intuitively*, A does not depend on $\mathbf{x}$—$\mathbf{x}$ is not free. So we can plug t into $\mathbf{x}$ by doing nothing: We do not change A.

(b) In the case before the last we are told that to form $A[\mathbf{x} := t]$ we just form $B[\mathbf{x} := t]$ first, and then we add the quantifier $(\forall \mathbf{y})$ up in front—*as long as no variable in t gets captured (p. 132) by this* $(\forall \mathbf{y})$. Of course, the only variable that can ever belong to $(\forall \mathbf{y})$—cf. 4.1.22—is $\mathbf{y}$ itself, so the condition "$\mathbf{y}(\text{not } \mathbf{x})$ does not occur in t" is *sufficient* (but not necessary) to avoid capture. The condition "$\mathbf{x}$ is not free in B" is also sufficient by itself as then $\big((\forall \mathbf{y})B\big)[\mathbf{x} := t]$ is just $(\forall \mathbf{y})B$—same as before the substitution. See Exercise 4.1.33 below.

(c) In the last case, there will be capture—if we go ahead with the substitution— since at least one occurrence of $\mathbf{x}$ is free in B. Thus, at least one $\mathbf{x}$ of B will be replaced by t causing a $\mathbf{y}$ in t to bind with the $(\forall \mathbf{y})$. We have agreed that in this case we must abort (cf. the discussion following 4.1.27).

Thus we say that the substitution as *requested cannot be performed*. For short, the result of the operation "$A[\mathbf{x} := t]$" in this case is unavailable; it is *undefined*.

We conclude our syntactic preliminaries by expanding Definition 1.3.15 to first-order languages. As we said before, we need something like "replace all occurrences of $\mathbf{p}$ in a formula A by the formula B" in order to make the Leibniz rule formulation friendly.

It turns out that we will need two concepts of formula substitution, one of which will be denoted analogously with $A[\mathbf{x} := t]$ as $A[\mathbf{p} := B]$, and, also analogously, it will *not* be allowed to proceed in case of variable capture; it will be undefined ("impossible") in that case. We may call this substitution *conditional*.

We will also benefit from the presence of an "unconditional" substitution, one that we allow to proceed regardless of any capture. This will be denoted by $A[\mathbf{p} \setminus B]$.

At this point, having two types of substitution into Boolean variables may look like an extra burden, but I promise that the usefulness of both will be evident before too long.

4.1.29 Definition. (Unconditional Substitution) The *unconditional* substitution of a formula into all occurrences of a Boolean variable $\mathbf{p}$ in a formula A is denoted by $A[\mathbf{p} \setminus B]$ and is defined almost exactly as in 1.3.15, by adjusting the atomic case and by adding a clause for "$(\forall \mathbf{x})$".

$$A[\mathbf{p} \setminus B] \text{ is } \begin{cases} B & \text{if } A \text{ is } \mathbf{p} \\ A & \text{if } A \text{ is in } \mathbf{AF} \text{ but is not } \mathbf{p} \\ \neg C[\mathbf{p} \setminus B] & \text{if } A \text{ is } \neg C \\ C[\mathbf{p} \setminus B] \circ D[\mathbf{p} \setminus B] & \text{if } A \text{ is } C \circ D \\ (\forall \mathbf{x})C[\mathbf{p} \setminus B] & \text{if } A \text{ is } (\forall \mathbf{x})C \end{cases}$$

$\square$

The above is straightforward. The inductive cases go through without restrictions, and what they do is to apply the substitution to the immediate subformulae of A, and then apply the appropriate connective (Boolean, or the quantifier, as the case may be). The immediate subformula of $(\forall \mathbf{x})C$ is C, i.e., the formula on which we applied "$(\forall \mathbf{x})$" to get $(\forall \mathbf{x})C$ during the relevant formula-calculation.

Case one is clear cut, but so is the second: The only way an atomic formula A can contain $\mathbf{p}$ is to *be* $\mathbf{p}$. So if it is not, plugging B to $\mathbf{p}$ is irrelevant; it does not change A.

The conditional substitution is defined similarly to the unconditional one, but the former will disallow the last case in the definition above if capture occurs. This will render the result of the operation $A[\mathbf{p} := B]$ undefined.

4.1.30 Definition. (Conditional Substitution) *Conditional* substitution of a formula into all occurrences of a Boolean variable $\mathbf{p}$ in a formula A is denoted by $A[\mathbf{p} := B]$.

The result of this substitution will be undefined if capture of a variable occurs at any step. Below we will use reduced-bracket notation:

$$A[\mathbf{p} := B] \text{ is } \begin{cases} B & \text{if } A \text{ is } \mathbf{p} \\ A & \text{if } A \text{ is in } \mathbf{AF} \text{ but is not } \mathbf{p} \\ \neg C[\mathbf{p} := B] & \text{if } A \text{ is } \neg C \\ C[\mathbf{p} := B] \circ D[\mathbf{p} := B] & \text{if } A \text{ is } C \circ D \\ (\forall \mathbf{x})C[\mathbf{p} := B] & \text{if } A \text{ is } (\forall \mathbf{x})C \text{ and } \mathbf{x} \text{ is not free in } B \\ & \text{else } \textit{undefined} \end{cases}$$

In each case above, the left-hand side, $A[\mathbf{p} := B]$, will be defined iff so are all the contributing substitutions in the right-hand side.   □

Intuitively, it is immediately evident that whenever $A[\mathbf{p} := B]$ is defined, then it is expanded (stands for) as the same string that $A[\mathbf{p} \setminus B]$ stands for. We can actually prove this by induction on the complexity of A.

For atomic A this is obvious, indeed the "whenever" is superfluous: Both $A[\mathbf{p} := B]$ and $A[\mathbf{p} \setminus B]$ are defined in this case, period.

We have three more cases, precisely the ones in 4.1.30 and 4.1.29:

(1) Suppose A is $\neg C$.

Let $A[\mathbf{p} := B]$ be defined. By 4.1.30 this is $\neg C[\mathbf{p} := B]$, *and* $C[\mathbf{p} := B]$ *is defined*. By the I.H.

$$C[\mathbf{p} := B] \text{ is the same as } C[\mathbf{p} \setminus B] \tag{$*$}$$

By 4.1.29, $A[\mathbf{p} \setminus B]$ is $\neg C[\mathbf{p} \setminus B]$. By $(*)$, $A[\mathbf{p} \setminus B]$ and $A[\mathbf{p} := B]$ are the same.

(2) Suppose A is $C \circ D$ ($\circ$ is any of $\wedge, \vee, \rightarrow, \equiv$). This case is similar to the previous, so I leave it to the reader.

(3) Suppose A is $(\forall \mathbf{x})C$. Let $A[\mathbf{p} := B]$ be defined.

By 4.1.30, this is $(\forall \mathbf{x})C[\mathbf{p} := B]$, and $C[\mathbf{p} := B]$ is defined, and $\mathbf{x}$ is not free in B.

Now, the I.H. applies to the less complex C (immediate subformula of A), hence by the middle conjunct of the preceding statement, again $(*)$ above holds. By Definition 4.1.29—which is totally indifferent to "and $\mathbf{x}$ is not free in B"—$A[\mathbf{p} \setminus B]$ is $(\forall \mathbf{x})C[\mathbf{p} \setminus B]$. By $(*)$, $A[\mathbf{p} \setminus B]$ and $A[\mathbf{p} := B]$ are the same once more. We are done.

4.1.31 Example. We calculate a few substitutions according to Definitions 4.1.28, 4.1.29, and 4.1.30.

(1) $(x = y)[y := x]$.

This (by 4.1.28) is $x[y := x] = y[y := x]$, which again by 4.1.28 is $x = x$.

(2) $\big((\forall x)x = y\big)[y := x]$. Using 4.1.28, this is *undefined* because it falls under the last case in the definition (part (2)).

(3) $(\forall x)(x = y)[y := x]$. According to our redundant-brackets convention, this is $(\forall x)\Big\{(x = y)[y := x]\Big\}$.

This *does not* ask that we form the substitution in (2) above, it rather asks (in a roundabout way, using (1) above), "Apply $(\forall x)$ to $x = x$", which is fine.[98] The final answer is $(\forall x)x = x$.

(4) $\Big((\forall x)(\forall y)\phi(x, y)\Big)[y := x]$.

According to 4.1.28, we are asked to do $\big((\forall y)\phi(x,y)\big)[y := x]$ and then apply $(\forall x)$. We can do this if y is not free in $(\forall y)\phi(x, y)$.

Luckily, it is not. So the final result is $(\forall x)(\forall y)\phi(x, y)$ since $\big((\forall y)\phi(x,y)\big)[y := x]$ is just $(\forall y)\phi(x, y)$ by the third case (from the bottom) in the definition of $A[\mathbf{x} := t]$.

(5) $\Big(z = a \vee (\forall x)x = y\Big)[y := x]$. This requires us to calculate the following two substitutions first:

(a) $(z = a)[y := x]$

(b) $\Big((\forall x)x = y\Big)[y := x]$

and then connect them with a "$\vee$" if both are defined. The substitution requested in (b) is undefined by (2) above; hence the whole thing is undefined even though part (a) is defined. For the record, part (a) is calculated as $z[y := x] = a[y := x]$, that is, $z = a$ (cf. (1) in Definition 4.1.28).

(6) $\Big((\forall x)p\Big)[p \setminus x = y]$ is $(\forall x)x = y$.

(7) $\Big((\forall x)p\Big)[p := x = y]$ is undefined (x in $x = y$ gets captured). $\square$

4.1.32 Exercise. Prove by induction on the complexity of A that $A[\mathbf{p} \setminus B]$ is[99] a formula, and that so are $A[\mathbf{x} := t]$ and $A[\mathbf{p} := B]$ whenever defined. $\square$

4.1.33 Exercise. Prove by induction on the complexity of A that if $\mathbf{x}$ is not free in A, then, for any term t, $A[\mathbf{x} := t]$ is A. $\square$

4.1.34 Exercise. Intuitively, $A[\mathbf{x} := \mathbf{x}]$ is A since we change nothing in A when we replace $\mathbf{x}$ by $\mathbf{x}$. But just for the sake of doing yet another proof by induction, prove by induction on the complexity of A that indeed $A[\mathbf{x} := \mathbf{x}]$ expands as A. $\square$

4.1.35 Example. This is an important example, as we will quote it in the crucial "dummy renaming metatheorem" later on (6.4.4).

Assume that $\mathbf{z}$ does not occur in A (as either a free or a bound variable). As we say, $\mathbf{z}$ is *fresh*. That $A[\mathbf{x} := \mathbf{z}][\mathbf{z} := \mathbf{x}]$ should be just A is pretty much "obvious", right?

Pause. The term *obvious* is very useful in mathematics. It leads to the technique of *proof by intimidation*. You just say, "It is obvious", and no one who does not want his intelligence questioned will dare ask you "why"!

---

[98] This is because $(x = y)[y := x]$ is $x = x$ by part (1) above.
[99] *Is* here is argot for *expands as*.

You change $\mathbf{x}$ into $\mathbf{z}$ and then $\mathbf{z}$ back into $\mathbf{x}$. This *should* get you back to where you started, namely, A. Shouldn't it?

Well, let A be $(\forall \mathbf{z})\mathbf{x} = \mathbf{z}$. Then $A[\mathbf{x} := \mathbf{z}]$ is undefined (i.e., not a formula) and therefore so is $A[\mathbf{x} := \mathbf{z}][\mathbf{z} := \mathbf{x}]$. Or, let A be $\mathbf{x} = \mathbf{z}$. Then $A[\mathbf{x} := \mathbf{z}]$ is $\mathbf{z} = \mathbf{z}$ and $A[\mathbf{x} := \mathbf{z}][\mathbf{z} := \mathbf{x}]$ is $\mathbf{x} = \mathbf{x}$—which is not A, unless $\mathbf{x}$ and $\mathbf{z}$ denote the same variable. *Thus the stated condition is necessary.*

But is it sufficient? Yes, as we see by induction on the complexity of A—*under the stated condition*—i.e.,

$$A[\mathbf{x} := \mathbf{z}][\mathbf{z} := \mathbf{x}] \text{ expands as } A \tag{1}$$

Implicit in (1) is that both substitutions $A[\mathbf{x} := \mathbf{z}]$ and $\{A[\mathbf{x} := \mathbf{z}]\}[\mathbf{z} := \mathbf{x}]$ are defined, where $\{\ \}$ are metabrackets indicating here a completed operation.

We look only at the case of distinct variables $\mathbf{x}$ and $\mathbf{z}$, since otherwise we are looking at $A[\mathbf{x} := \mathbf{x}][\mathbf{x} := \mathbf{x}]$, which is A, because $A[\mathbf{x} := \mathbf{x}]$ is also A (4.1.34).

To handle the atomic case we need to handle the case of terms, since, e.g., $t = s$ is one of the atomic cases. So we prove first, by induction on the complexity of terms t, that

$$t[\mathbf{x} := \mathbf{z}][\mathbf{z} := \mathbf{x}] \text{ is } t \text{ as long as } \mathbf{z} \text{ does not occur in } t \tag{2}$$

There are three cases (Definition 4.1.28, part (1)):

Case 1: t is $\mathbf{x}$. Then $t[\mathbf{x} := \mathbf{z}]$ is $\mathbf{z}$ and $\mathbf{z}[\mathbf{z} := \mathbf{x}]$ is $\mathbf{x}$; thus $t[\mathbf{x} := \mathbf{z}][\mathbf{z} := \mathbf{x}]$ is $\mathbf{x}$, that is, t.

Case 2: t is $\mathbf{y}$, other than $\mathbf{x}$—and also other than $\mathbf{z}$ by hypothesis—or t is a constant a. Then $t[\mathbf{x} := \mathbf{z}]$ is t ($\mathbf{y}$ or a), and since $\mathbf{y}$ is not $\mathbf{z}$, $t[\mathbf{z} := \mathbf{x}]$—i.e., $t[\mathbf{x} := \mathbf{z}][\mathbf{z} := \mathbf{x}]$—is still t in either case ($\mathbf{y}$ or a).

Case 3: t is $f(s_1, \ldots, s_n)$. Now $t[\mathbf{x} := \mathbf{z}][\mathbf{z} := \mathbf{x}]$ is $f(s_1[\mathbf{x} := \mathbf{z}][\mathbf{z} := \mathbf{x}], \ldots, s_n[\mathbf{x} := \mathbf{z}][\mathbf{z} := \mathbf{x}])$, which by the I.H.—applicable since all the s_i are less complex than t—is $f(s_1, \ldots, s_n)$, that is, t.

Now that (2) is proved, we turn to (1). For A atomic—following part (2) of 4.1.28—we find that the claim is trivial in the cases where A is other than $t = s$ or $\phi(s_1, \ldots, s_n)$. Even these cases are immediate with (2) settled. For example, $(t = s)[\mathbf{x} := \mathbf{z}][\mathbf{z} := \mathbf{x}]$ is $t[\mathbf{x} := \mathbf{z}][\mathbf{z} := \mathbf{x}] = s[\mathbf{x} := \mathbf{z}][\mathbf{z} := \mathbf{x}]$, which by (2) above is $t = s$.

The claim also clearly "propagates" with the Boolean formation rules. That is, (cf. 4.1.10, parts (2) and (3)) if the immediate subformulae of A—i.e., B if A is $\neg B$; B and C if A is $B \circ C$, where $\circ$ is one of $\wedge, \vee, \rightarrow, \equiv$—satisfy the claim, then so does A itself.

Consider then the two subcases where A is either $(\forall \mathbf{x})B$ or $(\forall \mathbf{w})B$ where $\mathbf{w}$ and $\mathbf{x}$ are different. Note that under our assumptions, the subcase $(\forall \mathbf{z})B$ does not apply. In the first subcase $A[\mathbf{x} := \mathbf{z}]$ is A (cf. 4.1.28, 3rd case from the bottom). Since, by assumption, $\mathbf{z}$ is not free in A, we get that $A[\mathbf{x} := \mathbf{z}][\mathbf{z} := \mathbf{x}]$—that is, $A[\mathbf{z} := \mathbf{x}]$—is just A (Exercise 4.1.33).

The second subcase now is displayed in the calculation below:

$$A[\mathbf{x} := \mathbf{z}][\mathbf{z} := \mathbf{x}] \text{ is } \Big((\forall \mathbf{w})B\Big)[\mathbf{x} := \mathbf{z}][\mathbf{z} := \mathbf{x}]$$

$$\text{is } \Big((\forall \mathbf{w})B[\mathbf{x} := \mathbf{z}]\Big)[\mathbf{z} := \mathbf{x}] \text{ by } 4.1.28 + \text{I.H.}$$

$$\text{is } (\forall \mathbf{w})B[\mathbf{x} := \mathbf{z}][\mathbf{z} := \mathbf{x}] \quad \text{by } 4.1.28 + \text{I.H.}[100]$$

$$\text{is } (\forall \mathbf{w})B \qquad\qquad\qquad \text{by I.H.}$$

$$\text{is } A \qquad\qquad\qquad\qquad\qquad\qquad \square$$

One last useful syntactic concept that we need to formulate our axioms is that of *partial generalization*.

4.1.36 Definition. We say that B is a *partial generalization* of A if B is formed by prefixing A with zero or more expressions such as $(\forall \mathbf{x})$ for any choice of the bound variable $\mathbf{x}$. $\square$

4.1.37 Example. Here is a list of some partial generalizations of $x = z$:

$$x = z \quad (\forall w)x = z \qquad (\forall x)(\forall x)x = z \qquad\qquad\qquad (\forall x)(\forall z)x = z$$
$$(\forall z)(\forall x)x = z \quad (\forall z)(\forall z)(\forall z)(\forall x)(\forall z)x = z \qquad\qquad\qquad \square$$

The preceding definition may be rephrased recursively:

4.1.38 Alternative Definition. A is a partial generalization of A. If B is a partial generalization of A, then the same is true of $((\forall \mathbf{x})B)$ for any choice of the bound variable $\mathbf{x}$. $\square$

4.2   AXIOMS AND RULES OF FIRST-ORDER LOGIC

At the beginning of Section 1.4, I said "Boolean logic is a (crude) vehicle through which we formulate and explore mathematical truth." To this end, the axioms of Boolean logic determine how the Boolean connectives and the constants $\bot$ and $\top$ behave. They postulate the *a priori* truths from which we proceed to discover more and more truths in mathematics.

Note how I used the qualifier *crude* and elaborated further on (p. 113) that "I regret to say that [Boolean logic] is totally insufficient [toward discovering mathematical truth]" because it manipulates mathematical statements "by name" only, and is thus incapable of seeing inside the first-order structure of the statements; therefore it does not allow us to talk about objects, quantifiers, or equality.

---

[100]The I.H. relevance stems from the italicized remark following (1) on p. 137.

Enter first-order logic. Its axioms must still determine the behavior of Boolean connectives and constants, but they must also tell us how the quantifier "∀" and equality "=" behave. Naturally, then, the axiom set for first-order logic *includes* all the axioms of Boolean logic.

Just as in the case of Boolean logic, one chooses the axioms of predicate calculus very carefully so that they express "universally acceptable"—i.e., application-independent—principles. As such, an axiom like "$(\forall x)(\neg x + 1 = 0)$"—while valuable for the study of the application known as *number theory*, which is about the properties of natural numbers—has no place as an acceptable *universal* principle, and will not be included.[101]

4.2.1 Definition. (Logical Axiom Schemata of Predicate Logic) In what follows, A, B, C stand for arbitrary formulae, and x for an arbitrary variable.

The set of logical axioms of first-order logic consists of *all possible partial generalizations* of the formulae in the following groups, **Ax1–Ax6**:

Ax1. This group contains all tautologies (cf. 4.1.27).

For example, $p \vee \neg p$, $x = 0 \vee \neg x = 0$ and $r \to r \vee q$ are each included. So is $\neg x = 5 \equiv x = 5 \equiv \bot$.

 Although we present predicate calculus in a format suitable *for all applications*, we have already said that *in examples* we reserve the right to use symbols from *specific* applications, such as the "0"—or the "5", short for $SSSSS0$—of the language where number theory is spoken.

Ax2. This group contains all formulae of the *form* $(\forall x)A \to A[x := t]$.

It has the name *specialization axiom* but also *substitution axiom*.

Ax3. This group contains all formulae of the form $(\forall x)(A \to B) \to (\forall x)A \to (\forall x)B$ (I have used least parenthesized notation; cf. 4.1.19).

Ax4. This group contains all formulae of the form $A \to (\forall x)A$, *where* x *is not free in* A.

Thus $x = 0 \to (\forall y)x = 0$ is included but $z = 5 \to (\forall z)z = 5$ is not.

Ax5. This group contains all formulae of the form $x = x$. The name of this axiom group is "*identity* axiom (group)". Again we have infinitely many axioms in the group, even before the application of partial generalization, because there are infinitely many instances of the metavariable x.

Ax6. This group contains all formulae of the form $t = s \to (A[x := t] \equiv A[x := s])$. It is called the "Leibniz *axiom (group) for equality*".

---

[101] For example, $(\forall x)(\neg x + 1 = 0)$ is a false statement about the *real* numbers, for it would say that "for every real number x it is true that if you add 1 to it, the result will be different from zero". Yet $1 + (-1) = 0$. This axiom is actually included as a special or nonlogical axiom, if one wants to *just do number theory*. It is part of the so-called *Peano axioms* for number theory or arithmetic.

We will denote by Λ_1 the set of all logical first-order axioms defined here in order to distinguish it from the Λ of 1.4.4. Think of the subscript "1" as a reminder that we are talking about *1*st-order axioms here!   □

Of course, Λ is a proper (i.e., not equal) subset of group **Ax1**. Even though it may look like overkill to include all tautologies in group **Ax1**, doing so has both a technical and practical basis. This approach is encountered in the literature, e.g., [13, 33, 50, 51, 53]. The foundation here is that of [50, 51].

In a correctly founded first-order logic we can certainly prove Post's theorem. Thus, technically, whether or not we include all tautologies as axioms, we will have them as theorems, anyway. Practically, Post's theorem *already, in Part I, has given us the license* to invoke without proof, *within a formal proof*, whichever tautologies are known to us exactly in the same manner that we invoke axioms. Moreover, the technical demands of writing predicate calculus proofs justify our putting to rest the until-now meticulous syntactic proofs of tautologies and tautological implications that we practiced in Part I, and to concentrate instead on mastering the difficulties that are peculiar to the handling of quantifiers and other non-Boolean elements.

Rest assured that whichever such tautologies (or tautological implications) we invoke in practice are of a very trivial nature and are readily recognizable as such.

4.2.2 Remark. We argue here, *intuitively, of course*, that the axioms listed above are indeed *universal principles*, unbiased by considerations specific to particular theories.[102]

(1) That the axioms in group **Ax1** express universally true principles stems from our intuitive understanding of the term *tautology*.

(2) The name *specialization* of **Ax2** stems from what it says: "If a statement (here A) is true for all (values) of **x**, then it must be true of any *special* value t that we are *allowed* to plug into **x**." The principle in quotes is valid no matter what application of logic we may have in mind.

Which part says "that we are *allowed*"? The "$[\mathbf{x} := t]$" part, of course, as you will recall from 4.1.28.

Definition 4.1.28 keeps **Ax2** honest, and true, since substitution is *not* always allowed.

For example, take A to be $(\exists y)\neg x = y$. The formula in $(*)$ below is *not* included in group **Ax2**, since we are not allowed to perform the substitution "$((\exists y)\neg x = y)[x := y]$" to get "$(\exists y)\neg y = y$".

This is just as well, because $(*)$ states an invalid principle! It states, "If for every value of x there is a value of y that is different, then there is some value of y that is different from itself."

$$(\forall x)(\exists y)\neg x = y \rightarrow (\exists y)\neg y = y \qquad\qquad (*)$$

---

[102]*Theory* was defined on p. 116.

Why is $(*)$ invalid? Because the *if* part does have true instances—e.g., no matter which real number you pick, I can find one that is different—but the *then* part is never true!

(3) Let us see now why, *intuitively*, every instance of the schema

$$(\forall \mathbf{x})(A \rightarrow B) \rightarrow (\forall \mathbf{x})A \rightarrow (\forall \mathbf{x})B \qquad (\ddagger)$$

of group **Ax3** expresses a universally valid principle. Well, let us freeze for our discussion the formulae A and B and the variable $\mathbf{x}$. Statement $(\ddagger)$ says:

$$\text{If } (\forall \mathbf{x})(A \rightarrow B) \text{ holds} \qquad (i)$$

and

$$\text{if } (\forall \mathbf{x})A \text{ holds} \qquad (ii)$$

then

$$(\forall \mathbf{x})B \text{ holds} \qquad (iii)$$

Let us then suppose the *informal* statements (i) and (ii) and conclude (iii).

Statement (i) says, in words, "Any value of $\mathbf{x}$ that makes A true, also makes B true." Statement (ii) says, "A holds for all values of $\mathbf{x}$." Hence, by (i), B holds for all values of $\mathbf{x}$; we have got (iii)!

(4) The universal validity of the principle expressed by the schema in group **Ax4** is easy to verify: You see, if $\mathbf{x}$ is not free in A, this means, *intuitively*, that what A says is *independent of* $\mathbf{x}$. Therefore proclaiming "For all $\mathbf{x}$, A holds" or just "A holds" makes no difference.

Have I not just argued that $A \equiv (\forall \mathbf{x})A$ is a correct universal principle? Yes, I did!

Note, however, that mathematicians (and logicians) prefer to say as little as necessary in their axioms. This is why the group **Ax4** is formulated as above, with "$\rightarrow$" rather than with "$\equiv$". The $\leftarrow$ direction[103] need not be stated since it can be proved formally to follow from **Ax2**. For now, in outline, let me only indicate that such a "proof" would be based on the following observations: $(\forall \mathbf{x})A \rightarrow A[\mathbf{x} := \mathbf{x}]$ is in **Ax2** and we know that $A[\mathbf{x} := \mathbf{x}]$ is just A (cf. 4.1.34).

(5) The schema $\mathbf{x} = \mathbf{x}$ expresses an obviously valid universal principle: An object is the same (equal) as itself, no matter what this object may be.

(6) Schema **Ax6** also expresses a universally valid principle. It says, "If two objects t and s are the same, then for any property A, either both t and s have the property, or neither does." At the informal level, the intuitive concept of *property* is synonymous with that of *statement*. Indeed, a statement P determines a property, shared by all objects that make P true. Conversely, a property P determines the statement "x has P".

The origins of **Ax6** are traced back to Leibniz's characterization of equality *between objects*. He suggested that *two objects are equal iff they have exactly the same properties*. In symbols, Leibniz's statement is captured by

$$t = s \equiv (\forall P)(P[\mathbf{x} := t] \equiv P[\mathbf{x} := s]) \qquad (**)$$

---

[103]In a Ping-Pong argument; cf. 3.4.5.

Of course, the language of first-order logic does not allow us to write $(\forall\phi)$—let alone $(\forall P)$. So how can we translate $(**)$ into our first-order language? Our translation is forgetful. First, we forget the $\leftarrow$ direction of $(**)$ and are content with stating just

$$t = s \rightarrow (\forall P)(P[\mathbf{x} := t] \equiv P[\mathbf{x} := s]) \qquad (***)$$

Second, not being allowed to use "$(\forall P)$" to express "for all P", we do the next best thing: We exhaustively postulate all the infinitely many formulae of the *form*

$$t = s \rightarrow (P[\mathbf{x} := t] \equiv P[\mathbf{x} := s]) \qquad (\P)$$

For each instance of the syntactic variables $t, s, \mathbf{x}$, and P we obtain a formula of the form ($\P$), in essence replacing the intuitive concept of *property* or *statement* by the formal one of *formula*. But ($\P$) is precisely the schema **Ax6**!

Unfortunately, this latter compromise, going from $(***)$ to **Ax6**, is also forgetful. It is a fact that we will not establish here (it uses a bit of a set theory argument) that there are far more properties than there are *first-order formulae*; thus our "representation", or "coding", of the former by the latter is far too coarse. Thus, saying, on one hand, *for all properties P*, which is the meaning of $(***)$, and, on the other hand, *for all formulae P*, which is the meaning of schema ($\P$), i.e., **Ax6**, are two very different things!

Oh well, not to worry about this. As it turns out, and this is rather surprising after all this hacking at $(**)$, we have retained enough in **Ax6** to still be able to prove the usual properties of equality, such as symmetry $(\mathbf{x} = \mathbf{y} \rightarrow \mathbf{y} = \mathbf{x})$ and transitivity $(\mathbf{x} = \mathbf{y} \wedge \mathbf{y} = \mathbf{z} \rightarrow \mathbf{x} = \mathbf{z})$. □

 (1) Just as in our discussion of **Ax2** we observe that the definition of substitution (4.1.28), which has an "undefined" or "don't do it!" case, keeps **Ax6** honest: By disallowing substitution in certain cases we disallow incorrect instances of the axiom schema. For, imagine that A is $(\exists y)\neg x = y$.

We can try **Ax6** on this formula, taking t and $\mathbf{x}$ to be the variable x and s to be the variable y:

$$x = y \rightarrow \left(\big((\exists y)\neg x = y\big)[x := x] \equiv \big((\exists y)\neg x = y\big)[x := y] \right) \qquad (2)$$

However, "$\big((\exists y)\neg x = y\big)[x := x]$" is just $(\exists y)\neg x = y$ (cf. 4.1.34), but "$\big((\exists y)\neg x = y\big)[x := y]$" is *undefined* by 4.1.28. This means that (2) does *not* translate into

$$x = y \rightarrow \big((\exists y)\neg x = y \equiv (\exists y)\neg y = y\big) \qquad (3)$$

in other words, we cannot proclaim (3) as a member of axiom group **Ax6**. *This is just as well*, because (3) is unacceptable!

Let us see why, arguing *intuitively* (intelligently, but loosely). Now "$(\exists y)\neg y = y$" says, "There is a value of y that is not equal to itself." This is a falsehood.

Thus, (3) simplifies into

$$x = y \rightarrow \neg(\exists y)\neg x = y \qquad (4)$$

a simplification that uses principle (4) in the list of universal principles given in 1.4.4, and thus also included in **Ax1**.

Pause. You surely have noticed that when I argue nontechnically—when I am not writing a proof, that is—I avoid saying "axioms" and rather say "principles".

Recall that $\exists$ is just an abbreviation (4.1.17), so (4) really says[104]

$$x = y \rightarrow (\forall y)x = y \tag{5}$$

Do you *believe* (5)? If you do, then you must also accept any special case of it, such as $0 = 0 \rightarrow (\forall y)0 = y$. Unfortunately, to the left of "$\rightarrow$" I have a "true" statement but to the right I have a "false" one. So the special case topples, bringing down with it the general case (5).

(2) **Ax6** must *not* be confused with the *Leibniz Rule* (**Inf1** of 1.4.2) of Part I (which we will reintroduce shortly for predicate calculus).

Ax6 is about *object equality, "="*. *In Part I we had no objects to talk about*.

Theorem (SFL) of Section 3.4, which [17] nicknamed the *Leibniz axiom*, has no connection with axiom schema **Ax6** beyond a superficial structural similarity—if, that is, one forgets the fundamental difference between "$\equiv$" and "$=$". Moreover, while (SFL) is provable, as we saw, within propositional calculus, schema **Ax6** is known *not to follow* from the other axioms of predicate calculus.

We next turn to the primary rules of inference for predicate logic. We have augmented the logical axiom set of Boolean logic (1.4.4) by adding axiom groups **Ax2**–**Ax6**—and that mysterious "partial generalization" whose purpose will be clear soon.

We have added just enough axioms to the Boolean case so that all the properties of objects, quantifiers, and equality can be reasoned about without introducing any new rules of inference! We will use in first-order logic the very same rules we introduced in 1.4.2, namely **Inf1** and **Inf2**.

Now this must be interpreted to mean that we apply rules **Inf1** and **Inf2** to the *Boolean abstractions* (4.1.25) of first-order formulae—after all, these are Boolean or *propositional* rules, that is, they apply to Boolean formulae. Therefore we must view first-order formulae as Boolean ones in order to make sense of applying these rules to them. This is not hard to do:

The form of **Inf2** remains the same in the first-order case, namely,

$$\frac{A, A \equiv B}{B} \tag{Eqn}$$

since the abstraction of a first-order formula $A \equiv B$ will still look like $A \equiv B$—no change in shape—because abstractions do not eliminate Boolean connectives, unless these are in the scope of a quantifier. The "$\equiv$" of $A \equiv B$ is not in any such scope

---

[104]A tiny leap of faith, this. In slow motion, "$\neg(\exists y)\neg x = y$" is short for "$\neg\neg(\forall y)\neg\neg x = y$", which, losing the double negations, yields "$(\forall y)x = y$". Intuitively, anyway!

(Why?). In other words: *We do not need to explicitly convert first-order formulae to their abstractions in order to apply* **Inf2** *(equanimity). We apply it directly.*

We must be more careful with **Inf1**:

$$\frac{A \equiv B}{C[\mathbf{p} := A] \equiv C[\mathbf{p} := B]}$$

Since this rule is applied to Boolean abstractions of first-order formulae, any "**p**" that occurs in C must be "visible" in the abstraction; that is, it must not occur within the scope of a quantifier in C. Thus the rule, without any reference to abstractions, is stated as

$$\frac{A \equiv B}{C[\mathbf{p} := A] \equiv C[\mathbf{p} := B]},$$ (BL)

provided that **p** is not in the scope of a quantifier in C

We will call this rule *BL* for *Boolean Leibniz.*

4.2.3 Remark. (1) The qualifier "Boolean" in BL is important as, on one hand, it reminds us of its origin and in particular the restriction on **p**, and, on the other hand, it distinguishes it from *derived* Leibniz rules that we will soon prove and use. These derived rules *do* allow us to replace a Boolean variable **p** that *does* occur in the scope of a quantifier in C, so they act beyond the Boolean structure of formulae.

(2) Why is it a good thing *not* to add *new* (primary) rules of inference? Because since the Boolean axioms are a part (subset) of the first-order axioms, keeping the rule set invariant means that anything we can prove in Boolean logic we can still prove in first-order logic as long as the concept of *proof* remains the same (it does). We will keep all our Boolean theorems!   □

In summary,

4.2.4 Definition. (First-Order Rules of Inference) The *primary* first-order logic's rules of inference are Eqn and BL above.   □

4.2.5 Remark. We can rewrite BL as

$$\frac{A \equiv B}{C[\mathbf{p} \setminus A] \equiv C[\mathbf{p} \setminus B]},$$ (BL)

provided that **p** is not in the scope of a quantifier in C

the reason being that the restriction stated makes capture of a free variable by a quantifier impossible during the two substitutions in the conclusion part of the rule. We have seen (p. 135) that—under no-capture conditions—the formulae $C[\mathbf{p} := A]$ and $C[\mathbf{p} \setminus A]$ are identical strings.

Nevertheless, we will retain the original notation for BL for a reason we will explain later.   □

We are ready to calculate (theorems) once again! As promised earlier on in our discussion here, the concepts of *theorem-calculation* (proof) and *theorem* remain exactly the same as in Part I. We repeat the definitions here, for the record, word for word:

4.2.6 Definition. (Theorem-Calculations—or Proofs) Let Γ be an arbitrary, given, set of formulae.[105]

A *theorem-calculation* (or *proof*) *from* Γ is any finite (ordered) sequence of formulae that we may write respecting the following two requirements:

. In any stage we may write down

Pr1 Any member of Λ_1 or Γ. Any member of Γ *that is not also in* Λ_1 we will call a *nonlogical* axiom.

Pr2 Any formula that appears in the denominator of *an instance* of a rule **Inf1–Inf2** as long as *all the formulae* in the numerator of the same instance of the (same) rule have already been written down at an earlier stage.

We may call a *proof from* Γ by the alternative name Γ- *proof.* □

4.2.7 Definition. (Theorems) Any formula A that appears in a Γ-proof is called a Γ-theorem. We write $\Gamma \vdash A$ to indicate this. If Γ is empty ($\Gamma = \emptyset$)—i.e., we have no special assumptions—then we simply write $\vdash A$ and call A just "a theorem".
Caution! We may also do this out of laziness and call a Γ-theorem just "a theorem", if the context makes clear which $\Gamma \neq \emptyset$ we have in mind.
We say that A is *an absolute*, or *logical* theorem whenever Γ is empty. □

Note that, once again (cf. practice in Part I), in the configuration "$\vdash A$" we take Λ_1 for granted and do not mention it to the left of $\vdash$.

4.2.8 Exercise. It is a good place here to redo Exercise 1.4.11, so that you can check your understanding of the concepts *proof* and *theorem*. Once again, for exactly the same reason as in Part I (What was the reason?) we have (verify!):
(1) $A \vdash A$, for any A.
(2) $\Gamma \vdash A$, if $A \in \Gamma$.
(3) $\vdash B$, for any axiom B. □

4.2.9 Remark. This remark retells 1.4.8. A Γ-proof of a formula A is, by definition, a sequence of formulae

$$B_1, \ldots, B_n, A, C_1, \ldots, C_m$$

with proprieties as stated in 4.2.6. It is trivial that if we discard the part "$C_1, \ldots, C_m$" of the sequence, then

$$B_1, \ldots, B_n, A \tag{1}$$

[105]In Part II, *formulae* means *first-order formulae* unless otherwise indicated.

is still a proof. The reason is that every formula in a proof is either legitimized outright—without reference to any other formulae—or is legitimized by reference to formulae to its *left*.

Thus (1) too proves A, since A occurs in it. This technicality allows us to stop a proof once we have written down the formula that we want to prove. ☐

So, 4.2.7 tells us what kind of theorems we have:

(1) Anything in $\Lambda_1 \cup \Gamma$.[106]

(2) For any formula C and variable $\mathbf{p}$—not in the scope of any quantifier of C—$C[\mathbf{p} := A] \equiv C[\mathbf{p} := B]$, *provided* $A \equiv B$ was written down already, therefore is a (Γ-) theorem.

(3) B (any B), *provided* $A \equiv B$ *and* A were written down already and therefore are both (Γ-) theorems.

The above is a recursive definition of (Γ-) theorems, and is worth recording (compare with Definition 4.1.14).

4.2.10 Definition. (Theorems, Inductively) A formula E is a Γ-theorem iff it fulfills one of the following:

Th1 E is in $\Lambda_1 \cup \Gamma$.

Th2 For some formula C and variable $\mathbf{p}$—not in the scope of any quantifier of C—E is $C[\mathbf{p} := A] \equiv C[\mathbf{p} := B]$, *and* (we know that) $A \equiv B$ is a (Γ-) theorem.

Th3 (We know that) $A \equiv E$ *and* A are (Γ-) theorems. ☐

The concept of *proof* defined in 4.2.6 is that of *Hilbert-style* proof, which we will arrange vertically—just as in Part I—with annotations. All that we said in Section 1.4 carries over here unchanged. In particular note Example 1.4.13.

Note that *wherever in the proofs that we wrote in Part I we said "by Leibniz", or just "Leib", we will here replace it with "by BL", or just "BL", and the proof will remain valid.* In particular, part (d) in 1.4.13 establishes $\vdash A \equiv A$. Even though that proof carries over unchanged here, we have in the new setting a more immediate proof: For any A, $A \equiv A$ is a member of **Ax1**.

Moving to the results of Chapter 1, again, everything we did there carries over unchanged. In particular 2.1.1, 2.1.4 and Corollary 2.1.6—which allows us to use already-proved theorems in proofs—hold, as well as "the other Eqn" (2.1.11).

Just for the record, "Eqn + Leib Merged" (2.1.16) is still valid and still equivalent (as powerful as) Eqn and BL combined. This rule, transcribed in first-order logic, will now have the condition "where $\mathbf{p}$ does not occur in the scope of any quantifier of C".[107]

---

[106]We explained the notation "$\Lambda \cup \Gamma$" in 1.3.14 on p. 36, item (3).

[107]It will turn out that the restriction "where $\mathbf{p}$ does not occur in the scope of any quantifier of C" can be removed from BL to obtain a valid—for first-order logic—derived Leibniz rule. But I am ahead of myself.

Redundant true (2.1.21) and the extremely important metatheorem 2.1.23 remain, of course, valid—the former for a now-trivial reason (the relevant schema is a member of **Ax1**).

We do not need to add anything new to what we already said about the equational proof methodology (Section 2.2) nor need we reiterate that all the results of Sections 2.4 and 2.5 remain valid.

Note that the proof of modus ponens, which we gave in Boolean logic (2.5.3), is also valid in first-order logic.

The deduction theorem (2.6.1) holds in first-order logic exactly as stated originally (2.6.1). To see this, note that the proof of the "main lemma" (2.6.2) goes through pretty much unchanged (remember to say "BL" where we said "Leib" before).

We need only to rephrase the basis case as follows:

Basis. D has complexity 0: So it is one of:

(1) **p**:   Then we must show $A \to (B \equiv C) \vdash A \to (B \equiv C)$ and we are done by 1.4.11.

(2) **q** (other than **p**), or $\top$ or $\bot$, or $\phi(t_1, \ldots, t_n)$ or $t = s$:   By reference to 4.1.30 we see that we need to establish that $A \to (B \equiv C) \vdash A \to (D \equiv D)$. Well, start with $\vdash D \equiv D$ (**Ax1**). By (3) in 2.5.1, and 2.1.4, $\vdash A \to (D \equiv D)$. We are done by 2.1.1.

It is worth noting that a really simple alternative proof of the deduction theorem can be given for first-order logic using the tools of the next chapter.

It is extremely important to emphasize, especially for the reader who encountered alternative, very complex versions in the literature that: *The deduction theorem in our version of first-order logic is exactly the same as in the Boolean (propositional) case.*

Are those very *complex versions* wrong. then? No. .

The small print here is that it *is* possible to found first-order logic somewhat differently: with fewer axioms but with the addition of a primary rule called *unconstrained* or *strong* generalization:

$$\frac{A}{(\forall \mathbf{x})A} \tag{1}$$

In a so-founded first-order logic, the deduction theorem is ugly—that is, it has cumbersome restrictions (cf. [35, 45, 52, 53]). The approach in the present volume is analogous to that in [2, 13, 50, 51], where the primary rules are, essentially, propositional. The trick is, rather than adopting rule (1), to encode it in the axioms (**Ax3, Ax4**) and to have these axioms and the only two primary rules, Eqn and BL, simulate *a somewhat weaker version of (1)*.

The side benefit: We can have a user-friendly deduction theorem that reads and applies exactly as in the Boolean case! Needless to say, the related results, 2.6.6 and 2.6.7, hold in first-order logic.

The most crucial result that carries over from Boolean to predicate calculus is Post's Theorem: *If $\Gamma \models_{taut} A$, then $\Gamma \vdash A$.*

In our particular foundation of first-order logic, the special case of $\Gamma = \emptyset$ is direct from **Ax1**:

$$\text{If } \models_{taut} A, \text{ then } A \text{ belongs to } \mathbf{Ax1}, \text{ hence } \vdash A$$

Even the case of nonempty, *but finite*, Γ follows easily from **Ax1** and MP, as we see in the next chapter—hence so does 3.3.1.

The (meta)proof of the infinite ("general") case given in Section 3.2 caries over unchanged as long as one remembers to work throughout with *Boolean abstractions* (4.1.25) of first-order formulae. In particular, the amended alphabet (3) of p. 94 is still correct, if one were to repeat the proof, since a Boolean abstraction contains none of the symbols that cannot occur in a Boolean formula (such as $=, \forall, x''_{13}, \phi'$, etc.). Corollary 3.3.1 is important for the user, and we will continue to rely on it heavily.

 Very Important! Although we have 3.2.1 in predicate calculus, we must not expect to have the very same soundness result of 3.1.3. In fact, note that $\vdash x = x$ (Definitions 4.2.1 and 4.2.7), but $\not\models_{taut} x = x$. Indeed, the Boolean abstraction of "$x = x$" is just some Boolean variable, say, p. A Boolean variable, of course, is not a tautology.

So, is predicate logic *not* sound?

In fact, it *is*. That is, its rules preserve truth and its axioms are true. However, the *truth* concept in predicate calculus is *narrower* than that of tautology and tautological implication, as is to be expected: Tautologies (and tautological implications) speak of the *broadest possible*, most abstract concept of truth, one that is oblivious to the presence of (object) variables, constants, predicates, functions, quantifiers, and equality (between objects) and therefore cannot describe mathematical truth in its full variety.

A different definition of truth (and "preservation of truth"—so-called *logical implication*) is necessary for predicate calculus. More on this in the chapter on first-order semantics (Chapter 8).

4.3 ADDITIONAL EXERCISES

1. In the formula

$$(\forall x)\Big((\forall x)(\forall y)x < y \vee x > z\Big) \to (\forall y)y = x$$

find to which "$(\forall x)$", if any, each occurrence of x belongs. Here "$<$" and "$>$" are just some nonlogical symbols (predicates of arity 2) of the alphabet.

2. Consider the following (not fully parenthesized) formula, in the first-order language of arithmetic where 0 is a nonlogical constant symbol:

$$(\forall x)x = 0 \vee \neg(\forall x)x = 0 \tag{1}$$

(a) Identify *all* the prime subformulae of the above, and display them by boxing as in 4.1.25.

(b) Using boxing indicate the Boolean abstraction of the formula.

(c) Can you prove formula (1) in predicate logic without the benefit of any nonlogical axioms that speak of "0"?

3. Find the results of the following substitutions. For item (a), work it out according to the inductive definition 4.1.28 step by step; for the others, just give the answer and explain the reasons if the result is undefined:

(a) $g(f(x), f(y))[x := 7]$ (where f is a unary function symbol, g is a binary function symbol, and 7 is a nonlogical constant symbol)

(b) $(f(x) < 7)[x := 7]$ (as above, plus we have a nonlogical binary predicate symbol $<$, in connection with which we are using infix notation)

(c) $((\forall x)(f(x) < 7))[x := 7]$

(d) $((\forall y)(f(x) < 7))[x := 7]$

(e) $((\forall x)(\forall y)(f(7) < g(x, y)))[z := g(y, 7)]$

(f) $((\forall x)(\forall y)(f(z) < g(x, y)))[z := g(y, 7)]$

(g) $((\forall x)(\forall y)(\forall z)(f(z) < g(x, y)))[z := g(y, 7)]$

4. Consider the language (i.e., set of strings) over alphabet 4.1.2 that is defined as follows:

- A *P-formula-calculation* consists of a finite sequence of steps. In each step we may write:

  (a) An atomic formula of 4.1.8

  (b) A formula of the form $((\forall \mathbf{x})A)$ for any A in the **WFF** of 4.1.13

  (c) $(\neg A)$, provided A has already been written

  (d) $(A \circ B)$—for any $\circ \in \{\wedge, \vee, \rightarrow, \equiv\}$—provided A and B have already been written

- We define a *P-formula* as any string that appears in a P-formula-calculation.

- We define the *complexity of a P-formula* as the total number of connectives (counting repetitions) from $\{\neg, \wedge, \vee, \rightarrow, \equiv\}$ that appear in the formula but not in any subformula of type (b).

Prove that the set of all P-formulae over the alphabet of 4.1.2 is the same as the set **WFF** of 4.1.13 over the same alphabet.

Hint. First, verify that the definition of P-formula-calculation differs from 4.1.10 in one essential aspect. Then, by induction on the complexity of P-formulae, prove that every such formula is in **WFF**. By induction on the complexity of formulae of **WFF**, prove that every such formula is a P-formula.

5. Prove—by induction on terms—that for any terms t and s, if s is a *prefix* of t, then the strings t and s *must* be identical.

6. Prove that any nonempty proper prefix of a first-order formula must have an excess of left brackets.

7. Prove the *unique readability* of first-order formulae; that is, for every such formula its immediate predecessors are uniquely defined.

8. Is $(\forall x)(\forall y)x = y \rightarrow (\forall y)y = y$ an instance of **Ax2**? Why?

9. Give a proof of $\vdash (\forall x)(\forall y)x = y \rightarrow (\forall y)y = y$.

CHAPTER 5

TWO EQUIVALENT LOGICS

This brief chapter aims to develop a few more useful tools that we will employ, among others, in the proof of the crucial 6.1.1.

Recall Remark 2.1.17. In exactly the same manner, two different foundations of predicate logic *over the same first-order language*—two "different logics"—are *equivalent* iff they have *the same absolute theorems*.

 With half a minute's reflection, and using the deduction theorem, we see that equivalent logics also have the same "relative theorems". That is, fixing a set of assumptions Γ, one logic proves a formula A from the assumptions Γ iff the other does.

Pause. How could we *have* two logics over the same language?

Well, just think of the ingredients: We can choose different logical axioms, or different *primary* rules of inference, or both!

Here are the two logics over the language of Section 4.1 that we want to compare:
(1) Our logic as given by 4.2.1 and 4.2.4 (cf. [50, 51])
(2) The logic given exactly as in (1) above, except that *its only primary rule is MP* (cf. [13], although in Enderton's exposition the Boolean variables and constants are absent)

We note that 2.1.1, 2.1.4 and 2.1.6 depend only on the general concept of proof and *not* on the choice of logical axioms or rules of inference. Thus

   Both logics above, (1) and (2), satisfy 2.1.1, 2.1.4 and 2.1.6.

5.0.1 Lemma. *Post's theorem (3.2.1) holds in logic (2) for finite* Γ.

Proof. Indeed, let

$$A_1, A_2, A_3, \ldots, A_n \models_{\text{taut}} B \qquad\qquad (i)$$

We will show

$$A_1, A_2, A_3, \ldots, A_n \vdash B \qquad\qquad (ii)$$

By 1.3.14(2), (i) yields

$$\models_{\text{taut}} A_1 \to A_2 \to A_3 \to \ldots \to A_n \to B \qquad\qquad (iii)$$

Thus $A_1 \to A_2 \to A_3 \to \ldots \to A_n \to B$ is an axiom of logic (2). We can therefore write the following Hilbert proof—*in the logic (2)*—of B from the hypotheses $A_1, A_2, A_3, \ldots, A_n$:

(1)	A_1	$\langle$hypothesis$\rangle$
(2)	A_2	$\langle$hypothesis$\rangle$
(3)	A_3	$\langle$hypothesis$\rangle$
$\vdots$		$\vdots$
(n)	A_n	$\langle$hypothesis$\rangle$
(n + 1)	$A_1 \to A_2 \to A_3 \to \ldots \to A_n \to B$	$\langle$axiom$\rangle$
(n + 2)	$A_2 \to A_3 \to \ldots \to A_n \to B$	$\langle (1, n+1) + MP \rangle$
(n + 3)	$A_3 \to \ldots \to A_n \to B$	$\langle (2, n+2) + MP \rangle$
(n + 4)	$A_4 \to \ldots \to A_n \to B$	$\langle (3, n+3) + MP \rangle$
$\vdots$		$\vdots$
(n + n)	$A_n \to B$	$\langle (n-1, n+n-1) + MP \rangle$
(n + n + 1)	B	$\langle (n, n+n) + MP \rangle$ $\square$

5.0.2 Lemma. *Logic (2) has BL and Eqn as derived rules of inference.*

Proof. Indeed, we know from Part I (3.1.1) that $A, A \equiv B \models_{\text{taut}} B$ and $A \equiv B \models_{\text{taut}}$ $C[\mathbf{p} := A] \equiv C[\mathbf{p} := B]$, where in the latter case A, B and C are abstractions of first-order formulae.

   By the previous lemma we can replace $\models_{\text{taut}}$ by $\vdash$; that is, BL and Eqn are derived rules of logic (2). $\square$

5.0.3 Metatheorem. *Logics (1) and (2) are equivalent.*

Proof. The two have the same axioms. Now, the axioms plus the rules BL and Eqn of the first can simulate the rule MP of the second. Conversely, the axioms plus the rule MP of the second can simulate the rules BL and Eqn of the first (5.0.2).   □

Why did we bother to prove the above metatheorem? How is it useful? Because the metatheory of logic (2) is simpler, i.e., logic (2) is easier to talk and argue *about*. This was the main reason. We can gauge its usefulness in Chapter 6 where we prove the generalization (derived) rule.

An immediate taste of the facilitation provided by 5.0.3 is offered by the next exercise.

5.0.4 Exercise. Invoking 5.0.3—and thus arguing about logic (2) rather than logic (1)—give an easy, short, and *2.6.2-independent* proof of the deduction theorem (exactly as stated in 2.6.1) for first-order logic. Needless to say, this will not be circular: We did not use the (first-order) deduction theorem toward any of the results of this chapter.   □

CHAPTER 6

GENERALIZATION AND ADDITIONAL LEIBNIZ RULES

In this chapter we finally bring quantifiers, in particular "∀", to the fore; thus we *really* start doing predicate logic!

 By *predicate logic* we understand either logic (1) or (2) of the previous chapter. It does not matter which one, as they are equivalent! (5.0.3).

We will be sure to use logic (2) when we prove facts about logic.

6.1 INSERTING AND REMOVING "(∀x)"

6.1.1 Metatheorem. (Weak Generalization) *Let* $\Gamma \vdash A$ *and let moreover* **x** *not occur free in any formula found in the set* Γ. *Then* $\Gamma \vdash (\forall \mathbf{x})A$ *as well.*

Proof. The (meta)proof is, essentially, a *construction* that takes a Γ-proof of A in *logic (2)* and transforms it step by step into a valid Γ-proof of $(\forall \mathbf{x})A$—still in logic (2).

The meta-tool employed is once again induction on natural numbers, indeed, induction on the length of a Γ-proof that includes A.

Basis (shortest possible proof). A occurs in a proof of length 1. Then A is one of the following:

(1) Axiom. Then—since all partial generalizations of all formulae found in groups **Ax1**–**Ax6** are also axioms—$(\forall \mathbf{x})A$ is also an axiom. But then (cf. 4.2.6 and 4.2.7) $(\forall \mathbf{x})A$ is a Γ-theorem.

(2) In Γ. By hypothesis, A has no free $\mathbf{x}$ occurrences. Thus $A \rightarrow (\forall \mathbf{x})A$ is an axiom (group **Ax4**). Since $\Gamma \vdash A$ and $\Gamma \vdash A \rightarrow (\forall \mathbf{x})A$ (cf. 4.2.6 and 4.2.7), we have $\Gamma \vdash (\forall \mathbf{x})A$ by an application of MP.

We now take as I.H. the truth of the claim for any A that appears in a proof of length n or less.

Let us establish the case where A appears in a proof of length $n + 1$. We have three cases:

(i) A is not the last formula of the proof. Then we are done by the I.H.

(ii) A *is* the last formula in the proof, but is an axiom or in Γ. Then we are done by the Basis argument.

(iii) None of the above. So A was obtained by MP, that is, B occurs in the proof, and so does $B \rightarrow A$ for some formula B. Since both B and $B \rightarrow A$ occur in Γ-proofs of lengths n or less, the I.H. kicks in and we have

$$\Gamma \vdash (\forall \mathbf{x})B \qquad\qquad (*)$$

and

$$\Gamma \vdash (\forall \mathbf{x})(B \rightarrow A) \qquad\qquad (**)$$

Now

$$\Gamma \vdash (\forall \mathbf{x})(B \rightarrow A) \rightarrow (\forall \mathbf{x})B \rightarrow (\forall \mathbf{x})A \qquad\qquad (***)$$

by **Ax3** and 4.2.6 and 4.2.7, thus, by MP (twice)—from $(*)$, $(**)$, and $(***)$— $\Gamma \vdash (\forall \mathbf{x})A$. $\qquad\square$

(1) The proof we just gave justifies the—at first sight mysterious—requirement that along with the formulae in the groups **Ax1**–**Ax6** we include all their partial generalizations, too, as logical axioms. This was used in step 1 of the basis.

(2) The above metatheorem was proved under a premise that at first sight might appear awfully restrictive: "No formula in Γ has a free $\mathbf{x}$." Surely this restriction is close to impossible to meet—rendering the metatheorem "mostly inapplicable"—when one has infinitely many formulae in Γ. How are we to be *sure* that some particular variable is not free in *every single one* of them?[108] It is irrelevant; see below.

---

[108]There are significant practical cases where Γ is infinite; e.g., Peano arithmetic and Zermelo/Fraenkel set theory have infinitely many nonlogical axioms. In the case of these two theories (cf. p. 116) we have some other tricks that neutralize the problem of an infinite Γ. Although infinitely many, we can choose the nonlogical axioms so that they do not have free variables at all.

6.1.2 Corollary. *If there is a proof of A from Γ, where all the formulae from Γ used in the proof have no free* **x**, *then* $\Gamma \vdash (\forall \mathbf{x})A$.

Proof. Let $B_1, \ldots, B_n$ be all the formulae from Γ used in the proof.
(1) By assumption, none of them has **x** free.
(2) By 4.2.6 and 4.2.7, $B_1, \ldots, B_n \vdash A$.
By (1) and (2) and Metatheorem 6.1.1, $B_1, \ldots, B_n \vdash (\forall \mathbf{x})A$. By 2.1.1, $\Gamma \vdash (\forall \mathbf{x})A$ as well. $\square$

Thus, *without loss of generality*, we can always pretend that the Γ in 6.1.1 is finite, for even if it is not, in every single proof only a finite part of it is used, anyway.

6.1.3 Corollary. *If* $\vdash A$, *then* $\vdash (\forall \mathbf{x})A$.

Proof. This is immediate by taking $\Gamma = \emptyset$. Surely, you can't find an A in this Γ with a free **x** in it! $\square$

6.1.4 Remark. Two important remarks need to be made here:
(1) Metatheorem 6.1.1, and in particular Corollary 6.1.2, allow us to insert the formula $(\forall \mathbf{x})A$ at any point *after* A was written in a Γ-proof as long as whatever formulae of Γ were invoked in the proof contain no free **x**.

This is all right simply because in a Γ-proof we can insert any Γ-theorem we happen to know of—cf. 2.1.6. We do know that $(\forall \mathbf{x})A$ is a Γ-theorem as soon as we learn that A *is*! (6.1.1)

Similarly, the metatheorem's second corollary, 6.1.3, allows us to insert $(\forall \mathbf{x})A$ in any proof, as long as we *already know* that A is an absolute theorem (this may have been established by the proof we are working on, or by some previous proof—a lemma, for example).

(2) One must be careful *not to confuse* the derived rule "if $\vdash A$, then $\vdash (\forall \mathbf{x})A$" with "$A \vdash (\forall \mathbf{x})A$". The former rule, *weak generalization, imposes a constraint on the premise A in its use*: Once we have A we may write down $(\forall \mathbf{x})A$, but this step is *constrained by how A was obtained*. A cannot be arbitrary; rather it must be an absolute theorem.

The latter rule, called *strong* or *unconstrained* generalization, allows us to write $(\forall \mathbf{x})A$ once we have A (written down, that is) *regardless of how A was obtained*. There are no conditions! A could have very well been an arbitrary hypothesis.

Is this distinction between the two rules "real"? Yes! We just saw that our logic ((1) or (2) on p. 151—it does not matter which) does allow *weak* generalization. We will see later on that it does *not* allow *strong* generalization. The reason is that if we had $A \vdash (\forall \mathbf{x})A$ in our logic, then we would also have—by the deduction theorem—$\vdash A \rightarrow (\forall \mathbf{x})A$. Semantic considerations in Chapter 8 will show this to be impossible (cf. also the discussion of formula (5) on p. 143). $\square$

A closely related and extremely useful derived rule is trivially derived from first principles:

6.1.5 Metatheorem. (Specialization Rule) $(\forall \mathbf{x})A \vdash A[\mathbf{x} := t]$

Proof. Of course, if the expression "$A[\mathbf{x} := t]$" ends up being undefined, then we have nothing to prove. Here is a Hilbert-style proof:

$$
\begin{array}{lll}
(1) & (\forall \mathbf{x})A & \langle \text{hypothesis} \rangle \\
(2) & (\forall \mathbf{x})A \to A[\mathbf{x} := t] & \langle \textbf{Ax2} \rangle \\
(3) & A[\mathbf{x} := t] & \langle (1,2) + MP \rangle \qquad \square
\end{array}
$$

6.1.6 Corollary. $(\forall \mathbf{x})A \vdash A$

Proof. Take t to be $\mathbf{x}$ in 6.1.5 (cf. 4.1.34). $\qquad \square$

Corollary 6.1.6 and Metatheorem 6.1.1 (or Corollary 6.1.3)—*which we will refer to as* spec *and* gen *respectively in annotations*—are a team that makes life extremely easy in Hilbert proofs when it comes to dealing with the quantifier $\forall$.

Corollary 6.1.6 lets us *unconditionally remove* the leftmost quantifier in the course of a proof. This action uncovers whatever Boolean connectives were buried in the scope of the removed quantifier[109] so that we can—in the subsequent steps of the proof—apply the purely Boolean techniques of Part I.

Before the end of the proof, if the targeted theorem requires us to do so, we can reintroduce the removed quantifier using 6.1.1 (or 6.1.3 as appropriate)—conditions for insertion, of course, apply.

This comment may appear too general now, but it will gain complete clarity as we progress, presenting several examples where the comment is put to use.

Here is the first such example:

6.1.7 Theorem. (Distributivity of $\forall$ over $\wedge$) $\vdash (\forall \mathbf{x})(A \wedge B) \equiv (\forall \mathbf{x})A \wedge (\forall \mathbf{x})B$

Proof. We use a Ping-Pong argument (cf. 3.4.5) to prove $\vdash (\forall \mathbf{x})(A \wedge B) \to (\forall \mathbf{x})A \wedge (\forall \mathbf{x})B$ and $\vdash (\forall \mathbf{x})A \wedge (\forall \mathbf{x})B \to (\forall \mathbf{x})(A \wedge B)$. Below we label "($\to$)" and "($\leftarrow$)" the two directions that we have to do:

($\to$)

$$
\begin{array}{lll}
(1) & (\forall \mathbf{x})(A \wedge B) & \langle \text{hypothesis} \rangle \\
(2) & A \wedge B & \langle (1) + \text{spec } (6.1.6) \rangle \\
(3) & A & \langle (2) \models_{\text{taut}} (3) + 3.3.1 \rangle \\
(4) & B & \langle (2) + \text{tautological implication} \rangle \\
(5) & (\forall \mathbf{x})A & \langle (3) + \text{gen } (6.1.1; \text{ Okay: Hypothesis has no free } \mathbf{x}) \rangle \\
(6) & (\forall \mathbf{x})B & \langle (4) + \text{gen } (6.1.1; \text{ Okay: Hypothesis has no free } \mathbf{x}) \rangle \\
(7) & (\forall \mathbf{x})A \wedge (\forall \mathbf{x})B & \langle (5,6) + \text{tautological implication} \rangle
\end{array}
$$

---

[109] Recall that the Boolean abstraction of a formula of the form $(\forall x)A$ is a Boolean variable, say p. Thus all the Boolean connectives of A are invisible in the abstraction.

Note. Annotation of use of Post's theorem (usually invoked in the form of 3.3.1) might take the form given in step (3), "(2) $\models_{\text{taut}}$ (3)", but the ones in steps (4) and (7) are equally acceptable.

$(\leftarrow)$

(1)	$(\forall x)A \wedge (\forall x)B$	$\langle$hypothesis$\rangle$
(2)	$(\forall x)A$	$\langle$(1) + tautological implication$\rangle$
(3)	$(\forall x)B$	$\langle$(1) + tautological implication$\rangle$
(4)	A	$\langle$(2) + spec$\rangle$
(5)	B	$\langle$(3) + spec$\rangle$
(6)	$A \wedge B$	$\langle$(4, 5) + tautological implication$\rangle$
(7)	$(\forall x)(A \wedge B)$	$\langle$(6) + gen (6.1.1); Okay: Line (1) has no free $x\rangle$ $\quad \square$

Easy and natural, was it not?

Worth Stating: We applied Post's theorem in the proof above, and we will be increasingly doing so in future proofs. But how do we *know* when

$$A_1, \ldots, A_n \models_{\text{taut}} B \tag{$*$}$$

holds, in order to apply Post's theorem—which allows us to state and use the derived rule $A_1, \ldots, A_n \vdash B$?

This is a practical, not a theoretical question, and it has a simple answer: *In practice*, $(*)$ should be trivial to verify using semantic ideas, or trivial for us to "see" outright—as, e.g., certainly $A \equiv B \models_{\text{taut}} A \rightarrow B$ is.

In any case where such verification is not outright obvious, $(*)$ should be justified separately from—i.e., outside of—the proof where it is used.

How? Semantically ($\models_{\text{taut}}$) or syntactically ($\vdash$)—it is all the same by Post's theorem—*use whichever approach is easier to you.*

6.1.8 Theorem. $\vdash (\forall x)(\forall y)A \equiv (\forall y)(\forall x)A$

Proof. Another Ping-Pong argument:

$(\rightarrow)$

(1)	$(\forall x)(\forall y)A$	$\langle$hypothesis$\rangle$
(2)	$(\forall y)A$	$\langle$(1) + spec (6.1.6)$\rangle$
(3)	A	$\langle$(2) + spec$\rangle$
(4)	$(\forall x)A$	$\langle$(3) + 6.1.1—line (1) has no free $x\rangle$
(5)	$(\forall y)(\forall x)A$	$\langle$(4) + 6.1.1—line (1) has no free $y\rangle$

$(\leftarrow)$

The proof can be reversed, (5) to (1), with a straightforward change of annotation. Exercise! $\quad \square$

6.1.9 Metatheorem. (∀-Monotonicity) *If* $\Gamma \vdash A \rightarrow B$, *then* $\Gamma \vdash (\forall \mathbf{x})A \rightarrow (\forall \mathbf{x})B$, *provided that no formula in* Γ *has a free* $\mathbf{x}$.

Proof.

(1) $A \rightarrow B$ ⟨theorem from Γ; we start where its proof finished⟩
(2) $(\forall \mathbf{x})(A \rightarrow B)$ ⟨(1) + gen (6.1.1)—restriction on Γ makes this okay⟩
(3) $(\forall \mathbf{x})(A \rightarrow B) \rightarrow$
      $(\forall \mathbf{x})A \rightarrow (\forall \mathbf{x})B$ ⟨axiom⟩
(4) $(\forall \mathbf{x})A \rightarrow (\forall \mathbf{x})B$ ⟨(2,3) + MP⟩   □

Why "monotonicity"?

Well, one may view "$\rightarrow$" as an analogue of "$\leq$". You see, $\leq$ satisfies "$x \leq y$ iff $\max(x, y) = y$" while $\rightarrow$ satisfies (**Ax1**) "$A \rightarrow B \equiv A \vee B \equiv B$".

If you next think of "$\vee$" as analogous to "$\max$"—not a terribly far fetched suggestion, since many reasonable people think of **t** as 1 and **f** as 0 (certainly the designers of the languages PL/1 and C do)— then you see my point.

But then the metatheorem says that if A is "less than or equal to" B, then $(\forall \mathbf{x})A$ is "less than or equal to" $(\forall \mathbf{x})B$.

If you finally *intuitively* think of "$(\forall \mathbf{x})$" as a "function" (in the algebraic sense) of the "argument" A, then the jargon "monotonicity of $\forall$" makes sense.

So there. A silly comment, justifying a silly terminology. *In annotations* A-mon *will stand for* ∀-monotonicity.

6.1.10 Corollary. *If* $\vdash A \rightarrow B$, *then* $\vdash (\forall \mathbf{x})A \rightarrow (\forall \mathbf{x})B$.

Proof. Just take $\Gamma = \emptyset$.   □

6.1.11 Corollary. *If* $\Gamma \vdash A \equiv B$, *then also* $\Gamma \vdash (\forall \mathbf{x})A \equiv (\forall \mathbf{x})B$, *as long as* Γ *has no formulae with free* $\mathbf{x}$.

Proof.

(1) $A \equiv B$ ⟨proved from Γ; now continue the proof⟩
(2) $A \rightarrow B$ ⟨(1) $\models_{\text{taut}}$ (2) + 3.3.1⟩
(3) $B \rightarrow A$ ⟨(1) $\models_{\text{taut}}$ (3) + 3.3.1⟩
(4) $(\forall \mathbf{x})A \rightarrow (\forall \mathbf{x})B$ ⟨(2) + A-mon (6.1.9)⟩
(5) $(\forall \mathbf{x})B \rightarrow (\forall \mathbf{x})A$ ⟨(3) + A-mon (6.1.9)⟩
(6) $(\forall \mathbf{x})A \equiv (\forall \mathbf{x})B$ ⟨(4, 5) $\models_{\text{taut}}$ (6) + 3.3.1⟩   □

6.1.12 Corollary. *If* $\vdash A \equiv B$, *then also* $\vdash (\forall \mathbf{x})A \equiv (\forall \mathbf{x})B$.

Proof. Just take $\Gamma = \emptyset$.   □

Corollaries 6.1.11 and 6.1.12 are forms of the Leibniz rule that, for the first time so far, allow us to replace a Boolean variable that occurs *inside the scope of a quantifier*, and *moreover* they allow us to do so not caring if capture occurs!

For example, think of 6.1.12 as "if $\vdash A \equiv B$, then $\vdash C[\mathbf{p} \setminus A] \equiv C[\mathbf{p} \setminus B]$—where C is $(\forall \mathbf{x})\mathbf{p}$".

Can we extend 6.1.11 and 6.1.12 to hold for *any* formula C? You bet! We do so in the next section, but first let me make a few more comments on the generalization rule(s).

6.1.13 Remark. (1) Is the name *generalization* for the rule described in 6.1.1 apt? Yes. The rule is used in everyday math practice, where to prove that "a formula A that depends on the variable x holds for all values of x" we do instead the following:

We prove that "A holds for an *arbitrary* value of x—in other words, a value that we have made no special assumptions about". We then *generalize* and conclude—since the value of x that we used was unbiased—that A holds for all values of x.

Which part in 6.1.1 formally says that when we proved A we did not take into account any special value of $\mathbf{x}$? It is the part that says that our *assumptions* (Γ) do not even *talk* about $\mathbf{x}$, that is, $\mathbf{x}$ is not referenced in them at all, because $\mathbf{x}$ is not free.

(2) Is $(\forall \mathbf{x})A$ the "same as" the formula below?

$$A(0) \wedge A(1) \wedge A(2) \wedge A(3) \wedge \cdots \qquad (i)$$

where I am using here the shorthand $A(k)$ for $A[\mathbf{x} := k]$.

No, for some obvious and for some more esoteric reasons.

First the obvious:

Obvious 1. $0, 1, 2, \ldots$ are *nonlogical* symbols. A *logical* expression such as $(\forall \mathbf{x})A$ that is meant to make sense for *all* applications of logic—not just for number theory and other theories that speak of the integers—has no business being defined by (or being "equivalent" to) an expression that refers to such nonlogical symbols.

Obvious 2. Classical logic, that is, the logic we develop and use in this volume—which, by the way, is the logic used by mathematicians, computer scientists, etc.—does not allow infinitely long formulae like (i).

There are esoteric reasons according to which even if we remove the two objections above by cleverly tweaking question (2), the answer remains "no".

Okay, let us tweak the original question and ask instead: Would "$(\forall \mathbf{x})A$" *when its use is restricted to number theory* mean (i) above? This bypasses objection 1 above by restricting the question.

Let us next remove objection 2 above (**Obvious 2**) and rephrase, asking instead—without reference to infinitely long formulae—the *intuitively* equivalent question:

Is it true *in number theory* that we have both (ii) and (iii) below?

$$(\forall \mathbf{x})A \vdash A(k), \text{ for all } k \geq 0 \qquad (ii)$$

$$A(0), A(1), A(2), \ldots \vdash (\forall \mathbf{x})A \tag{iii}$$

Well, we do have (ii) by 6.1.5. However we do *not* have (iii): A result of Kurt Gödel ([16]) known as "the first incompleteness theorem" has as a corollary that there are formulae A of number theory for which all the premises in (iii) are theorems of number theory, but $(\forall \mathbf{x})A$ is not.   □

For the reader who will not easily let go: "But," you say, "if the nonlogical axioms of (Peano) number theory characterize $\mathbb{N}$ and the various operations and relations on it (e.g., $+, \times, <$) *uniquely*, then surely (iii) ought to hold since the hypothesis part says that A is true *of each* natural number. Surely, $(\forall \mathbf{x})A$ says the same thing?"

This is precisely where the problem lies! You see, it is known that first-order logic (over the language of Peano arithmetic and equipped with Peano's axioms) is incapable of *uniquely* characterising $\mathbb{N}$ equipped with its operations and relations. That is, there are supersets of $\mathbb{N}$ that contain infinitely many *other* "numbers"—but where $S, +, \times, <$ and all standard operations on numbers still make sense. More concretely, these sets equipped with analogues of the operations and relations $S, +, \times, <$ also satisfy all the Peano axioms of number theory!

Clearly, on such sets, saying that the left-hand side of $\vdash$ in (iii) holds is *not enough guarantee* that the right-hand side also holds, because the left-hand side of $\vdash$ does *not* say that A holds for *all* numbers (it says so only for the *natural* numbers, and is silent about the additional numbers).

The team of spec and gen also allows us to work with "simultaneous substitution": For example, if I have proved A that has x and y (and perhaps other variables that I do not care about) free, and if t and s are any terms, can I then conclude that after I substitute t and s into x and y *simultaneously* I will obtain a theorem?

6.1.14 Example. First, I should be clear what this "simultaneous" means: Suppose that A is $x = y$. Let "t" be y and "s" be x. Then,

$$(x = y)[x := t][y := s] \text{ is } x = x$$

while

$$(x = y)[y := s][x := t] \text{ is } y = y$$

The result depends on the order of the two substitutions.

On the other hand, if we drop t and s into the x and y slots simultaneously, so that neither of the two substitutions has the time to mess up the other, then we get $y = x$.

In effect, the "simultaneous substitution" is (or can be simulated by) a sequence of consecutive substitutions performed with the help of fresh variables: Pick two new variables, z and w. Then do

$$(x = y)[x := z][y := w][z := t][w := s]$$

or

$$(x = y)[x := z][y := w][w := s][z := t]$$

Both work, i.e., yield $y = x$. Verify!

In the general case (of two simultaneous substitutions)—that involves any formula A, any substitution slots $\mathbf{x}$ and $\mathbf{y}$, and terms t and s—the new (distinct) variables $\mathbf{z}$ and $\mathbf{w}$ are chosen to be fresh with respect to *all* of A, t, s. Then, a bit of reflection (or an induction on the complexity of A) shows that $A[\mathbf{x} := \mathbf{z}][\mathbf{y} := \mathbf{w}][\mathbf{z} := t][\mathbf{w} := s]$ and $A[\mathbf{x} := \mathbf{z}][\mathbf{y} := \mathbf{w}][\mathbf{w} := s][\mathbf{z} := t]$ yield the same string,[110] corresponding to the intuitive understanding of "simultaneous substitution into $\mathbf{x}$ and $\mathbf{y}$". $\square$

6.1.15 Definition. (Simultaneous Substitution) The expression

$$A[\mathbf{x}_1, \ldots, \mathbf{x}_r := t_1, \ldots, t_r] \tag{1}$$

denotes *simultaneous substitution* of the terms $t_1, \ldots, t_r$ into the variables $\mathbf{x}_1, \ldots, \mathbf{x}_r$ in the following sense:

Let $\mathbf{z}_1, \ldots, \mathbf{z}_r$ be distinct new variables that do not occur at all (either as free or bound) in any of $A, t_1, \ldots, t_r$. Then (1) is short for

$$A[\mathbf{x}_1 := \mathbf{z}_1] \ldots [\mathbf{x}_r := \mathbf{z}_r][\mathbf{z}_1 := t_1] \ldots [\mathbf{z}_r := t_r] \tag{2}$$

where in the interest of generality our notation employed the metavariables $\mathbf{x}_i$ and $\mathbf{z}_i$. $\square$

This "simultaneity" that the definition is talking about is *not* in the physical time sense, but it rather means that effecting the substitutions *sequentially*, we are nevertheless guaranteed that *none of the t_i that have already replaced an $\mathbf{x}_i$—in two steps, via $\mathbf{z}_i$—are subsequently altered* by virtue of some t_j being substituted into one of t_i's variables. This cannot happen since none of the t_i contains any $\mathbf{z}_j$.

In effect, we have—through 6.1.15—simulated what we intuitively understand as "simultaneous substitution": The substitution of the t_i into the $\mathbf{x}_i$ is *order independent*.

6.1.16 Exercise. Given a formula A, if $\mathbf{z}$ is fresh, then $A[\mathbf{x} := \mathbf{z}]$ is defined (cf. 4.1.28).
Hint. Induction on A. $\square$

6.1.17 Exercise. Suppose that $A[\mathbf{x} := t]$ is defined.
If $\mathbf{z}$ is fresh, then $A[\mathbf{x} := \mathbf{z}][\mathbf{z} := t]$ is also defined, and has the same result as $A[\mathbf{x} := t]$.
Hint. Induction on A. $\square$

6.1.18 Exercise. Definition 6.1.15 yields the same string, unaffected by permutations within the groups "$[\mathbf{x}_1 := \mathbf{z}_1] \ldots [\mathbf{x}_r := \mathbf{z}_r]$" and "$[\mathbf{z}_1 := t_1] \ldots [\mathbf{z}_r := t_r]$" in (2). The definition is also unaffected by the choice of fresh variables $\mathbf{z}_1, \ldots, \mathbf{z}_r$.
Hint. Induction on A. $\square$

---

[110]And so do $A[\mathbf{y} := \mathbf{w}][\mathbf{x} := \mathbf{z}][\mathbf{z} := t][\mathbf{w} := s]$ and $A[\mathbf{y} := \mathbf{w}][\mathbf{x} := \mathbf{z}][\mathbf{w} := s][\mathbf{z} := t]$.

6.1.19 Metatheorem. (Substitution Theorem) *If $\vdash A$ and $t_1, \ldots, t_r$ are any terms, then $\vdash A[\mathbf{x}_1, \ldots, \mathbf{x}_r := t_1, \ldots, t_r]$ as well.*

Of course, if the substitution is not defined, then there is nothing to prove.

Proof. By 6.1.15 we need to prove that if $\mathbf{z}_1, \ldots, \mathbf{z}_r$ are fresh with respect to $A, t_1, \ldots, t_r$, then we have

$$\vdash A[\mathbf{x}_1 := \mathbf{z}_1] \ldots [\mathbf{x}_r := \mathbf{z}_r][\mathbf{z}_1 := t_1] \ldots [\mathbf{z}_r := t_r]$$

The above follows by applying the metatheorem below $2r$ times:

If $\vdash A$ and $A[\mathbf{x} := t]$ is defined, then $\vdash A[\mathbf{x} := t]$, where t is any term

The above has a trivial proof (Hilbert style, but not written down rigidly):
I have $\vdash (\forall \mathbf{x})A$ by 6.1.3. An application of 6.1.5 yields $\vdash A[\mathbf{x} := t]$. □

6.1.20 Corollary. *If $\Gamma \vdash A$ and there is a proof of this fact that uses formulae from Γ that have no free $\mathbf{x}_1, \ldots, \mathbf{x}_r$, then $\Gamma \vdash A[\mathbf{x}_1, \ldots, \mathbf{x}_r := t_1, \ldots, t_r]$ as well.*

Proof. Let us fix attention to one such proof that uses hypotheses from Γ where none of $\mathbf{x}_1, \ldots, \mathbf{x}_r$ occur free. Say, $B_1, \ldots, B_m$ are the hypotheses used. Thus, by definition of proof (4.2.6)

$$B_1, \ldots, B_m \vdash A \tag{1}$$

Applying the deduction theorem to (1) m times, we get

$$\vdash B_1 \to \ldots \to B_m \to A \tag{2}$$

By 6.1.19 we get

$$\vdash \Big(B_1 \to \ldots \to B_m \to A\Big)[\mathbf{x}_1 := \mathbf{z}_1] \ldots [\mathbf{x}_r := \mathbf{z}_r][\mathbf{z}_1 := t_1] \ldots [\mathbf{z}_r := t_r] \tag{3}$$

where the $\mathbf{z}_i$ are fresh.

Since none of the $\mathbf{x}_i$ or $\mathbf{z}_i$ appear free in the B_j, (3) can be rewritten as (remember the priority of the "$[\mathbf{x} := \ldots]$" notation!)

$$\vdash B_1 \to \ldots \to B_m \to A[\mathbf{x}_1 := \mathbf{z}_1] \ldots [\mathbf{x}_r := \mathbf{z}_r][\mathbf{z}_1 := t_1] \ldots [\mathbf{z}_r := t_r] \tag{4}$$

Applying MP to (4), m times, yields

$$B_1, \ldots, B_m \vdash A[\mathbf{x}_1 := \mathbf{z}_1] \ldots [\mathbf{x}_r := \mathbf{z}_r][\mathbf{z}_1 := t_1] \ldots [\mathbf{z}_r := t_r]$$

The above yields what we want, by hypothesis strengthening (2.1.1):

$$\Gamma \vdash A[\mathbf{x}_1 := \mathbf{z}_1] \ldots [\mathbf{x}_r := \mathbf{z}_r][\mathbf{z}_1 := t_1] \ldots [\mathbf{z}_r := t_r]$$ □

6.2 LEIBNIZ RULES THAT AFFECT QUANTIFIER SCOPES

6.2.1 Metatheorem. (Weak Leibniz—"WL") *If* $\vdash A \equiv B$, *then* $\vdash C[\mathbf{p} \setminus A] \equiv C[\mathbf{p} \setminus B]$.

Proof. Knowing that the result holds in the "simple case" of 6.1.12—cf. also the remarks following the proof of 6.1.12—we are motivated to provide a proof by induction on the complexity of C.

Basis. In the case of formula complexity 0 we have two subcases of interest:

(1) C is $\mathbf{p}$. Then we must show "if $\vdash A \equiv B$, then $\vdash A \equiv B$". There is nothing to do!

(2) C is not $\mathbf{p}$, i.e., it is one of the following: $\mathbf{q}$ ($\neq \mathbf{p}$), $t = s$, $\phi(t_1, \ldots, t_n)$, $\top$, $\bot$. Then we must show "if $\vdash A \equiv B$, then $\vdash C \equiv C$". Since $\vdash C \equiv C$ holds anyway (**Ax1**), the *if* part is redundant and we are done.

The complex cases:

(i) C is $\neg D$. By I.H. $\vdash D[\mathbf{p} \setminus A] \equiv D[\mathbf{p} \setminus B]$; hence $\vdash \neg D[\mathbf{p} \setminus A] \equiv \neg D[\mathbf{p} \setminus B]$ by tautological implication (3.3.1) and thus $\vdash (\neg D)[\mathbf{p} \setminus A] \equiv (\neg D)[\mathbf{p} \setminus B]$ (cf. 4.1.29).

(ii) C is $D \circ E$, where $\circ \in \{\wedge, \vee, \rightarrow, \equiv\}$. By I.H. $\vdash D[\mathbf{p} \setminus A] \equiv D[\mathbf{p} \setminus B]$ and $\vdash E[\mathbf{p} \setminus A] \equiv E[\mathbf{p} \setminus B]$; hence $\vdash D[\mathbf{p} \setminus A] \circ E[\mathbf{p} \setminus A] \equiv D[\mathbf{p} \setminus B] \circ E[\mathbf{p} \setminus B]$ by tautological implication and thus $\vdash (D \circ E)[\mathbf{p} \setminus A] \equiv (D \circ E)[\mathbf{p} \setminus B]$ by 4.1.29.

(iii) C is $(\forall \mathbf{x})D$. This is "the interesting case".

By I.H. $\vdash D[\mathbf{p} \setminus A] \equiv D[\mathbf{p} \setminus B]$. By 6.1.12, $\vdash (\forall \mathbf{x})D[\mathbf{p} \setminus A] \equiv (\forall \mathbf{x})D[\mathbf{p} \setminus B]$, which Definition 4.1.29 allows us to rewrite as $\vdash ((\forall \mathbf{x})D)[\mathbf{p} \setminus A] \equiv ((\forall \mathbf{x})D)[\mathbf{p} \setminus B]$ $\square$

In [51] I called the WL-rule "WLUS" (Weak Leibniz with unconditional substitution). I now prefer the simpler "WL".

But why "weak"? Because unlike BL, which allows us to apply it regardless of where, or how, we got the hypothesis $A \equiv B$, we will *not* apply WL unless we know that the hypothesis is an absolute theorem.

Bad things will happen if we ignore this restriction: We can end up contradicting things we know. Ignoring the restriction means that *no matter why we were allowed to write down $A \equiv B$ in a proof*, we may next write, for any C, $C[\mathbf{p} \setminus A] \equiv C[\mathbf{p} \setminus B]$.

In other words, that

$$A \equiv B \vdash C[\mathbf{p} \setminus A] \equiv C[\mathbf{p} \setminus B] \tag{i}$$

is a derived rule.

But if that is so, then I can also derive the "rule" below

$$A \vdash (\forall \mathbf{x})A \tag{ii}$$

which I know is impossible in our logic (cf. 6.1.4).

Here is how to get (ii) from (i): First note that $\vdash (\forall \mathbf{x})\top \equiv \top$. Indeed, a trivial Ping-Pong argument (cf. 3.4.5) suffices, noting that $(\forall \mathbf{x})\top \rightarrow \top$ is in **Ax2** while—as $\top$ has no free $\mathbf{x}$—$\top \rightarrow (\forall \mathbf{x})\top$ is in **Ax4**. Secondly,

(1)	A	$\langle$hypothesis$\rangle$
(2)	$A \equiv \top$	$\langle$(1) + tautological implication$\rangle$
(3)	$(\forall \mathbf{x})A \equiv (\forall \mathbf{x})\top$	$\langle$(2) + rule (i); "C-part" is $(\forall \mathbf{x})\mathbf{p}\rangle$
(4)	$(\forall \mathbf{x})A \equiv \top$	$\langle$(3) + $\vdash (\forall \mathbf{x})\top \equiv \top$ + tautological implication$\rangle$
(5)	$(\forall \mathbf{x})A$	$\langle$(4) + tautological implication$\rangle$

WL is perhaps the most useful Leibniz rule in predicate logic, as it allows total freedom in substituting "equivalents for equivalents". The price to pay for this freedom is, of course, the restriction on how the premise is obtained.

By analyzing the proof of WL one readily sees that we can be a bit less restrictive in the assumption, as follows:

6.2.2 Corollary. (A More Generous WL) *If $\Gamma \vdash A \equiv B$ and if none of the bound variables of C occur free in the formulae of Γ, then $\Gamma \vdash C[\mathbf{p} \setminus A] \equiv C[\mathbf{p} \setminus B]$ as well.*

Proof. The proof is exactly the same as that of WL on p. 165. We change all "$\vdash$" there by "$\Gamma \vdash$" here and note that there are no changes in the Basis part. All the induction step cases regarding Boolean connectives are handled by tautological implication once more.

The interesting case is when C is once again $(\forall \mathbf{x})D$. The I.H. here yields $\Gamma \vdash D[\mathbf{p} \setminus A] \equiv D[\mathbf{p} \setminus B]$. Now the assumption guarantees that Γ-formulae have no free $\mathbf{x}$, so 6.1.11 applies:

$$\Gamma \vdash (\forall \mathbf{x})D[\mathbf{p} \setminus A] \equiv (\forall \mathbf{x})D[\mathbf{p} \setminus B]$$

In view of 4.1.29, this is what we want. $\square$

In practice, one finds 6.2.1 used more often than 6.2.2.

We next strengthen BL. This is a "strong" Leibniz in that the hypothesis $A \equiv B$ can be plucked out of the blue without any restriction on its origin. On the other hand, we have an annoying side condition that $\mathbf{p}$ must not be in the scope of a quantifier of C.

Well, *we can drop the side condition* from BL, but we have to continue using "conditional substitution" (4.1.30) as the discussion following the proof of 6.2.1 made clear.

6.2.3 Metatheorem. (Strong Leibniz—"SL") $A \equiv B \vdash C[\mathbf{p} := A] \equiv C[\mathbf{p} := B]$

Again, it goes without saying that if the right hand side of $\vdash$ is *undefined* then we have nothing to prove as the expression "$C[\mathbf{p} := A] \equiv C[\mathbf{p} := B]$" does not denote any formula. I will not remind us again of this understanding of use of the metanotations "$[\mathbf{p} := A]$" and "$[\mathbf{x} := t]$" since it has been already established enough times.

Proof. The proof is by induction on the complexity of C. It is similar, but not identical, to the proof of WL (6.2.1).

Basis. In the case of formula complexity 0 we have two subcases of interest:

(1) C is $\mathbf{p}$. Then we must show $A \equiv B \vdash A \equiv B$, which holds by 1.4.11.

(2) C is not $\mathbf{p}$, i.e., it is one of the following: $\mathbf{q}$ ($\neq \mathbf{p}$), $t = s$, $\phi(t_1, \ldots, t_n)$, $\top$, $\bot$. Then we must show $A \equiv B \vdash C \equiv C$. This follows from $\vdash C \equiv C$ (**Ax1**) and 2.1.1.

The complex cases:

(i) C is $\neg D$. By I.H. $A \equiv B \vdash D[\mathbf{p} := A] \equiv D[\mathbf{p} := B]$; hence $A \equiv B \vdash \neg D[\mathbf{p} := A] \equiv \neg D[\mathbf{p} := B]$ by tautological implication, and thus $A \equiv B \vdash (\neg D)[\mathbf{p} := A] \equiv (\neg D)[\mathbf{p} := B]$ (cf. 4.1.30).

(ii) C is $D \circ E$. By I.H. $A \equiv B \vdash D[\mathbf{p} := A] \equiv D[\mathbf{p} := B]$ and $A \equiv B \vdash E[\mathbf{p} := A] \equiv E[\mathbf{p} := B]$; hence $A \equiv B \vdash D[\mathbf{p} := A] \circ E[\mathbf{p} := A] \equiv D[\mathbf{p} := B] \circ E[\mathbf{p} := B]$ by tautological implication and thus $A \equiv B \vdash (D \circ E)[\mathbf{p} := A] \equiv (D \circ E)[\mathbf{p} := B]$ by 4.1.30.

(iii) C is $(\forall \mathbf{x})D$. This is "the interesting case". By I.H. $A \equiv B \vdash D[\mathbf{p} := A] \equiv D[\mathbf{p} := B]$. Since $C[\mathbf{p} := A]$ and $C[\mathbf{p} := B]$ are defined, Definition 4.1.30 implies that $\mathbf{x}$ is not free in either A or B. Therefore (cf. 4.1.21) it is not free in $A \equiv B$. By 6.1.11, $A \equiv B \vdash (\forall \mathbf{x})D[\mathbf{p} := A] \equiv (\forall \mathbf{x})D[\mathbf{p} := B]$, which 4.1.30 allows us to rewrite as $A \equiv B \vdash ((\forall \mathbf{x})D)[\mathbf{p} := A] \equiv ((\forall \mathbf{x})D)[\mathbf{p} := B]$. $\square$

In [51] I called this rule "SLCS" (Strong Leibniz with conditional substitution). I now prefer the simpler "SL".

Note that we may from now on, conveniently, forget BL and—as far as rules of type Leibniz are concerned—to use only WL and SL since the latter subsumes BL. This last observation justifies the concluding sentence in Remark 4.2.5.

6.2.4 Corollary. $D \to (A \equiv B) \vdash D \to (C[\mathbf{p} := A] \equiv C[\mathbf{p} := B])$

Proof. We use the deduction theorem to prove instead

$$D \to (A \equiv B), D \vdash C[\mathbf{p} := A] \equiv C[\mathbf{p} := B]$$

Indeed

(1) D ⟨hypothesis⟩

(2) $D \to (A \equiv B)$ ⟨hypothesis⟩

(3) $A \equiv B$ $\langle (1, 2) + \text{MP} \rangle$

(4) $C[\mathbf{p} := A] \equiv C[\mathbf{p} := B]$ $\langle (3) + \text{SL} \rangle$

Note how SL (rather than WL) is applicable, because line (3) does not necessarily contain an absolute theorem. □

6.2.5 Remark. (1) The previous proof is not circular, as one might carelessly assume by the proved statement's similarity to 2.6.2. Indeed, Exercise 5.0.4 promises— within predicate logic—a direct, short and easy, and oblivious to 2.6.2, proof of the deduction theorem.

Note as well that 6.2.4 extends 2.6.2 since the latter can deal only with "**p**" that are not in the scope of some quantifier that occurs in C (Boolean logic knows nothing about such scopes).

(2) An alternative proof of 6.2.4 extends the statement to

$$D \circ (A \equiv B) \vdash D \circ (C[\mathbf{p} := A] \equiv C[\mathbf{p} := B]) \qquad (*)$$

where $\circ$ is any of $\vee, \wedge, \equiv$: Note that SL and the deduction theorem yield

$$\vdash (A \equiv B) \to (C[\mathbf{p} := A] \equiv C[\mathbf{p} := B])$$

By tautological implication we get

$$\vdash D \circ (A \equiv B) \to D \circ (C[\mathbf{p} := A] \equiv C[\mathbf{p} := B])$$

By MP we get $(*)$. □

6.3 THE LEIBNIZ RULES "8.12"

This section is meant as no more than extra homework since we will never use the two tools that we discuss here.

The tag "8.12" in the section title refers to the one used for the "twin" rules Leibniz in [17], p.148. These, reproduced verbatim from loc. cit. but recast in standard quantifier notation and using the metavariables **x** and **p**, are

$$\frac{A \equiv B}{(\forall \mathbf{x})(C[\mathbf{p} := A] \to D) \equiv (\forall \mathbf{x})(C[\mathbf{p} := B] \to D)} \qquad (8.12a)$$

and

$$\frac{D \to (A \equiv B)}{(\forall \mathbf{x})(D \to C[\mathbf{p} := A]) \equiv (\forall \mathbf{x})(D \to C[\mathbf{p} := B])} \qquad (8.12b)$$

This section will *prove* valid forms of these two rules.

6.3.1 Remark. (1) Refer to 4.1.18 once more if you plan to go to the source [17].

(2) The rules (8.12a) and (8.12b) are axiomatically given in [17] as the *primary* rules for quantifiers. As we will see here this is totally unnecessary for us to imitate,

because—with a correction—these are provable rules, that is, they are derived rules, just like "gen", "WL" and "SL" are in our logic.

(3) Are there any constraints on the premises of these rules? If so, what might they be?

Indeed, there *must* be constraints; otherwise either rule can derive "$A \vdash (\forall x)A$", which we know is unprovable in our logic.   □

6.3.2 Exercise. (I) Show that unless the premise in (8.12a) has restrictions on how it is obtained, (8.12a) implies strong generalization, and is therefore invalid.

Hint. Experiment, taking B to be the formula $\top$, D to be the formula $\bot$, and C to be the formula $\neg\mathbf{p}$.

(II) Show that unless the premise in (8.12b) has restrictions on how it is obtained, (8.12b) implies strong generalization, and is therefore invalid.

Hint. Experiment, taking B to be the formula $\top$, D to be the formula $\top$, and C to be the formula $\mathbf{p}$.

Using the hints, you will produce, in each question (I) and (II), a Hilbert-style proof

$$
\begin{array}{lll}
(1) & A & \langle\text{hypothesis}\rangle \\
& \vdots & \\
(n) & (\forall x)A & \langle\text{some appropriate reason}\rangle
\end{array}
$$ □

6.3.3 Metatheorem. (The Valid "8.12a") *If* $\Gamma \vdash A \equiv B$, *then* $\Gamma \vdash (\forall x)(C[\mathbf{p} := A] \rightarrow D) \equiv (\forall x)(C[\mathbf{p} := B] \rightarrow D)$, *provided no formula of* Γ *has a free* x.

Proof.

$$
\begin{array}{lll}
(1) & A \equiv B & \langle\text{from } \Gamma; \text{ we now continue the proof}\rangle \\
(2) & (C[\mathbf{p} := A] \rightarrow D) \equiv & \\
 & (C[\mathbf{p} := B] \rightarrow D) & \langle(1) \text{ and SL} + 3.3.1 \text{ to add "} \rightarrow D\text{"}\rangle \\
(3) & (\forall x)(C[\mathbf{p} := A] \rightarrow D) \equiv & \\
 & (\forall x)(C[\mathbf{p} := B] \rightarrow D) & \langle(2) \text{ and } 6.1.11; \text{ restriction on } \Gamma \text{ used}\rangle
\end{array}
$$ □

Note how in step (2) above SL—unlike WL—affords us the luxury not to worry whether $A \equiv B$ is an absolute theorem. Here, in general, it is not.

6.3.4 Corollary. *If* $\vdash A \equiv B$, *then* $\vdash (\forall x)(C[\mathbf{p} := A] \rightarrow D) \equiv (\forall x)(C[\mathbf{p} := B] \rightarrow D)$.

Proof. Take $\Gamma = \emptyset$.   □

6.3.5 Metatheorem. (The Valid "8.12b") *If* $\Gamma \vdash D \rightarrow (A \equiv B)$, *then* $\Gamma \vdash (\forall x)(D \rightarrow C[\mathbf{p} := A]) \equiv (\forall x)(D \rightarrow C[\mathbf{p} := B])$, *provided no formula of* Γ *has a free* x.

Proof. The hypothesis yields—via 6.2.4 followed by 3.3.1:

$$\Gamma \vdash (D \to C[\mathbf{p} := A]) \equiv (D \to C[\mathbf{p} := B])$$

We are done by an application of 6.1.11. □

6.3.6 Corollary. *If* $\vdash D \to (A \equiv B)$, *then* $\vdash (\forall \mathbf{x})(D \to C[\mathbf{p} := A]) \equiv (\forall \mathbf{x})(D \to C[\mathbf{p} := B])$.

Proof. Take $\Gamma = \emptyset$. □

6.4 ADDITIONAL USEFUL TOOLS

The authors of [17] introduce several theorem-schemata for quantifiers in their Chapter 8, which nevertheless they present as "axioms" and offer without proof. We will prove in this section that all these "axioms" are actually provable in our setting. All are useful additions to our toolbox and embody standard techniques of predicate logic such as the "variant theorem" (also known as the *dummy renaming* theorem) used to rename bound variables, and the so-called *prenex operations* used to bubble quantifiers from the interior of a formula all the way to its left boundary (cf. for example, 6.4.1, 6.4.2, 6.4.3).

The inquisitive reader who will want to explore the detailed presentation of these metatheorems in the cited source will benefit from a word of caution: First, all these so-called axioms about "generalized quantifiers" are provable from first principles. Second, some of them—those that speak about the "other" quantifiers, $\sum$ and $\prod$, that is, the generalized sum and product (of natural numbers)—do not even belong to pure logic but belong to number theory.[111]

Thus the "unified" notation "$*$" intended by [17] to express properties of all "quantifiers", $\exists, \forall, \sum, \prod$—*within pure logic*—can do so only for $\exists$ and $\forall$.

The *why* is straightforward: In predicate logic—when we *are not* employing it within a specific application such as number theory—statements involving $\sum, \prod$ *that also rely on special properties of these symbols* have no place.

First, statements involving *properties* of the nonlogical symbols such as $\sum, \prod$ *cannot* be *logical* axioms, nor can they be absolute theorems, by definition.

Second, all these statements regarding these two ($\sum, \prod$) symbols happen to be *provable*, but this is achievable only after one has introduced the Peano axioms, which, incidentally, do not even *refer* to $\sum, \prod$ as these are not primitive symbols of the language of arithmetic! The symbols $\sum$ and $\prod$—after a lot of work and acrobatics—*can* be *defined* in (Peano) number theory as secondary nonlogical symbols, and then all their properties as stated in Chapter 8 of [17] can be proved. However, I promise not to do any of this work here since this volume is about the methods and tools of *nonapplied* (i.e., pure) logic.[112]

---

[111]Chapter 8 of [17] also conflates the "quantifiers" $\bigcup$ and $\bigcap$ along with the others. The properties of these are provable in axiomatic set theory.

[112]The ambitious reader who wants to see how all this is done—in detail—may look up [53].

Well, then, for the sake of exercise, but also to enrich further our logical toolbox, let us prove all these results in the balance of this section.

It is fair to warn that in the context of predicate logic Hilbert-style proofs have the edge over equational-style proofs. The former are amenable to the methodology of removing/inserting quantifiers so that between the removal and insertion one can use Boolean techniques, notably the all-powerful Post's theorem (3.2.1—usually in the form 3.3.1).

The technique of removing/inserting quantifiers is not well suited to equational proofs partly because "A" and "$(\forall \mathbf{x})A$" cannot be connected with "$\equiv$"—indeed not even with "$\rightarrow$" in one direction—in general.

Of course, as always we will apply "the most natural proof style" for the problem at hand, but be aware that "most natural" is a subjective assessment!

Before we embark on proofs of the results advertised at the beginning of this section, I would like to expand a bit on equational methodology.

Our equational proofs so far have been based on a sequence of equivalences

$$A_1 \equiv A_2, A_2 \equiv A_3, \ldots, A_{n-1} \equiv A_n$$

that we know are, each, Γ-theorems (p. 62). We showed that we have 2.2.1 and its corollaries, which allow an equational proof to establish $\Gamma \vdash A_1 \equiv A_n$, and thus $\Gamma \vdash A_1$ iff $\Gamma \vdash A_n$.

Our extension allows any one of the $\equiv$ symbols to be replaced by $\rightarrow$. In this case, by Post's theorem (3.2.1), we have just $\Gamma \vdash A_1 \rightarrow A_n$. If we also know that $\Gamma \vdash A_1$, then MP shows that $\Gamma \vdash A_n$ as well.

The layout is

$$A_1$$
$$\circ \ \langle \text{annotation} \rangle$$
$$A_2$$
$$\circ \ \langle \text{annotation} \rangle$$
$$\vdots$$
$$A_{n-1}$$
$$\circ \ \langle \text{annotation} \rangle$$
$$A_n$$
$$\circ \ \langle \text{annotation} \rangle$$
$$A_{n+1}$$

where "$\circ$" in each instance is—independently of the other instances—one of $\Leftrightarrow$ or $\Rightarrow$. The symbol $\Rightarrow$ in this layout *occurs only on the left margin*; it is an alias for "$\rightarrow$" but it is *conjunctional*. That is,

$$A$$
$$\Rightarrow \langle\text{annotation}\rangle$$
$$B$$
$$\Rightarrow \langle\text{annotation}\rangle$$
$$C$$

means "$A \rightarrow B, B \rightarrow C$"—which are two separate formulae—*not* "$A \rightarrow B \rightarrow C$". The latter is, of course, short for $(A \rightarrow (B \rightarrow C))$.

We now embark on our task.

6.4.1 Theorem. $\vdash (\forall \mathbf{x})(A \rightarrow B) \equiv (A \rightarrow (\forall \mathbf{x})B)$, *provided* $\mathbf{x}$ *is not free in* A.

Proof. We use a Ping-Pong argument (cf. 3.4.5) in conjunction with the deduction theorem.

($\rightarrow$) I want

$$\vdash (\forall \mathbf{x})(A \rightarrow B) \rightarrow (A \rightarrow (\forall \mathbf{x})B)$$

but I'd rather (cf. 2.6.1) prove

$$(\forall \mathbf{x})(A \rightarrow B) \vdash A \rightarrow (\forall \mathbf{x})B$$

and, indeed, I'd rather (cf. 2.6.1 again!) prove

$$(\forall \mathbf{x})(A \rightarrow B), A \vdash (\forall \mathbf{x})B$$

Okay, let's do it! (Never forget the discussion following 6.1.6 on p. 158.)

(1)	$(\forall \mathbf{x})(A \rightarrow B)$	$\langle$hypothesis$\rangle$
(2)	A	$\langle$hypothesis$\rangle$
(3)	$A \rightarrow B$	$\langle(1) + \text{spec } (6.1.6)\rangle$
(4)	B	$\langle(2, 3) + \text{MP}\rangle$
(5)	$(\forall \mathbf{x})B$	$\langle(4) + \text{gen } (6.1.1);$ Okay since $(1, 2)$ have no free $\mathbf{x}\rangle$

($\leftarrow$) I want

$$\vdash (A \rightarrow (\forall \mathbf{x})B) \rightarrow (\forall \mathbf{x})(A \rightarrow B)$$

I might as well do (cf. 2.6.1) the easier

$$A \rightarrow (\forall \mathbf{x})B \vdash (\forall \mathbf{x})(A \rightarrow B) \tag{1}$$

Seeing that $A \rightarrow (\forall \mathbf{x})B$ has no free $\mathbf{x}$, I can prove the still easier (no quantifiers in right-hand side!)

$$A \rightarrow (\forall \mathbf{x})B \vdash A \rightarrow B \tag{2}$$

and then apply 6.1.1 to conclude. The deduction theorem allows me to prove something even simpler:

$$A \to (\forall \mathbf{x})B, A \vdash B \tag{3}$$

Here it goes (proof of (3)):

$$
\begin{array}{lll}
(1) & A \to (\forall \mathbf{x})B & \langle \text{hypothesis} \rangle \\
(2) & A & \langle \text{hypothesis} \rangle \\
(3) & (\forall \mathbf{x})B & \langle (1, 2) + \text{MP} \rangle \\
(4) & B & \langle (3) + \text{spec} \rangle \qquad\qquad\quad \square
\end{array}
$$

We can also give an equational proof of the ($\to$) direction—extended in this section to allow both the conjunctional $\Leftrightarrow$ and $\Rightarrow$ in the left margin—as follows:

($\to$)

$(\forall \mathbf{x})(A \to B)$

$\Rightarrow \langle \mathbf{Ax3} \rangle$

$(\forall \mathbf{x})A \to (\forall \mathbf{x})B$

$\Leftrightarrow \langle \text{SL: "Numerator:" } \mathbf{Ax4} + (\forall \mathbf{x})A \to A \text{ (Ping-Pong); "}C\text{-part" is } \mathbf{p} \to (\forall \mathbf{x})B \rangle$

$A \to (\forall \mathbf{x})B$

6.4.2 Corollary. $\vdash (\forall \mathbf{x})(A \vee B) \equiv A \vee (\forall \mathbf{x})B$, *provided* $\mathbf{x}$ *is not free in A.*

Proof. (Equational)

$(\forall \mathbf{x})(A \vee B)$

$\Leftrightarrow \langle \text{WL and } \vdash A \vee B \equiv \neg A \to B \text{ (tautology!); "}C\text{-part" is } (\forall \mathbf{x})\mathbf{p} \rangle$

$(\forall \mathbf{x})(\neg A \to B)$

$\Leftrightarrow \langle 6.4.1 \rangle$

$\neg A \to (\forall \mathbf{x})B$

$\Leftrightarrow \langle \text{tautology, hence a theorem} \rangle$

$A \vee (\forall \mathbf{x})B$ $\qquad\qquad\qquad\qquad\qquad\qquad\qquad \square$

Most of the results we prove here have interesting *duals* where $\forall$ and $\exists$ are interchanged (and so are $\wedge$ and $\vee$). The first one we prove with a routine (equational) calculation, but we will leave the proof of the majority of these duals to the reader.

Recall the definition of the *informal* (logical) symbol $\exists$ (4.1.17):

$$(\exists \mathbf{x})A \text{ is short for } \neg(\forall \mathbf{x})\neg A$$

Now, $\vdash \neg(\forall \mathbf{x})\neg A \equiv \neg(\forall \mathbf{x})\neg A$ (member of **Ax1**). We may choose to use the $\exists$-abbreviation in, say, the left-hand side of $\equiv$ to obtain

$$\vdash (\exists \mathbf{x})A \equiv \neg(\forall \mathbf{x})\neg A$$

We will frequently use this absolute theorem in what follows, often in connection with WL, and will nickname it "definition of $\exists$".

6.4.3 Corollary. $\vdash (\exists\mathbf{x})(A \wedge B) \equiv A \wedge (\exists\mathbf{x})B$, *provided* $\mathbf{x}$ *is not free in A.*

Proof. Below we use WL when capture may in general happen. Of course, in such a case, the hypothesis of the rule has to be an absolute theorem "$\vdash X \equiv Y$".

$$(\exists\mathbf{x})(A \wedge B)$$
$$\Leftrightarrow \langle \text{definition of } \exists \rangle$$
$$\neg(\forall\mathbf{x})\neg(A \wedge B)$$
$$\Leftrightarrow \langle \text{WL and deMorgan; "}C\text{-part" is } \neg(\forall\mathbf{x})\mathbf{p} \rangle$$
$$\neg(\forall\mathbf{x})(\neg A \vee \neg B)$$
$$\Leftrightarrow \langle \text{SL and 6.4.2—no free } \mathbf{x} \text{ in } \neg A; \text{"}C\text{-part" is } \neg\mathbf{p} \rangle$$
$$\neg(\neg A \vee (\forall\mathbf{x})\neg B)$$
$$\Leftrightarrow \langle \text{deMorgan} \rangle$$
$$\neg\neg A \wedge \neg(\forall\mathbf{x})\neg B$$
$$\Leftrightarrow \langle \text{tautology} \rangle$$
$$A \wedge \neg(\forall\mathbf{x})\neg B$$
$$\Leftrightarrow \langle \text{SL and definition of } \exists; \text{"}C\text{-part" is } A \wedge \mathbf{p} \rangle$$
$$A \wedge (\exists\mathbf{x})B \qquad\qquad \square$$

Here are the remaining results we promised (in quotes are the nicknames that the authors of [17] give to some of these results):

1. "Empty range". $\vdash (\forall\mathbf{x})(\bot \rightarrow A) \equiv \top$. By redundant true we just prove $\vdash (\forall\mathbf{x})(\bot \rightarrow A)$. Since $\vdash \bot \rightarrow A$ (in **Ax1**) we are done by an application of 6.1.3.

2. "One point rule". Provided that $\mathbf{x}$ is not free in the term t, $\vdash (\forall\mathbf{x})(x = t \rightarrow A) \equiv A[\mathbf{x} := t]$. We employ a Ping-Pong argument.

   $(\rightarrow)$ Note that since there is no free $\mathbf{x}$ in t,

   $$(\mathbf{x} = t \rightarrow A)[\mathbf{x} := t] \quad \text{is} \quad t = t \rightarrow A[\mathbf{x} := t]$$

   Thus

(1)	$(\forall\mathbf{x})(x = t \rightarrow A) \rightarrow t = t \rightarrow A[\mathbf{x} := t]$	$\langle\textbf{Ax2}\rangle$
(2)	$(\forall\mathbf{x})(\mathbf{x} = \mathbf{x})$	$\langle\text{partial generalization of }\textbf{Ax5}\rangle$
(3)	$t = t$	$\langle(2) + 6.1.5\rangle$
(4)	$(\forall\mathbf{x})(x = t \rightarrow A) \rightarrow A[x := t]$	$\langle(1, 3) + 3.3.1\rangle$

   $(\leftarrow)$

(1)	$\mathbf{x} = t \rightarrow (A \equiv A[\mathbf{x} := t])$	$\langle\textbf{Ax6}\rangle$

(2) $\quad A[\mathbf{x} := t] \rightarrow \mathbf{x} = t \rightarrow A$ $\qquad \langle(1) + 3.3.1\rangle$

(3) $\quad (\forall \mathbf{x})A[\mathbf{x} := t] \rightarrow (\forall \mathbf{x})(\mathbf{x} = t \rightarrow A)$ $\quad \langle(2) + 6.1.10\text{---}(2) \text{ is absolute}\rangle$

(4) $\quad A[\mathbf{x} := t] \rightarrow (\forall \mathbf{x})(\mathbf{x} = t \rightarrow A)$ $\qquad \langle(3) + \mathbf{Ax4} + 3.3.1\rangle$

Note that **Ax4** is applicable in step (4) since there is no free $\mathbf{x}$ in $A[\mathbf{x} := t]$

3. "One point rule—∃-version". $\vdash (\exists \mathbf{x})(\mathbf{x} = t \wedge A) \equiv A[\mathbf{x} := t]$ if $\mathbf{x}$ is not free in t.

   Exercise! (*Hint.* Use an equational calculation and the ∀-version of the one point rule.)

4. "Distributivity of ∀ over ∧". $\vdash (\forall \mathbf{x})(A \rightarrow B) \wedge (\forall \mathbf{x})(A \rightarrow C) \equiv (\forall \mathbf{x})(A \rightarrow B \wedge C)$.

   We calculate a proof as follows:

   $$(\forall \mathbf{x})(A \rightarrow B) \wedge (\forall \mathbf{x})(A \rightarrow C)$$
   $$\Leftrightarrow \langle 6.1.7\rangle$$
   $$(\forall \mathbf{x})\Big((A \rightarrow B) \wedge (A \rightarrow C)\Big)$$
   $$\Leftrightarrow \langle 6.1.11 + \text{tautology } (\mathbf{Ax1})\ (A \rightarrow B) \wedge (A \rightarrow C) \equiv (A \rightarrow B \wedge C)\rangle$$
   $$(\forall \mathbf{x})(A \rightarrow B \wedge C)$$

   We could also invoke WL itself above, but 6.1.11 is simpler and invoking it involves writing less annotation.

 The metatheorem that we just proved generalizes 6.1.7 to the case of the "bounded" or "relativized" quantifier ∀. The authors of [17] write the metatheorem thus

$$(\forall \mathbf{x}|A : B) \wedge (\forall \mathbf{x}|A : C) \equiv (\forall \mathbf{x}|A : B \wedge C)$$

while the corresponding notation in [2] is

$$(\forall \mathbf{x})_A B \wedge (\forall \mathbf{x})_A B \equiv (\forall \mathbf{x})_A (B \wedge C)$$

The general notation for bounded quantification is not much in use in the literature. However, special cases are used, such as "$(\forall \mathbf{x})_{\in \mathbf{y}} A$" and also "$(\forall \mathbf{x} \in \mathbf{y})A$", which mean $(\forall \mathbf{x})(\mathbf{x} \in \mathbf{y} \rightarrow A)$, and "$(\forall \mathbf{x})_{<\mathbf{y}} A$" and also "$(\forall \mathbf{x} < \mathbf{y})A$" that mean $(\forall \mathbf{x})(\mathbf{x} < \mathbf{y} \rightarrow A)$.

5. The proof of the dual of the above, "distributivity of ∃ over ∨", is left to the reader. It states: $\vdash (\exists \mathbf{x})(A \wedge B) \vee (\exists \mathbf{x})(A \wedge C) \equiv (\exists \mathbf{x})\Big(A \wedge (B \vee C)\Big)$. In the notation of [2] it is written as: $\vdash (\exists \mathbf{x})_A B \vee (\exists \mathbf{x})_A C \equiv (\exists \mathbf{x})_A (B \vee C)$.

6. "Range split", where *range* in [17] refers to the A of $(\forall \mathbf{x})_A B$.

$\vdash (\forall \mathbf{x})(A \vee B \rightarrow C) \equiv (\forall \mathbf{x})(A \rightarrow C) \wedge (\forall \mathbf{x})(B \rightarrow C)$, or, in Bourbaki's notation, $\vdash (\forall \mathbf{x})_{A \vee B} C \equiv (\forall \mathbf{x})_A C \wedge (\forall \mathbf{x})_B C$.

We calculate as follows:

$$(\forall \mathbf{x})(A \rightarrow C) \wedge (\forall \mathbf{x})(B \rightarrow C)$$
$$\Leftrightarrow \langle 6.1.7 \rangle$$
$$(\forall \mathbf{x})\Big((A \rightarrow C) \wedge (B \rightarrow C)\Big)$$
$$\Leftrightarrow \langle 6.1.11 \text{ and the tautology } (A \rightarrow C) \wedge (B \rightarrow C) \equiv ((A \vee B) \rightarrow C) \rangle$$
$$(\forall \mathbf{x})\Big((A \vee B) \rightarrow C\Big)$$

7. "Interchange of *dummies*"—as bound variables are called in [17]. This generalizes 6.1.8 to the case of bounded quantifiers. It states, $\vdash (\forall \mathbf{x})(A \rightarrow (\forall \mathbf{y})(B \rightarrow C)) \equiv (\forall \mathbf{y})(B \rightarrow (\forall \mathbf{x})(A \rightarrow C))$, *on the condition* that $\mathbf{y}$ is not free in A and $\mathbf{x}$ is not free in B. To highlight the relationship to 6.1.8 I also write the above in [2]-notation:

$$\vdash (\forall \mathbf{x})_A (\forall \mathbf{y})_B C \equiv (\forall \mathbf{y})_B (\forall \mathbf{x})_A C$$

Let us now calculate:

$$(\forall \mathbf{x})(A \rightarrow (\forall \mathbf{y})(B \rightarrow C))$$
$$\Leftrightarrow \langle 6.1.11 + 6.4.1 \text{—no free } \mathbf{y} \text{ in } A \rangle$$
$$(\forall \mathbf{x})(\forall \mathbf{y})(A \rightarrow (B \rightarrow C))$$
$$\Leftrightarrow \langle \text{WL + obvious tautology; "}C\text{-part" } (\forall \mathbf{x})(\forall \mathbf{y})\mathbf{p} \rangle \qquad (*)$$
$$(\forall \mathbf{x})(\forall \mathbf{y})(B \rightarrow (A \rightarrow C))$$
$$\Leftrightarrow \langle 6.1.8 \rangle$$
$$(\forall \mathbf{y})(\forall \mathbf{x})(B \rightarrow (A \rightarrow C))$$
$$\Leftrightarrow \langle 6.1.11 + 6.4.1 \text{—no free } \mathbf{x} \text{ in } B \rangle$$
$$(\forall \mathbf{y})(B \rightarrow (\forall \mathbf{x})(A \rightarrow C))$$

Note. Step $(*)$ uses WL rather than the simpler 6.1.11 since the latter deals with a single $\forall$ up in front.

8. The dual of the above is

$\vdash (\exists \mathbf{x})(A \wedge (\exists \mathbf{y})(B \wedge C)) \equiv (\exists \mathbf{y})(B \wedge (\exists \mathbf{x})(A \wedge C))$, *on the condition* that $\mathbf{y}$ is not free in A and $\mathbf{x}$ is not free in B.

It can be trivially proved using the "$\exists$-definition" and an equational argument. Exercise!

9. "Nesting". $\vdash (\forall \mathbf{x})(\forall \mathbf{y})(A \wedge B \rightarrow C) \equiv (\forall \mathbf{x})(A \rightarrow (\forall \mathbf{y})(B \rightarrow C))$, *on the condition* that $\mathbf{y}$ is not free in A.

$$(\forall \mathbf{x})(A \to (\forall \mathbf{y})(B \to C))$$
$$\Leftrightarrow \langle 6.1.11 + 6.4.1\text{—no free } \mathbf{y} \text{ in } A \rangle$$
$$(\forall \mathbf{x})(\forall \mathbf{y})(A \to (B \to C))$$
$$\Leftrightarrow \langle \text{WL + obvious tautology; } ``C\text{-part''} (\forall \mathbf{x})(\forall \mathbf{y})\mathbf{p} \rangle$$
$$(\forall \mathbf{x})(\forall \mathbf{y})(A \wedge B \to C)$$

10. The $\exists$-dual of the above is

$\vdash (\exists \mathbf{x})(\exists \mathbf{y})(A \wedge B \wedge C) \equiv (\exists \mathbf{x})(A \wedge (\exists \mathbf{y})(B \wedge C))$, *on the condition* that $\mathbf{y}$ is not free in A.

It is easy to prove equationally. Exercise!

The next metatheorem was given the nickname *dummy renaming* in [17] (where it is *axiom* (8.21)). Elsewhere in the literature (e.g., [45]), one refers to the metatheorem as the *variant* metatheorem.[113]

By either nickname it simply states something that we expect at the intuitive level, something we would be prepared to shrug off by saying, referring to the bound variable, "What's in a name?" After all, we know that $\sum_{i=1}^{n} i^2 = \sum_{k=1}^{n} k^2$.

Indeed, under some simple and not so restrictive conditions, a bound variable can be *provably renamed* without changing a formula's provability.

6.4.4 Theorem. (Dummy Renaming for $\forall$) *If $\mathbf{z}$ does not occur in A—i.e., neither free, nor bound—then* $\vdash (\forall \mathbf{x})A \equiv (\forall \mathbf{z})A[\mathbf{x} := \mathbf{z}]$.

The practical usefulness of the above is when $\mathbf{z}$ does not occur in $(\forall \mathbf{x})A$ either, i.e., when $\mathbf{z} \neq \mathbf{x}$ as well, for if $\mathbf{z}$ and $\mathbf{x}$ are the same, then the theorem becomes the trivial tautology $(\forall \mathbf{x})A \equiv (\forall \mathbf{x})A$ (cf. 4.1.34).

Proof. We go Ping-Pong:
$(\to)$

(1)	$(\forall \mathbf{x})A \to A[\mathbf{x} := \mathbf{z}]$	$\langle$**Ax2**: $\mathbf{z}$ is fresh for A; no capture: $A[\mathbf{x} := \mathbf{z}]$ defined$\rangle$
(2)	$(\forall \mathbf{z})(\forall \mathbf{x})A \to$	
	$(\forall \mathbf{z})A[\mathbf{x} := \mathbf{z}]$	$\langle(1) + \text{A-mon } (6.1.10)\rangle$
(3)	$(\forall \mathbf{x})A \to$	
	$(\forall \mathbf{z})A[\mathbf{x} := \mathbf{z}]$	$\langle(2) + \vdash (\forall \mathbf{x})A \to (\forall \mathbf{z})(\forall \mathbf{x})A$ from **Ax4** $+ 3.3.1\rangle$

---

[113]A variant of $(\forall \mathbf{x})A$ is $(\forall \mathbf{z})A[\mathbf{x} := \mathbf{z}]$ for some fresh $\mathbf{z}$.

$(\leftarrow)$

(1) $\quad(\forall\mathbf{z})A[\mathbf{x} := \mathbf{z}] \rightarrow$

$\quad\quad A[\mathbf{x} := \mathbf{z}][\mathbf{z} := \mathbf{x}] \quad\quad \langle\mathbf{Ax2}$—$A[\mathbf{x} := \mathbf{z}][\mathbf{z} := \mathbf{x}]$ defined, cf. 4.1.35$\rangle$

(2) $\quad(\forall\mathbf{z})A[\mathbf{x} := \mathbf{z}] \rightarrow A \quad \langle(1)$ rewritten—by 4.1.35, $A[\mathbf{x} := \mathbf{z}][\mathbf{z} := \mathbf{x}]$ is $A\rangle$

(3) $\quad(\forall\mathbf{x})(\forall\mathbf{z})A[\mathbf{x} := \mathbf{z}] \rightarrow$

$\quad\quad(\forall\mathbf{x})A \quad\quad\quad\quad \langle(2) +$ A-mon$\rangle$

(4) $\quad(\forall\mathbf{z})A[\mathbf{x} := \mathbf{z}] \rightarrow$

$\quad\quad(\forall\mathbf{x})A \quad\quad\quad\quad \langle(3) + \vdash (\forall\mathbf{z})A[\mathbf{x} := \mathbf{z}] \rightarrow (\forall\mathbf{x})(\forall\mathbf{z})A[\mathbf{x} := \mathbf{z}]$

$\quad\quad\quad\quad\quad\quad\quad\quad\quad$ from $\mathbf{Ax4}$; no free $\mathbf{x}$ in $(\forall\mathbf{z})A[\mathbf{x} := \mathbf{z}]); + 3.3.1\rangle \quad \square$

6.4.5 Corollary. (Dummy Renaming for $\exists$) *If $\mathbf{z}$ does not occur in A, then* $\vdash (\exists\mathbf{x})A$ $\equiv (\exists\mathbf{z})A[\mathbf{x} := \mathbf{z}]$.

Proof. Exercise! $\quad\quad\quad\quad\quad\quad\quad\quad\quad\quad\quad\quad\quad\quad\quad\quad\quad\quad\quad\quad\quad\quad\square$

6.4.6 Exercise. Show that $\vdash (\forall\mathbf{x})(\forall\mathbf{x})A \equiv (\forall\mathbf{x})A$. $\quad\quad\quad\quad\quad\quad\square$

6.4.7 Exercise. With two examples (within logic) show that the restriction according to which $\mathbf{z}$ in 6.4.4 must be neither free nor bound in A is *necessary*. $\quad\quad\square$

Equipped with the dummy renaming metatheorem we can stop worrying about "capture". For example, whenever we plan to do $A[\mathbf{x} := t]$ we can always settle for $A'[\mathbf{x} := t]$ instead, where A' is obtained from A by replacing all of the latter's bound variables by fresh ones (with respect to A, t). An induction on the complexity of A along with the pair $WL + 6.4.4$ yields $\vdash A \equiv A'$. This yields the metatheorem (cf. 6.1.19) "if $\vdash A$, then $\vdash A'[\mathbf{x} := t]$, if A' is chosen as above" (because under the circumstances, $\vdash A$ implies $\vdash A'$ to begin with; on the other hand, $A'[\mathbf{x} := t]$ is defined).

6.5 INSERTING AND REMOVING "$(\exists\mathbf{x})$"

Inserting and removing $(\exists\mathbf{x})$ is an analogous acrobatic to that of inserting and removing $(\forall\mathbf{x})$, performed for the same reason: to reduce a proof, as much as possible, to one where Post's theorem can be liberally applied. It is a technique that is very often used in everyday math, and is quite powerful.

First, to insert $\exists$ is rather trivial. We have the following tools:

6.5.1 Theorem. (Dual of Ax2) $\vdash A[\mathbf{x} := t] \rightarrow (\exists\mathbf{x})A$

Proof.

$$A[\mathbf{x} := t] \rightarrow (\exists\mathbf{x})A$$

$\Leftrightarrow$ $\langle$SL + "∃-def" (p. 173); "C-part" is $A[\mathbf{x} := t] \to \mathbf{p}\rangle$

$A[\mathbf{x} := t] \to \neg(\forall\mathbf{x})\neg A$

$\Leftrightarrow$ $\langle$tautology$\rangle$

$(\forall\mathbf{x})\neg A \to \neg A[\mathbf{x} := t]$ $\square$

6.5.2 Corollary. (Dual of the Specialization Rule) $A[\mathbf{x} := t] \vdash (\exists\mathbf{x})A$

Proof. 6.5.1 and MP. $\square$

6.5.3 Corollary. $A \vdash (\exists\mathbf{x})A$

Proof. By 6.5.2, taking t to be $\mathbf{x}$. $\square$

Now let us turn to removing ∃. We will need a few tools at first.

6.5.4 Metatheorem. (∀ Introduction) *If* $\mathbf{x}$ *does not occur free either in* Γ *or in* A, *then* $\Gamma \vdash A \to B$ *iff* $\Gamma \vdash A \to (\forall\mathbf{x})B$.

Proof. Only if direction. We are assuming $\Gamma \vdash A \to B$. 6.1.9 yields $\Gamma \vdash (\forall\mathbf{x})A \to (\forall\mathbf{x})B$. By the condition on A, I have $\vdash A \to (\forall\mathbf{x})A$ (**Ax4**). I am done by tautological implication.

Pause. This was a Hilbert proof written in *free-style*, just as a mathematician or computer scientist would have written it—not vertically, and not fully numbered. So is the following.

If direction: A tautological implication of $(\forall\mathbf{x})B \to B$ (**Ax2**) is

$$\vdash (A \to (\forall\mathbf{x})B) \to (A \to B) \tag{1}$$

Thus, if I have $\Gamma \vdash A \to (\forall\mathbf{x})B$, then (1) and MP yield $\Gamma \vdash A \to B$. $\square$

6.5.5 Corollary. (∃ Introduction) *If* $\mathbf{x}$ *does not occur free either in* Γ *or in* B, *then* $\Gamma \vdash A \to B$ *iff* $\Gamma \vdash (\exists\mathbf{x})A \to B$.

Note that the condition switched from A to B.

Proof. If direction: $\vdash A \to (\exists\mathbf{x})A$ by 6.5.1. This yields

$$\vdash ((\exists\mathbf{x})A \to B) \to (A \to B) \tag{2}$$

by tautological implication. If we now assume $\Gamma \vdash (\exists\mathbf{x})A \to B$, then (2) and MP yield $\Gamma \vdash A \to B$.

Only if direction:

(1)	$A \to B$	$\langle\Gamma$-proved; we continue the proof$\rangle$
(2)	$\neg B \to \neg A$	$\langle(1) +$ taut. implication (3.3.1)$\rangle$
(3)	$\neg B \to (\forall\mathbf{x})\neg A$	$\langle(2) +$ 6.5.4—conditions met$\rangle$
(4)	$\neg(\forall\mathbf{x})\neg A \to B$	$\langle(3) +$ taut. implication$\rangle$

Line (4) really says "$(\exists \mathbf{x})A \rightarrow B$".   □

Corollary 6.5.5 is really the ticket to the technique of removing $(\exists \mathbf{x})$:

6.5.6 Metatheorem. (Auxiliary Variable Metatheorem) *Assume that* $\Gamma \vdash (\exists \mathbf{x})A$. *Moreover, assume that* $\Gamma + A[\mathbf{x} := \mathbf{z}] \vdash B$, *where* $\mathbf{z}$ *is fresh with respect to* Γ, $(\exists \mathbf{x})A$ *and* B. *Then* $\Gamma \vdash B$ *as well.*

Proof. We can argue as follows: By the deduction theorem, $\Gamma \vdash A[\mathbf{x} := \mathbf{z}] \rightarrow B$. Thus, from 6.5.5, $\Gamma \vdash (\exists \mathbf{z})A[\mathbf{x} := \mathbf{z}] \rightarrow B$.

We can now calculate equationally (from hypotheses Γ):

$$(\exists \mathbf{z})A[\mathbf{x} := \mathbf{z}] \rightarrow B$$
$$\Leftrightarrow \langle \text{SL} + 6.4.5; \mathbf{z} \text{ is fresh for } A; \text{"C-part" is } \mathbf{p} \rightarrow B \rangle$$
$$(\exists \mathbf{x})A \rightarrow B$$
$$\Leftrightarrow \langle \text{SL} + 2.1.23; \text{"C-part" is } \mathbf{p} \rightarrow B \rangle$$
$$\top \rightarrow B$$
$$\Leftrightarrow \langle \text{tautology} \rangle$$
$$B$$ □

(1) The seemingly weaker hypothesis that "$\mathbf{z}$ is fresh with respect to Γ, A, and B" also lets the proof through as we immediately see from the first step (first $\Leftrightarrow$). Obtaining the first line of the equational proof only requires $\mathbf{z}$ not to occur free in $\Gamma \cup \{B\}$. However, note that

- Under the weaker hypothesis, if $\mathbf{z}$ *is* $\mathbf{x}$, then the theorem trivializes to 2.1.6, by the dual of "$\vdash A \equiv (\forall \mathbf{x})A$ when $\mathbf{x}$ not free in A" and 4.1.34. We learn nothing new.

- Thus, the case of practical importance is when there is a *nontrivial* $(\exists \mathbf{x})$-prefix that the metatheorem teaches us how to "remove"; i.e., a prefix that actually binds some free $\mathbf{x}$ in A. But then $\mathbf{z}$ is *not* $\mathbf{x}$ if the former does not occur in A. Therefore, in the "interesting case", the weaker restriction "$\mathbf{z}$ is fresh with respect to A" is the same as the one stated in the metatheorem: "$\mathbf{z}$ is fresh with respect to $(\exists \mathbf{x})A$".

(2) **Very Important!** There is nothing cryptic about the metatheorem, which is actually used a lot by folks who write in mathematics, in computer science, and in other fields where mathematical reasoning is called for. The *intuition* behind it is this:

If I know that $(\exists \mathbf{x})A$ holds, then this tells me, *intuitively*, that for some (value of) $\mathbf{x}$, $A(\mathbf{x})$ holds.[114]

---

[114]Where, by "$A(\mathbf{x})$" I simply want to draw attention, notationally, to A's dependence on $\mathbf{x}$.

So, *even though I do not know (or care) which value makes* $A(\mathbf{x})$ *hold*, I can name $\mathbf{z}$ any such a value and say, "Okay then, let $\mathbf{z}$ be one of those values of $\mathbf{x}$ that make $A(\mathbf{x})$ hold"; for short I *assume* $A(\mathbf{z})$!

This additional—auxiliary—assumption helps a lot toward proving B:

(a) Because *more* assumptions make proofs easier!

(b) This assumption is potentially easier to manipulate with Boolean techniques than $(\exists \mathbf{x})A$ is, because unlike the latter whose Boolean abstraction (4.1.25) is just a formula of the form $\neg \mathbf{p}$, $A[\mathbf{x} := \mathbf{z}]$ (or $A(\mathbf{z})$) may have Boolean connectives that I can profitably use in conjunction with Post's theorem.

The metatheorem guarantees that once this *auxiliary assumption*, $A[\mathbf{x} := \mathbf{z}]$, has served its purpose, *it drops out of the picture*, leaving us with just the fact $\Gamma \vdash B$.

Intuitively, this is because the assumption $A(\mathbf{z})$ is an alternative way to say "$(\exists \mathbf{x})A$", this way: "Some unspecified but *fixed* value $\mathbf{z}$ of $\mathbf{x}$ makes $A(\mathbf{x})$ hold"—so it does not really add anything new that Γ did not already know about. Recall that Γ can prove $(\exists \mathbf{x})A$.

Pause. How exactly does the metatheorem suggest the intuitive interpretation that $\mathbf{z}$ is *fixed*? By inserting "$A[\mathbf{x} := \mathbf{z}]$" in a proof of B as *an auxiliary hypothesis*. This hypothesis disallows us from using generalization ($\forall \mathbf{z}$) anywhere in the proof *below the point of insertion*—review 6.1.1. Therefore, $\mathbf{z}$ *behaves like a constant*, being *unavailable* for universal quantification (generalization)!

This metatheorem is a mathematical phenomenon entirely analogous to what happens with induction over $\mathbb{N}$: There we want to prove $\mathscr{P}(n)$ from some assumptions Γ, and for arbitrary n. Out of the blue we *add an assumption*, that $\mathscr{P}(k)$ holds for all $k < n$, and *use it* in our proof.

But when the dust settles we say that we have proved $\mathscr{P}(n)$ *only* from Γ, not from $\Gamma +$ "$\mathscr{P}(k)$ holds for all $k < n$". The additional assumption (I.H.) is gone!

6.5.7 Corollary. *Assume that* $\vdash (\exists \mathbf{x})A$. *Moreover, assume that* $A[\mathbf{x} := \mathbf{z}] \vdash B$, *where* $\mathbf{z}$ *is fresh with respect to* $(\exists \mathbf{x})A$ *and* B. *Then* $\vdash B$ *as well.*

Proof. Take $\Gamma = \emptyset$.   □

6.5.8 Corollary. *Assume that* $A[\mathbf{x} := \mathbf{z}] \vdash B$, *where* $\mathbf{z}$ *is fresh with respect to* $(\exists \mathbf{x})A$ *and* B. *Then* $(\exists \mathbf{x})A \vdash B$ *as well.*

Proof. Take $\Gamma = \{(\exists \mathbf{x})A\}$.   □

6.5.9 Remark. In the examples that follow toward illustrating the use of 6.5.6 and its corollaries, we eliminate the existential prefix from a formula $(\exists \mathbf{x})A$ that occurs in a proof from Γ, say at line (n), by introducing—anywhere below line (n)—the formula $A[\mathbf{x} := \mathbf{z}]$ with the terse annotation *auxiliary hypothesis associated with* (n); $\mathbf{z}$ *fresh*.

According to the requirements of 6.5.6, "$\mathbf{z}$ fresh" means all three below:

(i) $\mathbf{z}$ does not occur in any hypotheses (auxiliary or not) written *before this step*.

(ii) **z** does not occur in the already-written existential formula $(\exists x)A$ (the "associate" of $A[\mathbf{x} := \mathbf{z}]$).

(iii) **z** does not occur in the formula we want to prove.

In practice, if we interpret "**z** fresh" more strongly, as (iii) plus *it does not occur in any formula written in the proof before this step*, then we are covered.   □

6.5.10 Example. We prove here, just for practice, that $\vdash (\exists x)(\forall y)A \to (\forall y)(\exists x)A$.
I'll give you two proofs:

First proof: I use deduction theorem, so I prove $(\exists x)(\forall y)A \vdash (\forall y)(\exists x)A$
instead:

(1) $(\exists x)(\forall y)A$ $\langle$hypothesis$\rangle$

(2) $(\forall y)A[\mathbf{x} := \mathbf{z}]$ $\langle$auxiliary hypothesis associated with (1); **z** fresh$\rangle$

(3) $A[\mathbf{x} := \mathbf{z}]$ $\langle$(2) + spec$\rangle$

(4) $(\exists x)A$ $\langle$(3) + 6.5.2$\rangle$

(5) $(\forall y)(\exists x)A$ $\langle$(4) + 6.1.1—lines (1,2) [hypotheses] have no free **y**$\rangle$

By 6.5.6 (or 6.5.8; $\Gamma = \{(\exists x)(\forall y)A\}$), the "auxiliary" hypothesis (2) drops out, and we have that line (1), *alone*, proves line (5). In effect, what the proof does is so obvious as to be dull: It removes quantifiers—using appropriate tools—and reinserts them in a different order.

You must be sure to annotate the auxiliary *hypothesis* as such. It is dead wrong to say that line (2) *follows* from line (1).

Second proof: From $\vdash A \to (\exists x)A$ (i.e., 6.5.1) we get $\vdash (\forall y)A \to (\forall y)(\exists x)A$
by 6.1.10.
     Corollary 6.5.5 yields $\vdash (\exists x)(\forall y)A \to (\forall y)(\exists x)A$.   □

6.5.11 Example. We prove that $(\exists x)(A \to B), (\forall x)A \vdash (\exists x)B$.

(1) $(\exists x)(A \to B)$ $\langle$hypothesis$\rangle$

(2) $(\forall x)A$ $\langle$hypothesis$\rangle$

(3) $A[\mathbf{x} := \mathbf{z}] \to B[\mathbf{x} := \mathbf{z}]$ $\langle$aux. hypothesis associated with (1); **z** is fresh$\rangle$

(4) $A[\mathbf{x} := \mathbf{z}]$ $\langle$(2) + 6.1.5$\rangle$

(5) $B[\mathbf{x} := \mathbf{z}]$ $\langle$(3, 4) + MP$\rangle$

(6) $(\exists x)B$ $\langle$(5) + 6.5.2$\rangle$

Lines (1) + (2) prove (6) by 6.5.6. *I'll stop making this pedantic assertion at the end of proofs by auxiliary hypothesis/variable. We know that the auxiliary hypothesis drops out. No more reminders!*   □

6.5.12 Example. It is instructive to look at an alternative proof that uses proof by contradiction (cf. 2.6.7). So rather than establishing $(\exists \mathbf{x})(A \to B), (\forall \mathbf{x})A \vdash (\exists \mathbf{x})B$, we will attempt

$$(\exists \mathbf{x})(A \to B), (\forall \mathbf{x})A, \neg(\exists \mathbf{x})B \vdash \bot$$

instead:

(1)	$(\exists \mathbf{x})(A \to B)$	$\langle$hypothesis$\rangle$
(2)	$(\forall \mathbf{x})A$	$\langle$hypothesis$\rangle$
(3)	$\neg(\exists \mathbf{x})B$	$\langle$hypothesis$\rangle$
(4)	$\neg\neg(\forall \mathbf{x})\neg B$	$\langle$(3) + writing in full the abbreviation "$\exists$"$\rangle$
(5)	$(\forall \mathbf{x})\neg B$	$\langle$(4) + tautological implication (3.3.1)$\rangle$
(6)	A	$\langle$(2) + 6.1.6$\rangle$
(7)	$\neg B$	$\langle$(5) + 6.1.6$\rangle$
(8)	$\neg(A \to B)$	$\langle$(6, 7) + tautological implication$\rangle$
(9)	$(\forall \mathbf{x})\neg(A \to B)$	$\langle$(8) + gen; Okay since (1, 2, 3) have no free $x\rangle$
(10)	$\neg(\exists \mathbf{x})(A \to B)$	$\langle$(9) + $\exists$-abbrev. + tautol. implication$\rangle$
(11)	$\bot$	$\langle$(1, 10) + 2.5.7$\rangle$ $\qquad \Box$

6.5.13 Example. We establish $(\forall \mathbf{x})(A \to B), (\exists \mathbf{x})A \vdash (\exists \mathbf{x})B$.

(1)	$(\forall \mathbf{x})(A \to B)$	$\langle$hypothesis$\rangle$
(2)	$(\exists \mathbf{x})A$	$\langle$hypothesis$\rangle$
(3)	$A[\mathbf{x} := \mathbf{z}]$	$\langle$auxiliary hypothesis associated with (2); z fresh$\rangle$
(4)	$A[\mathbf{x} := \mathbf{z}] \to B[\mathbf{x} := \mathbf{z}]$	$\langle$(1) + spec (6.1.5)$\rangle$
(5)	$B[\mathbf{x} := \mathbf{z}]$	$\langle$(3, 4) + MP$\rangle$
(6)	$(\exists \mathbf{x})B$	$\langle$(5) + 6.5.2$\rangle$ $\qquad \Box$

6.5.14 Remark. An experienced mathematician or computer scientist who never took a course in formal logic(!) would probably argue the above almost identically, however they would most likely opt for intuitively acceptable *semantic* terminology. They would also probably prefer to write $A(\mathbf{x})$ and $B(\mathbf{x})$—over the terse A and B— to draw attention to our interest in the variable $\mathbf{x}$. The argument would go something like this:

> Assume that $(\forall \mathbf{x})\big(A(\mathbf{x}) \to B(\mathbf{x})\big)$ and $(\exists \mathbf{x})A(\mathbf{x})$ are true. The truth of the latter implies that for some value of $\mathbf{x}$, say c, $A(c)$ is true. The truth of the first assumption (for all the values of $\mathbf{x}$) implies—in particular—that $A(c) \to B(c)$ is true.
>
> Modus ponens yields the truth of $B(c)$. But then it is true to say $(\exists \mathbf{x})B$.

Our formal methods achieve the following:

(1) Ratify the above informal technique for handling $\exists$. We have already noted that the "auxiliary variable" behaves formally like a *constant* throughout the proof, so it is all right that our fictitious computer scientist used an "auxiliary *constant*" c—cf. **Pause** on p. 181.

(2) Make the technique available to the nonexpert—the experts can fend for themselves without the benefit of formal rules; the latter mostly benefit beginners.

(3) Avoid errors that might creep into a loose semantic approach (see next example!).

By the way, our fictitious mathematician wrote a Hilbert proof but did so in a rather condensed style, without numbering. *This condensed informal Hilbert style is very common in mathematical practice.* If a proof is longer, then numbers are inserted *only where needed* so that one can later refer back to earlier statements. $\square$

6.5.15 Example. Let us prove the schema $\vdash (\exists \mathbf{x}) A \wedge (\exists \mathbf{x}) B \rightarrow (\exists \mathbf{x})(A \wedge B)$.

We will do this posturing as "experienced computer scientists or mathematicians". Thus, we will attempt to imitate the condensed Hilbert style of proof given above, using semantic terminology rather than formal (i.e., syntactic) methods:

Assume that $(\exists \mathbf{x}) A(\mathbf{x})$ and $(\exists \mathbf{x}) B(\mathbf{x})$[115] *are true. Thus, for some value of $\mathbf{x}$, say c, we have $A(c)$ and $B(c)$ are true. But then so is $A(c) \wedge B(c)$. Suppressing reference to the specific c, it is true to state $(\exists \mathbf{x})(A(\mathbf{x}) \wedge B(\mathbf{x}))$.*

This is nice, crisp, and short.

And wrong! We will see in Chapter 8 that $(\exists \mathbf{x}) A \wedge (\exists \mathbf{x}) B \rightarrow (\exists \mathbf{x})(A \wedge B)$ is *not* a theorem schema.

What went wrong? Well, an inexperienced person arguing semantically often makes this kind of error (I have seen it over and over again): "Thus, for some value of $\mathbf{x}$, say c, we have $A(c)$ and $B(c)$ are true." Surely, the truth of $(\exists \mathbf{x}) A(\mathbf{x})$ and $(\exists \mathbf{x}) B(\mathbf{x})$ does *not* imply that the *same* c makes both $A(c)$ and $B(c)$ true!

We should have said that some c and some d (possibly different) make $A(c)$ and $B(d)$ true.

Note how we cannot now conclude $(\exists \mathbf{x})(A(\mathbf{x}) \wedge B(\mathbf{x}))$ as we are saved by the "(possibly different)" qualification.

Would formal techniques be safer? Yes!

(1) $(\exists \mathbf{x}) A \wedge (\exists \mathbf{x}) B$ $\langle$hypothesis$\rangle$

(2) $(\exists \mathbf{x}) A$ $\langle$(1) + tautological implication$\rangle$

(3) $(\exists \mathbf{x}) B$ $\langle$(1) + tautological implication$\rangle$

(4) $A[\mathbf{x} := \mathbf{z}]$ $\langle$auxiliary hypothesis associated with (2); $\mathbf{z}$ fresh$\rangle$

(5) $B[\mathbf{x} := \mathbf{w}]$ $\langle$auxiliary hypothesis associated with (3); $\mathbf{w}$ fresh$\rangle$

---

[115]The experienced mathematician takes (at least) two things for granted and thus unworthy of explicit mention:

(1) The deduction theorem

(2) "Hypothesis splitting" (2.5.2), where a hypothesis $X \wedge Y$ splits into two hypotheses X and Y

Thus, by the requirement for "freshness" (cf. Remark 6.5.9), the "auxiliary variables" **z** and **w** are distinct. The variable **z** is the formal counterpart of c above, while **w** formalizes d.

Clearly, we cannot continue the formal argument in the fallacious way of our original informal argument.   □

6.5.16 Example. Our last example has a famous name attached to it: Bertrand Russell. We show that for any predicate ϕ of arity 2 (i.e., one that accepts two arguments)

$$\vdash \neg(\exists y)(\forall x)(\phi(x,y) \equiv \neg\phi(x,x)) \qquad (R)$$

By the tautology $\neg A \equiv A \to \bot$ it suffices that we show instead

$$\vdash (\exists y)(\forall x)(\phi(x,y) \equiv \neg\phi(x,x)) \to \bot$$

(1) $(\exists y)(\forall x)(\phi(x,y) \equiv \neg\phi(x,x))$ ⟨hypothesis⟩

(2) $(\forall x)(\phi(x,z) \equiv \neg\phi(x,x))$ ⟨aux. hypothesis for (1); z fresh⟩

(3) $\phi(z,z) \equiv \neg\phi(z,z)$ ⟨(2) + spec⟩

(4) $\bot$ ⟨(3) + tautological implication⟩

If we take $\phi(x,y)$ to be "$x \in y$"—where $\in$ is the "is a member of" predicate of set theory, then (R) that we just proved (for *any* predicate of arity 2) becomes

$$\vdash \neg(\exists y)(\forall x)(x \in y \equiv x \notin x) \qquad (R')$$

Translated in English (R') says that there is *no* set y that contains all those sets that satisfy $x \notin x$ (are not members of themselves).

Very remarkably, the nonexistence of such a set y—a situation known as "Russell's paradox"[116]—has just been proved within pure logic without using a single set theory axiom!   □

There is a result analogous to the "auxiliary variable metatheorem" (6.5.6), named the *auxiliary constant metatheorem*, to which we now turn our attention. It makes formally explicit the role of the auxiliary variable as a "constant" (cf. **Pause** on p. 181). We will not need the auxiliary constant metatheorem except in the Appendix to Part II.

6.5.17 Lemma. *Assume that no formula in Γ contains the constant c and that $\Gamma \vdash A$. Moreover, let* **x** *be a variable that does not occur in A. Then $\Gamma \vdash A'$, where A' is obtained from A by replacing all the occurrences of c in it by* **x**.

Proof. The proof, given by induction on the length of a proof of A from Γ for logic 2 of Chapter 5 is very similar to that of 6.1.1. *Throughout, the result of replacing c in a formula B by* **x** *will be denoted by B'.*

---

[116] The "paradox" exists in Cantor's informal—or as we say, naïve—set theory that leads us to believe that every collection of objects is a set.

For proofs of length one we have two cases:

(1) $A \in \Gamma$. Then A' is A since A cannot contain c. Thus $\Gamma \vdash A'$.

(2) $A \in \Lambda_1$ (4.2.1). There are six subcases where A is a partial generalization of one of the following:

  (i) Tautology B (group **Ax1**). The tautology is determined by how the Boolean connectives connect Boolean variables, constants, and *prime formulae* inside B. The former two remain invariant as c is replaced by $\mathbf{x}$. The prime formulae are of three types, $(\forall \mathbf{y})C$, $t = s$, or $\phi(t_1, \ldots, t_n)$. The replacement of c by by $\mathbf{x}$ leaves these *types* invariant, say, $(\forall \mathbf{y})C'$, $t' = s'$, or $\phi(t_1', \ldots, t_n')$. Thus we end up with a tautology B'.

  (ii) Formula $(\forall \mathbf{y})B \rightarrow B[\mathbf{y} := t]$ (group **Ax2**). Replacing c by $\mathbf{x}$ results in a formula still in group **Ax2**: $(\forall \mathbf{y})B' \rightarrow B'[\mathbf{y} := t']$. Note that $B'[\mathbf{y} := t']$ is still defined, for $\mathbf{x}$ cannot be captured (if it entered t'), being new for A.

  (iii) Formula $(\forall \mathbf{y})(B \rightarrow C) \rightarrow (\forall \mathbf{y})B \rightarrow (\forall \mathbf{y})C$ (group **Ax3**). Replacing c by $\mathbf{x}$ results in a formula still in group **Ax3**: $(\forall \mathbf{y})(B' \rightarrow C') \rightarrow (\forall \mathbf{y})B' \rightarrow (\forall \mathbf{y})C'$.

  (iv) Formula $B \rightarrow (\forall \mathbf{y})B$, when $\mathbf{y}$ is not free in B (group **Ax4**). Replacing c by $\mathbf{x}$ results in a formula $B' \rightarrow (\forall \mathbf{y})B'$ still in group **Ax4**, noting that $\mathbf{y}$, being different from $\mathbf{x}$, is still not free in B'.

  (v) Formula $\mathbf{y} = \mathbf{y}$ (group **Ax5**). This contains no c, so it remains invariant upon replacing c by $\mathbf{x}$.

  (vi) Formula $t = s \rightarrow (A[\mathbf{y} := t] \equiv A[\mathbf{y} := s])$ (group **Ax6**). Replacing c by $\mathbf{x}$ results in a formula still in group **Ax6**: $t' = s' \rightarrow (A'[\mathbf{y} := t'] \equiv A[\mathbf{y} := s'])$. Note that $A'[\mathbf{y} := t']$ and $A[\mathbf{y} := s']$ are still defined, for $\mathbf{x}$ cannot be captured (if it entered t' or s'), being new for A.

Assume the claim for proofs of length n where A appears. We now go to the case of length $n + 1$. If A appeared before the end, then we are done by the I.H. If it appears for the first time at the end and it is in $\Gamma \cup \Lambda_1$, then the case already has been argued. Let then A be the result of MP; that is, the formulae B and $B \rightarrow A$ (for some B) appear before it in the proof. By I.H., $\Gamma \vdash B'$ and $\Gamma \vdash B' \rightarrow A'$; thus $\Gamma \vdash A'$ by MP. □

6.5.18 Corollary. *Assume that $\Gamma \vdash A$ and that there is a proof certifying this, which uses no formula from Γ that contains the constant c. Moreover, let $\mathbf{x}$ be a variable that does not occur in A. Then $\Gamma \vdash A'$, where A' is obtained by A by replacing all the occurrences of c in it by $\mathbf{x}$.*

Proof. Let Δ be the set of all the formulae from Γ that appear in said proof. By 6.5.17, $\Delta \vdash A'$. We get $\Gamma \vdash A'$ by 2.1.1. □

6.5.19 Metatheorem. (Auxiliary Constant Metatheorem) *Let c be a constant that does not appear in the formulae A or B. Assume that* $\Gamma \vdash (\exists x)A$. *Moreover, let* $\Gamma + A[x := c] \vdash B$, *with a proof where the formulae invoked from* Γ *do not contain the constant c. Then* $\Gamma \vdash B$ *as well.*

Proof. Let Δ be the set of all formulae of Γ invoked in the certification of $\Gamma + A[x := c] \vdash B$ as in the statement of 6.5.19. Thus, $\Delta + A[x := c] \vdash B$. By the deduction theorem, $\Delta \vdash A[x := c] \to B$. Let z be new for $\Delta \cup \{A, B\}$. Thus, from 6.5.17, $\Delta \vdash A[x := z] \to B$. Hence, $\Delta \vdash (\exists z)A[x := z] \to B$ by 6.5.5 (this part needs that z is not free in Δ, nor in B). By 6.4.5 and an application of SL we get $\Delta \vdash (\exists x)A \to B$; hence $\Gamma \vdash (\exists x)A \to B$. By MP we have $\Gamma \vdash B$. $\square$

6.6 ADDITIONAL EXERCISES

1. Show that $\vdash (\forall x)(A \to B) \to (\exists x)A \to (\exists x)B$.

   Hint. Use 4.1.17 to eliminate "$\exists$".

2. Show that $\vdash (\forall x)(A \to (B \equiv C)) \to \big((\forall x)(A \to B) \equiv (\forall x)(A \to C)\big)$.

3. Show that $\vdash (\forall x)((A \vee B) \to C) \to (\forall x)(A \to C)$.

4. Show that $\vdash (\forall x)(A \to (B \wedge C)) \to (\forall x)(A \to B)$.

5. State and prove the $\exists$-dual of 6.1.8.

6. Prove the following version of the relativized $\forall$-monotonicity, which in the notation of [2] (cf. p. 175) is

$$\vdash (\forall x)_A(B \to C) \to (\forall x)_A B \to (\forall x)_A C$$

   while in standard notation it reads

$$\vdash (\forall x)(A \to B \to C) \to (\forall x)(A \to B) \to (\forall x)(A \to C)$$

7. Prove

$$\vdash (\exists x)_{A \wedge B} C \equiv (\exists x)_A (B \wedge C)$$

   Hint. Translate first to standard notation.

8. This relativizes 6.4.2. Prove

$$\vdash A \vee (\forall x)_B C \equiv (\forall x)_B (A \vee C), \text{ as long as } x \text{ not free in } A$$

   Hint. Translate first to standard notation.

9. This relativizes 6.4.3. Prove

$$\vdash A \wedge (\exists x)_B C \equiv (\exists x)_B (A \wedge C), \text{ as long as } x \text{ not free in } A$$

Hint. Translate first to standard notation.

10. Prove

$$\vdash (\exists \mathbf{x})_A B \vee (\exists \mathbf{x})_A C \equiv (\exists \mathbf{x})_A (B \vee C)$$

Hint. Translate first to standard notation.

11. Prove that if $\mathbf{x}$ is not free in A, then $\vdash A \equiv (\exists \mathbf{x})A$.

12. Prove the *one point rule*—∃-version: $\vdash (\exists \mathbf{x})(\mathbf{x} = t \wedge A) \equiv A[\mathbf{x} := t]$ if $\mathbf{x}$ is not free in t.

13. Prove $\vdash (\exists \mathbf{x})(A \wedge (\exists \mathbf{y})(B \wedge C)) \equiv (\exists \mathbf{y})(B \wedge (\exists \mathbf{x})(A \wedge C))$, *on the condition* that $\mathbf{y}$ is not free in A and $\mathbf{x}$ is not free in B.

14. This is the dual of result 6 on p. 175. Prove $\vdash (\exists \mathbf{x})_{A \vee B} C \equiv (\exists \mathbf{x})_A C \vee (\exists \mathbf{x})_B C$.

Hint. Translate to standard notation first.

15. Prove $\vdash (\exists \mathbf{x})(\exists \mathbf{y})(A \wedge B \wedge C) \equiv (\exists \mathbf{x})(A \wedge (\exists \mathbf{y})(B \wedge C))$, *on the condition* that $\mathbf{y}$ is not free in A.

16. Prove *dummy renaming* for ∃: If $\mathbf{z}$ does not occur in A, then $\vdash (\exists \mathbf{x})A \equiv (\exists \mathbf{z})A[\mathbf{x} := \mathbf{z}]$.

17. Here is a suggested proof of

$$\vdash (\forall \mathbf{x})(\exists \mathbf{y})A \rightarrow (\exists \mathbf{y})(\forall \mathbf{x})A \tag{$*$}$$

We split the $\rightarrow$ and go via the deduction theorem:

  (1) $(\forall \mathbf{x})(\exists \mathbf{y})A$ ⟨hypothesis⟩
  (2) $(\exists \mathbf{y})A$ ⟨(1) + spec⟩
  (3) $A[\mathbf{y} := \mathbf{z}]$ ⟨auxiliary hypothesis associated with (2); $\mathbf{z}$ is fresh⟩
  (4) $(\forall \mathbf{x})A[\mathbf{y} := \mathbf{z}]$ ⟨(3) + gen; Okay: $\mathbf{x}$ is not free in hypothesis line (1)⟩
  (5) $(\exists \mathbf{y})(\forall \mathbf{x})A$ ⟨(4) + 6.5.2⟩

Now, you should *not* believe $(*)$ (cf. 8.2.11).

However, you are asked not simply to dismiss the proof because of 8.2.11, but rather to find *precisely in which step* it went wrong, and *why* said step is wrong.

18. Prove *using the auxiliary variable metatheorem*: $\vdash (\exists \mathbf{x})(A \rightarrow B) \rightarrow (\forall \mathbf{x})A \rightarrow (\exists \mathbf{x})B$.

19. Prove *using the auxiliary variable metatheorem*: $\vdash (\exists \mathbf{x})B \rightarrow (\exists \mathbf{x})(A \vee B)$.

20. Prove the dual of Exercise 3: (In [2] notation) $\vdash (\exists \mathbf{x})_A C \rightarrow (\exists \mathbf{x})_{A \vee B} C$.

21. Let ϕ be a predicate of arity 2.

I claim $\phi(x, y) \vdash \phi(y, x)$ via 6.1.19.

- Am I right or wrong in my (one-line) claim + proof, and why exactly?
- If I am wrong in my proof, it *might* still be possible to provide a *different* correct proof.

  Well, either provide such a correct proof, or *definitively show* that $\phi(x, y) \vdash \phi(y, x)$ is *not* a theorem schema (if this is what you will end up doing, it will require tools from 8.2.3).

22. Prove $\vdash (\exists \mathbf{x})(A \rightarrow (\forall \mathbf{x})A)$.

23. Prove $\vdash \mathbf{x} = \mathbf{y} \wedge \mathbf{y} = \mathbf{z} \rightarrow \mathbf{x} = \mathbf{z}$.

24. Prove $\vdash (\forall \mathbf{x})B \rightarrow A \equiv (\exists \mathbf{x})(B \rightarrow A)$, *provided that* $\mathbf{x}$ *is not free in* A.

Hint. Be mindful of the priorities of the connectives. Use an equational proof.

25. Prove that the following is an absolute theorem schema *on the condition* that $\mathbf{x}$ is not free in B: $(\forall \mathbf{x})(A \rightarrow B) \equiv (\exists \mathbf{x})A \rightarrow B$.

Hint. Be mindful of the priorities of the connectives. Use an equational proof.

26. Prove $\vdash (\exists \mathbf{x})A \rightarrow \big((\exists \mathbf{x})_A(B \vee C) \equiv B \vee (\exists \mathbf{x})_A C\big)$, if $\mathbf{x}$ is not free in B.

Hint. Be mindful of connective priorities. Then translate in standard notation before embarking on a proof.

27. This relativizes Exercise 24. Prove $\vdash (\exists \mathbf{x})A \rightarrow \big((\forall \mathbf{x})_A B \rightarrow C \equiv (\exists \mathbf{x})_A(B \rightarrow C)\big)$, if $\mathbf{x}$ is not free in C.

Hint. Be mindful of connective priorities. Then translate in standard notation before embarking on a proof.

28. Professor N. A. Ïve has submitted the following claim for publication:

$$\vdash A \equiv (\forall \mathbf{x})A \tag{1}$$

He offered the following proof, and I quote:

"We know (generalization + specialization) that

$$\vdash A \text{ iff } \vdash (\forall \mathbf{x})A \tag{$*$}$$

By the metatheorem that says '*for any two (absolute) theorems, B and C, we have $\vdash B \equiv C$*', it follows from ($*$) that $\vdash A \equiv (\forall \mathbf{x})A$."

Hmm. We know from the text that (1) is wrong, so the question is: *Precisely which step* is wrong in Professor Ïve's proof, and why?

29. Let ϕ be a predicate of arity 2.

- Explain *why* $(\forall x)(\forall y)\phi(x, y) \rightarrow (\forall y)\phi(y, y)$ is *not* an instance of **Ax2**.

- Nevertheless, *prove* that $(\forall x)(\forall y)\phi(x,y) \rightarrow (\forall y)\phi(y,y)$ *is* an absolute theorem!

30. Let ϕ, ψ be predicates of one variable. Prove

$$\vdash \psi(x) \rightarrow (\forall y)\Big(\psi(y) \rightarrow (\forall z)\phi(z)\Big) \rightarrow (\forall x)\phi(x)$$

31. Let ϕ, ψ be any *unary* (of arity 1) predicates, and c an (object) constant. Prove

$$\vdash (\forall x)(\phi(x) \rightarrow \psi(x)), \ (\forall z)\phi(z) \vdash \psi(c)$$

32. (a) For any predicate ϕ of arity 2 prove

$$\vdash (\forall x)(\forall y)\phi(x,y) \equiv (\forall y)(\forall x)\phi(y,x)$$

(b) How is the above different from the known $\vdash (\forall \mathbf{x})(\forall \mathbf{y})A \equiv (\forall \mathbf{y})(\forall \mathbf{x})A$?

33. Prove or disprove the schema: "If $\mathbf{x}$ is not free in B, then $\vdash B \equiv (\forall \mathbf{x})_A B$".

34. Prove or disprove the schema: "If $\mathbf{x}$ is not free in B, then $\vdash B \equiv (\exists \mathbf{x})_A B$".

35. Consider the following formula, in a language with a nonlogical function symbol f of arity 1:
$$(\forall x)\big(x = f(x) \rightarrow f(x) = f(f(x))\big) \tag{1}$$

(a) What is the Boolean abstraction of this formula? (Indicate by boxing.)

(b) Is the abstraction a tautology? Why?

(c) Whether or not the *abstraction* is a tautology, can you prove (1) in first-order logic?

Hint. While this exercise is self-contained in its present context, you may benefit by peeking in the next chapter.

36. Assume that $\Gamma \vdash A$ and that there is a proof certifying this where no formula from Γ used in it contains the constant c. Then for some variable $\mathbf{x}$ we have $\Gamma \vdash (\forall \mathbf{x})A'$, where A' is obtained by A by replacing all the occurrences of c in it by $\mathbf{x}$.

37. Assume that $\vdash (\exists \mathbf{x})A$. Moreover, assume that $A[\mathbf{x} := c] \vdash B$, where c is a constant that does not occur in A or B. Then $\vdash B$ as well.

38. Assume that $A[\mathbf{x} := c] \vdash B$, where c is a constant that does not occur in A or B. Then $(\exists \mathbf{x})A \vdash B$ as well.

39. Revisit and formally re-prove all the examples in Section 6.5 that follow 6.5.8, but this time do so using a proof "by auxiliary constant" (cf. 6.5.19, and Exercises 37 and 38 above) to eliminate the leading existential quantifier.

CHAPTER 7

PROPERTIES OF EQUALITY

You will recall our brief discussion of "SFL" (Single-Formula Leibniz) in Section 3.4 of Part I. That was in the context of Boolean logic. Axiom 6 below (cf. 4.2.1)

$$t = s \rightarrow (A[\mathbf{z} := t] \equiv A[\mathbf{z} := s])$$

is a counterpart of SFL for equality of objects. We explore here some of the consequences of **Ax5–Ax6**, including another counterpart of SFL, where the types of expressions on either side of $\rightarrow$ are non-Boolean.

7.0.1 Lemma. $\vdash \mathbf{x} = \mathbf{y} \rightarrow \mathbf{y} = \mathbf{x}$ *and* $\vdash \mathbf{x} = \mathbf{y} \rightarrow \mathbf{y} = \mathbf{z} \rightarrow \mathbf{x} = \mathbf{z}$.

Proof. For $\vdash \mathbf{x} = \mathbf{y} \rightarrow \mathbf{y} = \mathbf{x}$ here is a Hilbert-style proof:

(1) $\mathbf{x} = \mathbf{y} \rightarrow (\mathbf{x} = \mathbf{x} \equiv \mathbf{y} = \mathbf{x})$ ⟨**Ax6**: "A" is $\mathbf{z} = \mathbf{x}$, "t" is $\mathbf{x}$ and "s" is $\mathbf{y}$⟩

(2) $\mathbf{x} = \mathbf{y} \rightarrow \mathbf{x} = \mathbf{x} \rightarrow \mathbf{y} = \mathbf{x}$ ⟨(1) + tautological implication (3.3.1)⟩

(3) $\mathbf{x} = \mathbf{x}$ ⟨**Ax5**⟩

(4) $\mathbf{x} = \mathbf{y} \rightarrow \mathbf{y} = \mathbf{x}$ ⟨(2, 3) + tautological implication (3.3.1)⟩

I leave the second proof to the reader as an easy exercise.   □

Mathematical Logic. By George Tourlakis
Copyright © 2008 John Wiley & Sons, Inc.

7.0.2 Exercise. Prove $\vdash x = y \rightarrow y = z \rightarrow x = z$. □

7.0.3 Lemma. *For any function symbol f of arity n,*

$$\vdash x = y \rightarrow f(z_1, \ldots, z_i, x, z_{i+2}, \ldots, z_n) = f(z_1, \ldots, z_i, y, z_{i+2}, \ldots, z_n)$$

Proof. This is a Hilbert proof in a relaxed style.

Let A stand for the formula

$$f(z_1, \ldots, z_i, x, z_{i+2}, \ldots, z_n) = f(z_1, \ldots, z_i, y, z_{i+2}, \ldots, z_n)$$

Then, by **Ax6**,

$$\vdash x = y \rightarrow \Big(f(z_1, \ldots, z_i, x, z_{i+2}, \ldots, z_n) = f(z_1, \ldots, z_i, y, z_{i+2}, \ldots, z_n) \equiv$$
$$f(z_1, \ldots, z_i, y, z_{i+2}, \ldots, z_n) = f(z_1, \ldots, z_i, y, z_{i+2}, \ldots, z_n) \Big)$$

The subformula "$f(z_1, \ldots, z_i, y, z_{i+2}, \ldots, z_n) = f(z_1, \ldots, z_i, y, z_{i+2}, \ldots, z_n)$" can be dropped. Why? By (**Ax5**) $w = w$ is an axiom. By 6.1.19 we get

$$\vdash f(z_1, \ldots, z_i, y, z_{i+2}, \ldots, z_n) = f(z_1, \ldots, z_i, y, z_{i+2}, \ldots, z_n)$$

"Redundant true" does the rest. □

7.0.4 Corollary. *For any function symbol f of arity n,*

$$x = y \vdash f(z_1, \ldots, z_i, x, z_{i+2}, \ldots, z_n) = f(z_1, \ldots, z_i, y, z_{i+2}, \ldots, z_n)$$

Proof. By MP and 7.0.3. □

7.0.5 Corollary. *For any function symbol f of arity n,*

$$\vdash x_1 = y_1 \rightarrow \cdots \rightarrow x_n = y_n \rightarrow f(x_1, \ldots, x_n) = f(y_1, \ldots, y_n)$$

Proof. (Sketch) Move all the $x_i = y_i$ to the left of "$\vdash$" and prove instead

$$x_1 = y_1, \ldots, x_n = y_n \vdash f(x_1, \ldots, x_n) = f(y_1, \ldots, y_n)$$

This is a legitimate approach by the deduction theorem. Then we deduce using Corollary 7.0.4

$$f(x_1, \ldots, x_n) = f(y_1, x_2, \ldots, x_n) \text{ from } x_1 = y_1$$

$$f(y_1, x_2, \ldots, x_n) = f(y_1, y_2, x_3, \ldots, x_n) \text{ from } x_2 = y_2$$

$$f(y_1, y_2, x_3, \ldots, x_n) = f(y_1, y_2, y_3, x_4, \ldots, x_n) \text{ from } x_3 = y_3$$

$$\vdots$$

Finally,

$$f(\mathbf{y}_1, \mathbf{y}_2, \ldots, \mathbf{y}_{n-1}, \mathbf{x}_n) = f(\mathbf{y}_1, \mathbf{y}_2, \ldots, \mathbf{y}_{n-1}, \mathbf{y}_n) \text{ from } \mathbf{x}_n = \mathbf{y}_n$$

Transitivity (Lemma 7.0.1) does the rest. $\qquad\square$

7.0.6 Corollary. *For any function symbol f of arity n, and any terms t_i and s_i, $i = 1, 2, \ldots, n$,*

$$\vdash t_1 = s_1 \to \cdots \to t_n = s_n \to \Big(f(t_1, \ldots, t_n) = f(s_1, \ldots, s_n) \Big)$$

Proof. By 7.0.5 and the substitution theorem (6.1.19). $\qquad\square$

7.0.7 Corollary. *For any function symbol f of arity n, and any terms t_i and s_i, $i = 1, 2, \ldots, n$,*

$$t_1 = s_1, \ldots, t_n = s_n \vdash f(t_1, \ldots, t_n) = f(s_1, \ldots, s_n)$$

Proof. By 7.0.6 and n applications of MP. $\qquad\square$

7.0.8 Theorem. *For any terms t, t', s, we have that $\vdash t = t' \to s[\mathbf{x} := t] = s[\mathbf{x} := t']$.*

Proof. We do induction on the complexity of the term s (cf. 4.1.6).

Basis 1. s is a constant or a variable other than $\mathbf{x}$. Then the claim reads

$$\vdash t = t' \to s = s$$

and follows by a tautological implication from $\vdash s = s$. The latter is correct by **Ax5** and an application of 6.1.19.

Basis 2. s is the variable $\mathbf{x}$. The claim now reads

$$\vdash t = t' \to t = t'$$

and is correct (**Ax1**).

Induction step. s is $f(t_1, \ldots, t_n)$, where f is a function symbol of arity n and $t_i, i = 1, \ldots, n$, are terms.

We proceed via the deduction theorem, so we add the hypothesis $t = t'$ and embark upon proving

$$f(t_1, \ldots, t_n)[\mathbf{x} := t] = f(t_1, \ldots, t_n)[\mathbf{x} := t']$$

that is (cf. 4.1.28)

$$f(t_1[\mathbf{x} := t], \ldots, t_n[\mathbf{x} := t]) = f(t_1[\mathbf{x} := t'], \ldots, t_n[\mathbf{x} := t'])$$

$$(0) \quad t = t' \qquad\qquad\qquad\qquad\qquad \langle \text{hypothesis} \rangle$$

$$(1) \quad t_1[\mathbf{x} := t] = t_1[\mathbf{x} := t'] \qquad \langle (0) + \text{I.H.} + \text{MP} \rangle$$
$$(2) \quad t_2[\mathbf{x} := t] = t_2[\mathbf{x} := t'] \qquad \langle (0) + \text{I.H.} + \text{MP} \rangle$$
$$\vdots \qquad\qquad\qquad\qquad\qquad \vdots$$
$$(n) \quad t_n[\mathbf{x} := t] = t_n[\mathbf{x} := t'] \qquad \langle (0) + \text{I.H.} + \text{MP} \rangle$$
$$(n+1) \quad f(t_1[\mathbf{x} := t], \ldots, t_n[\mathbf{x} := t]) =$$
$$f(t_1[\mathbf{x} := t'], \ldots, t_n[\mathbf{x} := t']) \quad \langle (1)\text{--}(n) + 7.0.7 \rangle \qquad \square$$

Theorem 7.0.8 is an SFL counterpart where both sides of $\rightarrow$ are of non-Boolean type.

CHAPTER 8

FIRST-ORDER SEMANTICS—VERY NAÏVELY

This chapter is on naïve semantics. That is, we see here what these abstract, "meaningless", strings—the first-order formulae—actually say. Specifically, we will see what it means, and how, to compute truth values of such formulae.

I would like to emphasize the qualifier *naïve*. In the more advanced literature one defines semantics rigorously: either defining the truth or falsehood of formulae within *informal mathematics*—that is, in the metatheory—a process originated by Tarski and nicknamed *Tarski semantics*, or defining the truth or falsehood of formulae of the original language *within some other formal theory, T, possibly over a different language*, in which case one speaks of *formal semantics*. In formal semantics, formulae of the original language are first translated into formulae over the language of T. Then one defines that a formula in the original language is "true" iff its translation is *provable* in T (cf. [45, 53, 54]).

Here we do neither, but instead *imitate* Tarski informal semantics, albeit in a very simple and purposely sloppy (8.1.2) manner. But this will do for our purpose, which is to learn to easily build counterexamples that expose fallacious statements in predicate logic.

Central to all this is the concept of an *interpretation*. Until now, we treated formulae as "meaningless strings of symbols" that we knew how to manipulate

Mathematical Logic. By George Tourlakis
Copyright © 2008 John Wiley & Sons, Inc.

syntactically—even to prove formally—with our logical axioms and rules of inference.

But how can we give mathematical meaning to these symbols and formulae?

8.1 INTERPRETATIONS

An *interpretation* of a formula is inherited from the interpretation of the symbols of the language where it belongs[117] via a process that we will describe below (8.1.2).

It is a tool that when applied to any formula in the language will produce an *interpreted* mathematical formula of the metatheory—the "meaning" or "semantics" of the original. As was the case in propositional logic, where the semantics of a formula (in that case, its truth value) is not unique but, in general, depends on the *chosen state*, an analogous situation holds in the first-order case: The "meaning" of a formula is not unique.

An interpretation of a first-order language is a pair of two components, a *non empty* set D—called the *domain* or *underlying set* of the interpretation—and a *translator* M, which is a mapping that assigns an appropriate mathematical object to each of the following elements of the language: nonlogical symbols, $\bot$ and $\top$, each object variable and each propositional variable.

Let me stress that the choices of both D and M are entirely up to us.

We usually denote the pair (D, M) with the same capital letter that names the domain, but in German calligraphic typesetting; that is, $\mathfrak{D}$. Clearly, the name $\mathfrak{D}$ does not uniquely determine an interpretation (as neither does D) since an interpretation also depends on M. However, the context will fend off possible ambiguities.

8.1.1 Definition. (Interpreting a Language—Step 1: Translating the Alphabet)
An interpretation $\mathfrak{D} = (D, M)$ that we choose gives meaning to $\bot$, $\top$, to all object and Boolean variables, and to all nonlogical symbols of the alphabet as follows, where the result of the *translation* "$M(\ldots)$" is written as "$\ldots^{\mathfrak{D}}$":

(1) For each free variable $\mathbf{x}$, $\mathbf{x}^{\mathfrak{D}}$—i.e., the translation $M(\mathbf{x})$—is some member of D.

(2) For each Boolean variable $\mathbf{p}$, $\mathbf{p}^{\mathfrak{D}}$ is some member of $\{\mathbf{t}, \mathbf{f}\}$.

(3) $\top^{\mathfrak{D}} = \mathbf{t}$ and $\bot^{\mathfrak{D}} = \mathbf{f}$.

(4) For each object constant c of the alphabet, its translation $c^{\mathfrak{D}}$ is some member of D.

(5) For each function f of the alphabet, the translation $f^{\mathfrak{D}}$ is a mathematical function of the metatheory with the same arity as the formal f. $f^{\mathfrak{D}}$ takes its inputs from D and has its output values in D.

---

[117]We recall, of course, that a first-order language consists of three significant sets: the alphabet, **Term**, and **WFF**.

(6) For each predicate ϕ of the alphabet, the translation $\phi^{\mathfrak{D}}$ is a relation of the same arity as ϕ that takes its inputs from D and has its output values in $\{\mathbf{t}, \mathbf{f}\}$.

Note that the Boolean connectives, the symbol "=", and the brackets are not translated into something else; they retain their standard fixed meaning and notation. □

The next definition describes the inheritance mechanism through which any formula of the language inherits an interpretation that was given to its alphabet.

8.1.2 Definition. (Interpreting a Language—Step 2: Translating a Formula)
Given a formula A over some first-order language, and an interpretation $\mathfrak{D} = (D, M)$ of the language. The interpretation or translation of A via $\mathfrak{D}$ is *a mathematical formula of the metatheory* that we denote by $A^{\mathfrak{D}}$, which is constructed as follows:

(i) *We replace* each occurrence of $\bot, \top$ in A by $\bot^{\mathfrak{D}}, \top^{\mathfrak{D}}$—i.e., $\mathbf{f}, \mathbf{t}$—respectively.

(ii) *We replace* each occurrence of a Boolean variable $\mathbf{p}$ in A with the specific truth value $\mathbf{p}^{\mathfrak{D}}$ given by the interpretation of the language.

(iii) *We replace* each occurrence of a *free* variable $\mathbf{x}$ in A with the specific value $\mathbf{x}^{\mathfrak{D}}$ from D.

(iv) *We replace* each occurrence of $(\forall \mathbf{x})$ in A by $(\forall \mathbf{x} \in D)$, which means "for all values of $\mathbf{x}$ in D".

Nonlogical elements of A:

(v) *We replace* each occurrence of an object constant c in A with the specific value $c^{\mathfrak{D}}$ from D.

(vi) *We replace* each occurrence of a function f in A with the specific function $f^{\mathfrak{D}}$, which has inputs from D and output values in D.

(vii) *We replace* each occurrence of predicate ϕ in A with the specific relation $\phi^{\mathfrak{D}}$, which has inputs from D and output in $\{\mathbf{t}, \mathbf{f}\}$.

(viii) We *emphasize* once more what was left unsaid in the transformations (i)–(vii) above: *Every Boolean connective, $=$, and brackets are translated as themselves.* □

(1) A translation $A^{\mathfrak{D}}$ does not have any free variables or any Boolean variables; thus we cannot make any substitutions into variables that may alter $A^{\mathfrak{D}}$. All its object variables are bound. Indeed—by construction—each free variable $\mathbf{x}$ and each Boolean variable $\mathbf{p}$ of the original ("meaningless") A has been everywhere replaced by the value $\mathbf{x}^{\mathfrak{D}}$ from D and $\mathbf{p}^{\mathfrak{D}}$ from $\{\mathbf{t}, \mathbf{f}\}$ respectively.

This is good, because it ensures that $A^{\mathfrak{D}}$ has a determinate truth value, $\mathbf{t}$ or $\mathbf{f}$. We will be writing $A^{\mathfrak{D}} = \mathbf{t}$ or $A^{\mathfrak{D}} = \mathbf{f}$ respectively.

(2) Definition 8.1.2 imitates Tarski semantics in that the translation results are meta-mathematical objects. It also differs in a significant way: The rigorous definition of Tarski semantics does not effect a text-editor-like "find/replace" textual substitution in A toward constructing its metamathematical analogue, the *formula $A^{\mathfrak{D}}$*. Rather, by induction on the formula A, it computes the truth value of $A^{\mathfrak{D}}$, which is induced by the language interpretation, directly (cf. [45, 13, 35, 29, 53]).

Thus our approach also borrows from, but is far from being the same as, the process that defines formal semantics. The latter does have an intermediate interpretation step that produces a *formal* formula, which one next will check for *provability in some theory*. Our intermediate interpretation as defined here instead produces an *informal* formula of metamathematics, which one next checks for its metamathematical truth.

I believe the reader understands at the intuitive level how to compute the truth value of simple mathematical formulae that have no free variables. Thus, in the user-friendly and natural 8.1.2 I am content only to show how we compute the metamathematical counterpart of a "meaningless" formula.

It is useful for the discussion in the next section to also have a concept of a *partially translated* formula that *may have some free variables over D*.

8.1.3 Definition. (Partial Translation of a Formula) Given a formula A over a first-order language and an interpretation, $\mathfrak{D}$, of the language, we may decide, in the application of the translation 8.1.2, to select some particular finite set of formal object variables $\mathbf{x}, \mathbf{y}, \mathbf{z}_3'', \ldots$ *of the language*, which we want to *exempt*—that is, *all their occurrences in A*—from step (iii) of 8.1.2, thus leaving them *untranslated* as we go from A to its metamathematical counterpart.

Of course, $\mathbf{x}, \mathbf{y}, \mathbf{z}_3'', \ldots$ may or may not actually all occur free in A. If, say, $\mathbf{x}$ does not occur free in A, then $\mathbf{x}$ does not occur free in the partially translated formula, either. Otherwise, it is a free variable in the latter.

In any case, the resulting mathematical formula of the metatheory where the $\mathbf{x}_1, \ldots, \mathbf{x}_n$ have not been translated—called *the partial translation of A by $\mathfrak{D}$ with respect to the variables* $\mathbf{x}_1, \ldots, \mathbf{x}_n$—will be denoted by $A^{\mathfrak{D}}_{\mathbf{x}_1, \ldots, \mathbf{x}_n}$ to indicate the *possible free occurrence* of $\mathbf{x}_1, \ldots, \mathbf{x}_n$ in it. We view the $\mathbf{x}_1, \ldots, \mathbf{x}_n$ in the formula $A^{\mathfrak{D}}_{\mathbf{x}_1, \ldots, \mathbf{x}_n}$ of the metatheory as variables that vary over D.

Omission of the qualifier partial *will mean the* full *interpretation/translation—a formula with no free variables—as per 8.1.2.*   □

8.1.4 Remark. It follows that $\big((\forall \mathbf{x})A\big)^{\mathfrak{D}}$ is $(\forall \mathbf{x} \in D)A^{\mathfrak{D}}_{\mathbf{x}}$ since the variable $\mathbf{x}$ of A is not translated as part of the translation process of $(\forall \mathbf{x})A$—it is not free in $(\forall \mathbf{x})A$.   □

8.1.5 Example. Consider the formula $\phi(x, x)$, where ϕ is a 2-ary predicate in some first-order language. Here are a few possible interpretations:

(a) $D = \mathbb{N}$ (the natural numbers, $\{0, 1, 2, \ldots\}$), $\phi^{\mathfrak{D}} = <$.

In this context I mean "$<$" as the "less than" relation on natural numbers.

Thus, $\big(\phi(x, x)\big)^{\mathfrak{D}}$ is this formula over $\mathbb{N}$: $x^{\mathfrak{D}} < x^{\mathfrak{D}}$. For example, if we happened to take $x^{\mathfrak{D}} = 42$, then $\big(\phi(x, x)\big)^{\mathfrak{D}}$ is specifically "$42 < 42$".

By the way, we see that $\left(\phi(x,x)\right)^{\mathcal{D}} = \mathbf{f}$ no matter how we chose $x^{\mathcal{D}}$.

Here are two *partial interpretations*, the first having exempted the variables y, z, the second having exempted x:

(i) $\left(\phi(x,x)\right)^{\mathcal{D}}_{y,z}$ is $x^{\mathcal{D}} < x^{\mathcal{D}}$. Exempting y and z from the translation (8.1.3) makes no difference to the result as neither y nor z occur free in $\phi(x,x)$.

(ii) $\left(\phi(x,x)\right)^{\mathcal{D}}_{x}$ is $x < x$.

(b) $D = \mathbb{N}$, $\phi^{\mathcal{D}} = \leq$ (the "less than or equal" relation on $\mathbb{N}$).

Thus, $\left(\phi(x,x)\right)^{\mathcal{D}}$ is this formula over $\mathbb{N}$: $x^{\mathcal{D}} \leq x^{\mathcal{D}}$. Clearly, no matter what the choice of $x^{\mathcal{D}}$, we have $\left(\phi(x,x)\right)^{\mathcal{D}} = \mathbf{t}$.

(c) $D = \{0, \{0\}\}$, $\phi^{\mathcal{D}} = \in$.

By "$\in$" here I mean the concrete "is a member of" relation of set theory.

Thus, $\left(\phi(x,x)\right)^{\mathcal{D}}$ is this formula over D: $x^{\mathcal{D}} \in x^{\mathcal{D}}$.

Note that *every* choice of $x^{\mathcal{D}}$ makes this false (**f**). Indeed, $0 \in 0$ is false because (the right copy of) 0 has no members (it is not a set), and $\{0\} \in \{0\}$ is false because (the right copy of) $\{0\}$ contains the element "0", *not* the element "$\{0\}$"— these two are different, one is of type "number" the other is of type "set".   □

 The symbols "$<, \leq, \in$" are nonlogical and have *no* fixed *intrinsic* meaning. That is why we *had* to say, when we chose them to interpret "ϕ" above, what we mean by them. For example, we said that we mean "$\in$" to be the "is a member of" relation of set theory.

8.1.6 Example. Consider the formula $f(x) = f(y) \rightarrow x = y$, where f is a 1-ary (unary) function in some language. Here are a few possible interpretations:

(1) $D = \mathbb{N}$, $f^{\mathcal{D}}(x) = x + 1$ for all values of x in D (i.e., in $\mathbb{N}$).

Thus, $\left(f(x) = f(y) \rightarrow x = y\right)^{\mathcal{D}}$ is this formula over $\mathbb{N}$:

$$x^{\mathcal{D}} + 1 = y^{\mathcal{D}} + 1 \rightarrow x^{\mathcal{D}} = y^{\mathcal{D}}$$

Note how *every* choice of $x^{\mathcal{D}}$ and $y^{\mathcal{D}}$ makes this formula true.

(2) $D = \mathbb{Z}$, where $\mathbb{Z}$ is the set of all integers, $\{\ldots, -2, -1, 0, 1, 2, \ldots\}$. We take here $f^{\mathcal{D}}(x) = x^2$ for all x in $\mathbb{Z}$.

Thus, $\left(f(x) = f(y) \rightarrow x = y\right)^{\mathcal{D}}$ is this formula over $\mathbb{Z}$:

$$(x^{\mathcal{D}})^2 = (y^{\mathcal{D}})^2 \rightarrow x^{\mathcal{D}} = y^{\mathcal{D}}$$

The above is true for some, and false for some other, choices of $x^{\mathcal{D}}$ and $y^{\mathcal{D}}$. For example, it is false for $x^{\mathcal{D}} = -2$ and $y^{\mathcal{D}} = 2$.

And here are two *partial interpretations*:

(i) $\left(f(x) = f(y) \rightarrow x = y\right)^{\mathcal{D}}_{x}$ is $x^2 = (y^{\mathcal{D}})^2 \rightarrow x = y^{\mathcal{D}}$.

(ii) $\left(f(x) = f(y) \rightarrow x = y\right)^{\mathcal{D}}_{x,y}$ is $x^2 = y^2 \rightarrow x = y$. ☐

The moral from the above is: *There are some interpretations of "$f(x) = f(y) \rightarrow x = y$" that are (i.e., have truth value) false* (**f**).

8.1.7 Example. In this example we will consider, in order, two different interpretations with different domains D, each a finite subset of $\mathbb{N}$.

Consider the formula

$$x = y \rightarrow (\forall x)x = y \tag{1}$$

Here are a few possible interpretations:

(1) $D = \{3\}$, $x^{\mathcal{D}} = 3$, $y^{\mathcal{D}} = 3$.

Since there is only one element in D, my only option is to set "$x^{\mathcal{D}} = 3$" and "$y^{\mathcal{D}} = 3$", as I did above.

Thus, formula (1) is interpreted as the following formula over D:

$$3 = 3 \rightarrow (\forall x \in D)x = 3 \tag{2}$$

By the way, formula (2) has value **t** because "$3 = 3$" is **t** and so is "$(\forall x \in D)x = 3$", since it says "all values x in D equal 3".

(2) This time I take $D = \{3, 5\}$, and again $x^{\mathcal{D}} = 3$ and $y^{\mathcal{D}} = 3$.

Thus, formula (1) is interpreted as the following formula over this D:

$$3 = 3 \rightarrow (\forall x \in D)x = 3 \tag{3}$$

Now, formula (3) is **f**, because "$3 = 3$" is **t** as before, but this time "$(\forall x \in D)x = 3$" is **f**, since it still says "all values x in D equal 3", which fails—D now has two elements: It is not true that both are equal to 3.

The moral is: There is some interpretation where formula (1) above is interpreted as false (**f**). ☐

8.1.8 Example. Let us look at a few interpretations of

$$(\forall x)(x \in y \equiv x \in z) \rightarrow y = z \tag{1}$$

(1) If we take D to be the "collection"[118] of all sets, and $\in^{\mathfrak{D}}$ to mean "belongs to" (which we still denote as "$\in$"), then we get

$$(\forall x \in D)(x \in y^{\mathfrak{D}} \equiv x \in z^{\mathfrak{D}}) \to y^{\mathfrak{D}} = z^{\mathfrak{D}}$$

which is set theory's *requirement* (so-called *axiom of extensionality*) that two sets, $y^{\mathfrak{D}}$ and $z^{\mathfrak{D}}$, are equal if they have the same elements. This interpretation of formula (1) is therefore true for any choice of sets $y^{\mathfrak{D}}$ and $z^{\mathfrak{D}}$.

(2) Take now $D = \mathbb{N}$ and $\in^{\mathfrak{D}} = <$, where once more "$<$" is the relation "less than" on $\mathbb{N}$. Then, formula (1) is interpreted into:

$$(\forall x \in \mathbb{N})(x < y^{\mathfrak{D}} \equiv x < z^{\mathfrak{D}}) \to y^{\mathfrak{D}} = z^{\mathfrak{D}}$$

which is obviously true no matter how we chose the numbers $y^{\mathfrak{D}}$ and $z^{\mathfrak{D}}$.

(3) Take $D = \mathbb{N}$ and $\in^{\mathfrak{D}} = \mid$, where by "$\mid$" we denote the relation "divides" (with remainder 0). For example, $2 \mid 3$ and $2 \mid 1$ are false, but $2 \mid 4$ and $2 \mid 0$ are true.

Then, formula (1) is interpreted into:

$$(\forall x \in \mathbb{N})(x \mid y^{\mathfrak{D}} \equiv x \mid z^{\mathfrak{D}}) \to y^{\mathfrak{D}} = z^{\mathfrak{D}}$$

which is also obviously true for all choices of the numbers $y^{\mathfrak{D}}, z^{\mathfrak{D}}$. It says: "Two natural numbers, $y^{\mathfrak{D}}$ and $z^{\mathfrak{D}}$, are equal if they have precisely the same divisors". However,

(4) Take $D = \mathbb{Z}$ and $\in^{\mathfrak{D}} = \mid$. Now, (1) is interpreted into:

$$(\forall x \in \mathbb{Z})(x \mid y^{\mathfrak{D}} \equiv x \mid z^{\mathfrak{D}}) \to y^{\mathfrak{D}} = z^{\mathfrak{D}}$$

We note that unlike the previous interpretations of (1), the current interpretation may be false. For example, this is so if $y^{\mathfrak{D}} = 2$ and $z^{\mathfrak{D}} = -2$. The interpretation is then the formula

$$(\forall x \in \mathbb{Z})(x \mid 2 \equiv x \mid -2) \to 2 = -2$$

which is false, for the hypothesis $(\forall x \in \mathbb{Z})(x \mid 2 \equiv x \mid -2)$ is true (2 and -2 do have the same divisors), but the conclusion $2 = -2$ is false. □

8.2 SOUNDNESS IN PREDICATE LOGIC

8.2.1 Definition. (Universally—or Logically, or Absolutely—Valid Formulae) If $A^{\mathfrak{D}} = \mathbf{t}$, for some A and $\mathfrak{D}$, we will say that *A is true in the interpretation $\mathfrak{D}$* or that *$\mathfrak{D}$ is a model of A*. We write this briefly as:

$$\models_{\mathfrak{D}} A \tag{1}$$

---

[118]**Small print:** We are *not* interested here in esoteric issues such as: "But this D is too 'large' and is thus not a set."

A first-order formula, A, is *universally valid*—or just *valid*—iff *it has as a model every interpretation of the language where the formula belongs*, i.e., (1) holds for all interpretations $\mathfrak{D}$ of the language of A.

We indicate that A is universally valid by dropping the $\mathfrak{D}$ subscript from (1), writing

$$\models A \tag{V}$$

□

Note the absence of the subscript "taut" from notation (V) above. This is for good reason: $\models x = x$ is correct, but it is *not* the case that $\models_{\text{taut}} x = x$. For the latter says of the abstraction, say p, of $x = x$ that $\models_{\text{taut}} p$—a clearly incorrect metamathematical statement, because I may take a state v with $v(p) = \mathbf{f}$. The former says that for any $\mathfrak{D}$ and any choice of the value $x^{\mathfrak{D}}$ in D I have $x^{\mathfrak{D}} = x^{\mathfrak{D}}$—clearly true!

8.2.2 Remark. All axioms in 4.2.1 are universally valid. We have already argued this claim early on at a very intuitive level in 4.2.2 and we are going to elaborate further here in the light of Definitions 8.1.2 and 8.1.3 without however getting into a 100% rigorous proof that requires a much more careful Definition 8.1.2 (we return to this topic and give such a definition, and a rigorous proof, in the Appendix, Section A.1).

Valid Axioms 1: Ax1. It is easy to see that all axioms in group **Ax1** are valid.

Indeed, more generally, we argue in outline that

$$\text{if } \models_{\text{taut}} A, \text{ then } \models A \tag{1}$$

That the converse of (1) is false was already discussed taking A to be $x = x$ as a counterexample.

Now, why is (1) a true metamathematical statement?

Well,

$$\text{let us assume } \models_{\text{taut}} A \tag{2}$$

Of course, in the context of predicate calculus, (2) refers to the abstraction of A (cf. 4.1.27).

In that abstraction—we are told by (2)—*any arbitrary assignment of values to the Boolean variables* and *prime subformulae* of A[119] will lead to a computed truth value of (the abstraction of) A equal to $\mathbf{t}$.

Let us now see what happens when we fix a $\mathfrak{D}$ and try to compute the truth value of $A^{\mathfrak{D}}$. Well, the abstraction of $A^{\mathfrak{D}}$ has *exactly the same Boolean structure as that of A, because all the Boolean connectives and brackets are translated as themselves*. Therefore the abstraction of $A^{\mathfrak{D}}$ is also a tautology!

[119]Prime subformulae were defined in 4.1.25.

We next note:

(a) A prime subformula of A has one of the forms $\phi(t_1, \ldots, t_n)$, $t = s$, or $((\forall \mathbf{x})B)$. Upon translation of A into $A^{\mathfrak{D}}$, these prime subformulae are transformed into $t^{\mathfrak{D}} = s^{\mathfrak{D}}$, $\phi^{\mathfrak{D}}(t_1^{\mathfrak{D}}, \ldots, t_n^{\mathfrak{D}})$ and $((\forall \mathbf{x} \in D)B_{\mathbf{x}}^{\mathfrak{D}})$ respectively, i.e., into prime subformulae of $A^{\mathfrak{D}}$.

On the other hand, a Boolean variable $\mathbf{p}$ of A becomes a subformula $\mathbf{p}^{\mathfrak{D}}$ of $A^{\mathfrak{D}}$ (that is, the metamathematical Boolean constant $\mathbf{t}$ or $\mathbf{f}$). We may think of it as a "virtual" $\mathbf{p}$ to which we decided (by virtue of our choice of $\mathfrak{D}$) to assign the value $\mathbf{p}^{\mathfrak{D}}$.

(b) Now, in checking whether A is a tautology, one *assigns* to the prime subformulae *and* to the Boolean variables of A *arbitrary* Boolean values and computes the result for A according to truth tables. By (2), the result invariably is $\mathbf{t}$.

(c) How does the computation of the truth value of $A^{\mathfrak{D}}$ relate to the description in (b)? Well, rather than *assigning* truth values to the prime subformulae of $A^{\mathfrak{D}}$—which are direct translations of those of A—one computes and uses their *intrinsic* truth values instead. This latter subcomputation is informed by our knowledge of the metatheory where these subformulae mean something mathematically tangible (cf. the examples of the preceding section).

On the other hand, $A^{\mathfrak{D}}$ has no Boolean variables, but at the precise spot in the formula structure where A had a $\mathbf{p}$, $A^{\mathfrak{D}}$ has a $\mathbf{t}$ or $\mathbf{f}$ ($\mathbf{p}^{\mathfrak{D}}$). As we said in (a) above, we may think that this is a "virtual" $\mathbf{p}$ of $A^{\mathfrak{D}}$ with the assigned value $\mathbf{p}^{\mathfrak{D}}$.

(d) Having noted that $A^{\mathfrak{D}}$ is a tautology as a Boolean formula of the metatheory that is built from metamathematical Boolean constants, prime subformulae, brackets, and connectives, it follows that its truth value under the computation described in (c) is independent—and equal to $\mathbf{t}$—of what values one might *choose to assign* to its prime subformulae. Thus it is also independent of the *intrinsic, computed value* of said subformulae.

This concludes the case for (1) above.

Valid Axioms 2: **Ax2.** $(\forall \mathbf{x})A \rightarrow A[\mathbf{x} := t]$ is universally valid. Indeed, given $\mathfrak{D}$, and fixing $A, \mathbf{x}, t$, we have that $\left((\forall \mathbf{x})A \rightarrow A[\mathbf{x} := t] \right)^{\mathfrak{D}}$ is, according to 8.1.2,

$$(\forall \mathbf{x} \in D)A_{\mathbf{x}}^{\mathfrak{D}} \rightarrow \left(A[\mathbf{x} := t] \right)^{\mathfrak{D}} \tag{1}$$

The trickiest part to agree on is that the part $\left(A[\mathbf{x} := t] \right)^{\mathfrak{D}}$ of (1) is $A_{\mathbf{x}}^{\mathfrak{D}}[\mathbf{x} := t^{\mathfrak{D}}]$ (cf. also 8.1.4). Indeed, we start with

$$A: \quad \ldots \boxed{\mathbf{x}} \ldots \boxed{\mathbf{x}} \ldots$$

where, for the sake of visualization and without loss of generality, I show two free occurrences of $\mathbf{x}$ (actually I may have zero or more). Then we get

$$A[\mathbf{x} := t]: \quad \ldots \boxed{t} \ldots \boxed{t} \ldots \tag{2}$$

and, applying 8.1.2, we get

$$\left(A[\mathbf{x} := t] \right)^{\mathfrak{D}}: \quad (\ldots)^{\mathfrak{D}} \boxed{t^{\mathfrak{D}}} (\ldots)^{\mathfrak{D}} \boxed{t^{\mathfrak{D}}} (\ldots)^{\mathfrak{D}} \tag{3}$$

But (3) is what becomes of

$$(\ldots)^{\mathfrak{D}} \boxed{\mathbf{x}} (\ldots)^{\mathfrak{D}} \boxed{\mathbf{x}} (\ldots)^{\mathfrak{D}}$$

i.e., $A_{\mathbf{x}}^{\mathfrak{D}}$, after we apply the substitution "$[\mathbf{x} := t^{\mathfrak{D}}]$"!

With this out of the way, we readily see that (1) is true (**t**): So, assume the left-hand side of $\rightarrow$ in (1), that is, that $A_i^{\mathfrak{D}}$ is true for all $i \in D$. But then $A_i^{\mathfrak{D}}$ is true when $i = t^{\mathfrak{D}}$ in particular.

Valid Axioms 3: Ax3 and **Ax4.** I do not have anything to add to the discussion in 4.2.2.

Valid Axioms 4: Ax5. $\mathbf{x} = \mathbf{x}$ is interpreted as "$\mathbf{x}^{\mathfrak{D}} = \mathbf{x}^{\mathfrak{D}}$" in any $\mathfrak{D}$, as we have just discussed. And this is true, no matter what the $\mathbf{x}$ and $\mathfrak{D}$.

Clearly, the first "$=$" is formal, while the second is metamathematical.

Valid Axioms 5: Ax6. $t = s \rightarrow (A[\mathbf{x} := t] \equiv A[\mathbf{x} := s])$ is universally valid. Indeed, let us fix $t, s, A, \mathbf{x}$ and look at the $\mathfrak{D}$-interpretation of this formula for some arbitrary $\mathfrak{D}$. As in the argument for **Ax2**, we want to find that the computed truth value of (4) is **t**.

$$t^{\mathfrak{D}} = s^{\mathfrak{D}} \rightarrow (A_{\mathbf{x}}^{\mathfrak{D}}[\mathbf{x} := t^{\mathfrak{D}}] \equiv A_{\mathbf{x}}^{\mathfrak{D}}[\mathbf{x} := s^{\mathfrak{D}}]) \tag{4}$$

So let $t^{\mathfrak{D}} = s^{\mathfrak{D}}$ in D. Let us set $i = t^{\mathfrak{D}}$. But then each of $A_{\mathbf{x}}^{\mathfrak{D}}[\mathbf{x} := t^{\mathfrak{D}}]$ and $A_{\mathbf{x}}^{\mathfrak{D}}[\mathbf{x} := s^{\mathfrak{D}}]$ are the same formula of the metatheory: $A_i^{\mathfrak{D}}$. Trivially, $A_i^{\mathfrak{D}} \equiv A_i^{\mathfrak{D}}$ is true. $\square$

We now have:

8.2.3 Metatheorem. (Soundness in First-Order Logic) *If* $\vdash A$*, then* $\models A$*.*

Proof. We argue this for the equivalent logic 2 of Chapter 5. This simplifies the argument due to the presence of a single primitive rule of inference, MP, in that logic.

As in the proof of 3.1.3, we do induction on the length of (absolute) proofs that contain A, and prove that $\models A$; that is, for every $\mathfrak{D}$ we have $A^{\mathfrak{D}} = \mathbf{t}$.

Basis. A appears in a proof of length 1. But then A is the only formula in the proof, and hence is in Λ_1. We are done by 8.2.2.

As an I.H. we assume the claim for proofs of lengths $\leq n$, which contain A.

For the induction step, let A appear in an absolute proof of length $n + 1$. We have two cases:

(1) A was written for the first time *before* the last step. Then, in view of 4.2.9 and the I.H., we are done.

(2) A was written for the first time *in* the last step.

If it is in Λ_1, then the case has already been argued in the Basis step. So let instead A be derived by MP, that is, for some B, the formulae B and $B \to A$ have already appeared in the proof.

Now pick any $\mathfrak{D}$. By the I.H. we have

$$B^{\mathfrak{D}} = \mathbf{t} \tag{$*$}$$

and $(B \to A)^{\mathfrak{D}} = \mathbf{t}$, i.e.,

$$B^{\mathfrak{D}} \to A^{\mathfrak{D}} = \mathbf{t} \tag{$**$}$$

By the truth table for $\to$, $(*)$ and $(**)$ yield $A^{\mathfrak{D}} = \mathbf{t}$. $\square$

We have already remarked that in predicate logic "if $\vdash A$, then $\models_{\text{taut}} A$" is false.

8.2.4 Metatheorem. (Gödel's Completeness Theorem) *If* $\models A$, *then* $\vdash A$.

A proof is presented in the Appendix of Part II.

Soundness, just as in the case of propositional calculus, serves the purpose of obtaining counterexamples—in first-order logic these are called *countermodels*. Thus, if for some formula A we do *not* believe that $\vdash A$, we need only to show that $\not\models A$, that is, to find an interpretation $\mathfrak{D}$—a countermodel—where $\not\models_{\mathfrak{D}} A$, i.e., $A^{\mathfrak{D}} = \mathbf{f}$.

8.2.5 Example. Question: Can our logic derive the "rule"

$$\text{If } \Gamma \vdash A, \text{ then } \Gamma \vdash (\forall \mathbf{x})A$$

without a condition on $\mathbf{x}$?

Well, if it could derive the above, it could also derive *strong generalization* below, by setting $\Gamma = \{A\}$.[120]

$$A \vdash (\forall \mathbf{x})A \tag{1}$$

Why not? Because (1) would yield via the deduction theorem

$$\vdash A \to (\forall \mathbf{x})A \tag{2}$$

---

[120]Then $A \vdash A$; hence $A \vdash (\forall \mathbf{x})A$.

By soundness, (2) yields

$$\models A \rightarrow (\forall \mathbf{x})A \tag{3}$$

Now, (3) is a statement schema. It is supposed to hold—if we think that (1) is all right—for no matter what particular formula we may use for A and no matter what variable for **x**. Well, then, (3) should hold when we choose A to be "$x = y$" and **x** to be "x". However, as we saw in Example 8.1.7(2) (cf. Definition 8.2.1)

$$\not\models x = y \rightarrow (\forall x)x = y$$

so (2) is no good, and (1) does not hold in our logic.   □

8.2.6 Example. Knowing that strong generalization is illegal in our logic[121] we can show that certain other suggested "rules" are impossible (*underivable*) by "reducing strong generalization to them", that is, by saying:

> *If I could have this rule, then I could also do (derive) strong generalization, but this is impossible.*

For example, we cannot derive

$$A \equiv B \vdash (\forall \mathbf{x})(C[\mathbf{p} := A] \rightarrow D) \equiv (\forall \mathbf{x})(C[\mathbf{p} := B] \rightarrow D) \tag{1}$$

because it yields the unprovable (in our logic)

$$A \vdash (\forall \mathbf{x})A \tag{2}$$

Hence (1) is unprovable too, lest we want a contradiction. Here is the calculation that shows that (1) derives (2):

Hypothesis is A:

$$(\forall \mathbf{x})A$$
$$\Leftrightarrow \langle 6.1.11 \text{ and } \models_{\text{taut}} X \equiv \neg X \rightarrow \bot \rangle$$
$$(\forall \mathbf{x})\big(\underbrace{(\neg \mathbf{p})[\mathbf{p} := A]}_{\text{this is } \neg A} \rightarrow \bot \big)$$
$$\Leftrightarrow \langle (1) \text{ and } A \vdash A \equiv \top \rangle$$
$$(\forall \mathbf{x})\big(\underbrace{(\neg \mathbf{p})[\mathbf{p} := \top]}_{\text{this is } \neg \top} \rightarrow \bot \big)$$
$$\Leftrightarrow \langle \text{drop } \forall, \text{ by } \vdash X \equiv (\forall \mathbf{y})X \text{ when } X \text{ has no free } \mathbf{y} \rangle$$
$$\neg \top \rightarrow \bot$$

The last formula is a tautology, and hence a theorem. Thus, the first line is a theorem *from* A as the assumption used (middle $\Leftrightarrow$) was A.

---

[121]We *have to* say "in *our* logic". In the logic of [35, 45, 53], $A \vdash (\forall \mathbf{x})A$ is perfectly legal.

In the same manner one shows that "8.12b" ((1) is "8.12a" of [17]) is *not* strong, i.e., the following is *not* valid (cf. Section 6.3).

$$D \rightarrow (A \equiv B) \vdash (\forall \mathbf{x})(D \rightarrow C[\mathbf{p} := A]) \equiv (\forall \mathbf{x})(D \rightarrow C[\mathbf{p} := B]) \qquad \square$$

8.2.7 Exercise. Show that the "rule 8.12b" above is not valid in our logic. $\square$

8.2.8 Exercise. Given a formula A over a first-order language and an interpretation $\mathfrak{D}$ of the language, we saw that $((\forall \mathbf{x})A)^{\mathfrak{D}}$ is $(\forall \mathbf{x} \in D)A_{\mathbf{x}}^{\mathfrak{D}}$.

Show that $((\exists \mathbf{x})A)^{\mathfrak{D}}$ is $(\exists \mathbf{x} \in D)A_{\mathbf{x}}^{\mathfrak{D}}$.

Hint. Recall that $(\forall \mathbf{x} \in D)A$ is short for $(\forall \mathbf{x})((\mathbf{x} \in D) \rightarrow A)$ and $(\exists \mathbf{x} \in D)A$ is short for $(\exists \mathbf{x})((\mathbf{x} \in D) \wedge A)$. $\square$

8.2.9 Example. Why do we insist on choosing a *nonempty* domain D in an interpretation $\mathfrak{D}$?

Take any formula A. Clearly $(\forall \mathbf{x})A \rightarrow (\exists \mathbf{x})A$ is false when interpreted on an *empty domain* D.

Why? "$(\forall \mathbf{x} \in D)A_{\mathbf{x}}^{\mathfrak{D}}$" is true, since there are *no* $\mathbf{x}$ values in D to use toward a counterexample. On the other hand, "$(\exists \mathbf{x} \in D)A_{\mathbf{x}}^{\mathfrak{D}}$" is false, for it says "there exists an $\mathbf{x}$ value that verifies $A_{\mathbf{x}}^{\mathfrak{D}}$", but there are *no* values in D to choose from.

"Big deal", you say. Why should we worry about that?

Because it also happens that

$$\vdash (\forall \mathbf{x})A \rightarrow (\exists \mathbf{x})A \tag{1}$$

If we allow empty domains D, the above argument shows

$$\not\models (\forall \mathbf{x})A \rightarrow (\exists \mathbf{x})A$$

contradicting soundness, something we will not allow! $\square$

8.2.10 Exercise. Prove (1). $\square$

8.2.11 Exercise. Prove that $(\forall \mathbf{y})(\exists \mathbf{x})A \rightarrow (\exists \mathbf{x})(\forall \mathbf{y})A$ is not a theorem schema. That is, show that there is a choice of $A, \mathbf{x}, \mathbf{y}$ such that $\not\vdash (\forall \mathbf{y})(\exists \mathbf{x})A \rightarrow (\exists \mathbf{x})(\forall \mathbf{y})A$.

By the techniques of this chapter (specifically, using soundness) you have to do this: Find an appropriate A so that $\not\models (\forall \mathbf{y})(\exists \mathbf{x})A \rightarrow (\exists \mathbf{x})(\forall \mathbf{y})A$.

Hint. Take A to be $\mathbf{y} < \mathbf{x}$ ($<$ is a nonlogical symbol, of course). Now interpret: Take $D = \mathbb{N}$ interpreting $<$ as the "less than" relation on $\mathbb{N}$. $\square$

8.2.12 Exercise. This was promised on p. 184. Prove that $(\exists \mathbf{x})A \wedge (\exists \mathbf{x})B \rightarrow (\exists \mathbf{x})(A \wedge B)$ is not a theorem schema. That is, show that there is a choice of $A, \mathbf{x}$ and B such that $\not\vdash (\exists \mathbf{x})A \wedge (\exists \mathbf{x})B \rightarrow (\exists \mathbf{x})(A \wedge B)$.

Once again, by the techniques of this chapter (soundness) you have to do this: Find appropriate A and B so that $\not\models (\exists \mathbf{x})A \wedge (\exists \mathbf{x})B \rightarrow (\exists \mathbf{x})(A \wedge B)$.

Hint. Take A and B to be atomic formulae $\phi(\mathbf{x})$ and $\psi(\mathbf{x})$ respectively, where ϕ, ψ are predicates of arity 1.

Next interpret: Take $D = \mathbb{N}$ and let $\phi^{\mathfrak{D}}(\mathbf{x})$ be "the number $\mathbf{x}$ is even" and $\psi^{\mathfrak{D}}(\mathbf{x})$ be "the number $\mathbf{x}$ is odd".   □

8.3 ADDITIONAL EXERCISES

1. Axiom 3 implies (4.2.7) that no matter for which choice of A and $\mathbf{x}$, we have

$$\vdash (\forall \mathbf{x})(A \rightarrow B) \rightarrow (\forall \mathbf{x})A \rightarrow (\forall \mathbf{x})B$$

Prove by an *appropriate countermodel argument* that the converse

$$((\forall \mathbf{x})A \rightarrow (\forall \mathbf{x})B) \rightarrow (\forall \mathbf{x})(A \rightarrow B) \tag{1}$$

is not a universally valid schema.

Conclude, with reason (one sentence), that (1) cannot be a theorem schema, either.

2. Is the following schema a derived rule of our logic (that is, of logic 1 or 2)?

$$A \rightarrow B \vdash A \rightarrow (\forall \mathbf{x})B, \text{ provided } \mathbf{x} \text{ is not free in } A \tag{2}$$

   - If you think that it *is*, then give a proof in our logic.
   - If you do *not* think so, then give a definitive reason as to *why*—for example, using a concrete interpretation, or by proving the invalid "strong generalization" using (2) as a lemma.

3. Redo Exercise 2, but for the schema

$$A \rightarrow B \vdash (\exists \mathbf{x})A \rightarrow B, \text{ provided } \mathbf{x} \text{ is not free in } B$$

4. Would your answer change in Exercise 3 if to the left of $\vdash$ we had $(\forall \mathbf{x})(A \rightarrow B)$ instead?

5. Is the following schema a derived rule of our logic (that is, of logic 1 or 2)?

$$A \rightarrow B \vdash (\forall \mathbf{x})A \rightarrow (\forall \mathbf{x})B \tag{3}$$

   - If you think that it *is*, then give a proof in our logic.
   - If you do *not* think so, then give a definitive reason as to *why*—for example, using a concrete interpretation, or by proving the invalid *strong generalization* from (3).

6. Redo Exercise 5, but for the schema

$$(\forall \mathbf{x})(A \rightarrow B) \vdash (\forall \mathbf{x})A \rightarrow (\forall \mathbf{x})B$$

instead.

7. Redo Exercise 5, but for the schema

$$(\forall \mathbf{x})A \rightarrow (\forall \mathbf{x})B \vdash (\forall \mathbf{x})(A \rightarrow B)$$

instead.

8. Which of the following is a derived rule?

   - $A \rightarrow B \vdash (\exists \mathbf{x})A \rightarrow (\exists \mathbf{x})B$
   - $(\forall \mathbf{x})(A \rightarrow B) \vdash (\exists \mathbf{x})A \rightarrow (\exists \mathbf{x})B$

   In each case, a positive answer needs proof; a negative answer needs a precise argument in connection with a carefully built countermodel.

9. Formulate and explore whether the relativization of the schema in the second bullet of Exercise 8 is provable in our logic. Give precise reasons (proof or countermodel) whichever way you choose to conjecture.

10. Is this $(\forall \mathbf{x})(A \vee B) \rightarrow (\forall \mathbf{x})A \vee (\forall \mathbf{x})B$ an absolute theorem schema?

    - If you think, "Yes, it is", then give a proof within our logic.
    - If you think, "No, it is not", then find specific A and B and an appropriate interpretation to carefully and completely make your case for "no".

11. Is this $(\forall \mathbf{x})A \vee (\forall \mathbf{x})B \rightarrow (\forall \mathbf{x})(A \vee B)$ an absolute theorem schema?

    - If you think, "Yes, it is", then give a proof within our logic.
    - If you think, "No, it is not", then find specific A and B and an appropriate interpretation to carefully and completely make your case for "no".

12. Is this $(\exists \mathbf{x})(A \wedge B) \rightarrow (\exists \mathbf{x})A \wedge (\exists \mathbf{x})B$ an absolute theorem schema?

    - If you think, "Yes, it is", then give a proof within our logic.
    - If you think, "No, it is not", then find specific A and B and an appropriate interpretation to carefully and completely make your case for "no".

13. Consider the proof below:

(1)	$(\exists \mathbf{x})A$	$\langle$hypothesis$\rangle$
(2)	$A[\mathbf{x} := \mathbf{z}]$	$\langle$auxiliary hypothesis associated with (1); $\mathbf{z}$ fresh$\rangle$
(3)	$(\forall \mathbf{z})A[\mathbf{x} := \mathbf{z}]$	$\langle$(2) plus generalization; (1) has no free $\mathbf{z}\rangle$
(4)	$(\forall \mathbf{x})A$	$\langle$(3) plus dummy renaming plus Eqn$\rangle$

    By the deduction theorem we have $\vdash (\exists \mathbf{x})A \rightarrow (\forall \mathbf{x})A$.

    You have two tasks:

(a) Definitively show that $(\exists x)A \rightarrow (\forall x)A$ is *not* an absolute theorem schema.

(b) Grade the above "proof". That is, find *exactly where* (i.e., at which step) it went wrong—and *precisely* how.

14. Is the following—on condition that x is not free in B—an absolute theorem schema? (Cf. Exercise 26 on p. 189.)

$$(\exists x)_A(B \vee C) \equiv B \vee (\exists x)_A C$$

If yes, then prove it; if no, then provide a carefully constructed countermodel.

15. Is the following—on condition that x is not free in C—an absolute theorem schema? (Cf. Exercise 27 on p. 189.)

$$(\forall x)_A B \rightarrow C \equiv (\exists x)_A(B \rightarrow C)$$

If yes, then prove it; if no, then provide a carefully constructed countermodel.

Appendix A

Gödel's Theorems and Computability

This appendix develops two cornerstone results of the metatheory of logic, both due to Gödel: his *completeness* and (first) *incompleteness* theorem.

The first is the counterpart of Post's theorem (3.2.1) for first-order logic and intuitively says that when it comes to the notion of "absolute truth", that is, truth as understood philosophically for the entire edifice of mathematics, then predicate logic speaks "the whole truth". The second came as a shock when first announced ([16]): When it comes to restricted or "relative" truth, that is, the truth of statements (that have no free variables) made in some "powerful"[1] theory such as Peano arithmetic or axiomatic set theory, the formal axiomatic method *cannot* speak the whole truth.

[1]"Powerful" in that these theories can express fairly complicated statements *about their behavior*. For example, either of them contains a variable-free formula that in essence says "this axiomatic system cannot prove me".

A.1 REVISITING TARSKI SEMANTICS

This section looks more carefully at Tarski semantics, which were introduced in 8.1. The extra care is needed so that we can now give a rigorous definition of *absolute truth* that can be mathematically discussed and manipulated toward proving, as Gödel originally did ([15]), that predicate logic is complete, that is, *every absolutely true formula has a formal proof in our logic—i.e., the calculus captures the* whole truth.

The semantics, just as we did in 8.1, will be shaped within an *interpretation* $\mathfrak{D} = (D, M)$, where as before D will be a nonempty set and M a translator of symbols.

As we recall from 8.1.2, the goal in assigning semantics to an abstract formula A was to translate it into a "concrete" mathematical statement $A^{\mathfrak{D}}$ with the ultimate goal of "computing" the latter's truth value, something that we can in principle do by putting our "knowledge of mathematics" to work. Since the process eliminates all free variables by replacing any free variable $\mathbf{x}$ by a value $\mathbf{x}^{\mathfrak{D}}$ from D (cf. 8.1.2), this truth value is unambiguously obtained.

The rigorous definition of Tarski semantics, as was already noted in the remark (2) that followed Definition 8.1.2, bypasses this translation into a concrete formula, thus avoiding the implication—which we cannot make mathematically precise!—that "*we know enough mathematics*" to evaluate the resulting concrete formula as to its truth. The definition is thus impersonal and instead defines truth of any formula A over a first-order language directly and from first principles, as long as we have chosen a translation of the language in the style of 8.1.1.

 The reader should note that in this section the meaning of the symbol $A^{\mathfrak{D}}$ (or $M(A)$) has changed: It now denotes a member of $\{\mathbf{t}, \mathbf{f}\}$ rather than a metamathematical formula (see Exercises A.1.7 and A.1.8).

Central to the definition will be our ability to "replace any free $\mathbf{x}$ by $\mathbf{x}^{\mathfrak{D}}$", *but we need to do such substitutions without exiting into the metamathematical realm*, since we are not building a "(meta)mathematical formula" this time around.

Pause. But how can we effect such substitutions? As soon as a formal object like $\mathbf{x}$ in A is replaced by a metamathematical object $\mathbf{x}^{\mathfrak{D}}$ from D, the resulting string will not be a well-formed formula anymore: $\mathbf{x}^{\mathfrak{D}}$ is not even in the acceptable alphabet!

A trick that originated with Leon Henkin and Abraham Robinson bypasses the abovementioned difficulty: *Just augment the chosen first-order alphabet (cf. 4.1.2) to include (names of) objects from the domain D that we have in mind!*

Thus, as in 8.1, we start by fixing a first-order language L. By the way, it focuses the mind if we think of the language as *the triple* of "ingredients "$(\mathcal{V}, \mathbf{Term}, \mathbf{WFF})$— where $\mathcal{V}$ denotes the chosen first-order alphabet—just as we view an interpretation as a pair of ingredients, D and M (cf. 8.1). In step two, we choose an interpretation $\mathfrak{D} = (D, M)$ for our language L. In step three, the last preparatory step prior to giving the definitive version of Definition 8.1.1 below, we import the names of all the members of D to the alphabet $\mathcal{V}$ as (names of) *new* constants. It is intended that these new constants will be translated—by M, to be reintroduced in A.1.5 below—*as themselves*; i.e., if i is such a new imported constant from D, its meaning under the interpretation will be i. I use the same *name*, say "i", for a given object of D, both metamathematically and formally (e.g., the name for the object "three", if this constant is imported, will be "3" both formally and informally).

A.1.1 Definition. For *any nonempty set* D, the first-order alphabet obtained by importing the members of D as new constants is denoted by $\mathcal{V}(D)$. That is, $\mathcal{V}(D) = \mathcal{V} \cup D$. We will use the terminology "D-formulae" and "D-terms" for formulae and terms of the *original* language L, where *some* free variables have been replaced by D values. $\qquad\square$

It is an easy exercise to establish that this is tantamount to saying:

A.1.2 Definition. Given a nonempty set D, a D-formula and D-term are a formula and term over the alphabet $\mathcal{V}(D)$, respectively. The set of all D-formulae and D-terms will be denoted by **WFF**(D) and **Term**(D) respectively. The augmented language is $L(D) = (\mathcal{V}(D), \textbf{Term}(D), \textbf{WFF}(D))$. $\qquad\square$

A.1.3 Exercise. Fix a language L and a nonempty set D. Then **Term** $\subseteq$ **Term**(D) and **WFF** $\subseteq$ **WFF**(D). $\qquad\square$

A.1.4 Exercise. Fix a language L and a nonempty set D. Then all D-terms and all the D-formulae according to Definition A.1.2 can be obtained by repeated application of substitutions such as $t[\mathbf{x} := i]$ and $A[\mathbf{x} := i]$ where both t and A are over the original alphabet $\mathcal{V}$, and $i \in D$. $\qquad\square$

A.1.5 Definition. (Tarski Semantics—Step 1: Translating the Alphabet) Given a first-order language $L = (\mathcal{V}, \textbf{Term}, \textbf{WFF})$. An interpretation $\mathfrak{D} = (D, M)$ *appropriate for L* (or just "*for L*") is a pair where D is a nonempty set and M is a "translator" or "interpretation mapping" that assigns concrete meaning to the symbols in $\mathcal{V}(D)$. We will often denote the result of the *translation* "$M(\ldots)$" as "$\ldots^{\mathfrak{D}}$".

(1) For each free variable $\mathbf{x}$, $\mathbf{x}^{\mathfrak{D}}$—i.e., the translation $M(\mathbf{x})$—is some member of D.

(2) For each Boolean variable $\mathbf{p}$, $\mathbf{p}^{\mathfrak{D}}$ is some member of $\{\mathbf{t}, \mathbf{f}\}$.

(3) $\top^{\mathfrak{D}} = \mathbf{t}$ and $\perp^{\mathfrak{D}} = \mathbf{f}$.

(4) For each object constant c in $\mathcal{V}$, its translation $c^{\mathfrak{D}}$ is some member of D.

(5) For each object constant i in D, its translation $i^{\mathfrak{D}}$ is i itself.

 The letters i, j, k, m, n will name constants imported from D, utilizing primes or subscripts if necessary.

(6) For each function f of the language L, the translation $f^{\mathfrak{D}}$ is a mathematical function of the metatheory with the same arity as the formal f. $f^{\mathfrak{D}}$ takes its inputs from D and has its output values in D.

(7) For each predicate ϕ of the language L, the translation $\phi^{\mathfrak{D}}$ is a relation of the same arity as ϕ that takes its inputs from D and has its output values in $\{\mathbf{t}, \mathbf{f}\}$.   □

We next extend M so as to give metamathematical meaning—and by "meaning" I mean a concrete value from $D \cup \{\mathbf{t}, \mathbf{f}\}$, not an "intermediate" mathematical formula—to arbitrary D-terms and to arbitrary D-formulae. By A.1.3, this extension will also give meaning to terms and formulae of L.

The reader may want to briefly glimpse at Definition 1.3.5, where the "meaning function" v was extended to be meaningful not only on atomic but on all Boolean formulae. We noted there (immediately prior to the definition) that the *extension* is different from the *original* and thus some logicians would prefer a different symbol for it, say, $\bar{v}$. We decided against this as it clutters notation and the context can readily fend off any ambiguity. For the same reason we should be content with using the same symbol, M, for both the original mapping of the symbols of the alphabet $\mathcal{V}(D)$ into concrete ones and the mapping that maps terms and formulae into their values.

A.1.6 Definition. (Tarski Semantics—Step 2: Extending M to all of $L(D)$) Let $\mathfrak{D} = (D, M)$ be an interpretation for L. We extend M to all D-terms and D-formulae—still calling it M—as follows:

D**-terms:** We define $M(t)$—i.e., $t^{\mathfrak{D}}$—for each $\mathfrak{D}$-term t:

(i) If t is an object variable, or a constant symbol of $\mathcal{V}(D)$, then M is as in A.1.5.

(ii) If t is $f(t_1, \ldots, t_n)$ where f is any n-ary function symbol and $t_1, \ldots, t_n$ are D-terms, then $t^{\mathfrak{D}} = f^{\mathfrak{D}}(t_1^{\mathfrak{D}}, \ldots, t_n^{\mathfrak{D}})$

D**-formulae:** By induction on the complexity (cf. 4.1.15) of such formulae A we define $A^{\mathfrak{D}}$:

I. If A is the Boolean constant $\bot$ or $\top$ or a Boolean variable $\mathbf{p}$, then $A^{\mathfrak{D}}$ has been defined in A.1.5.

II. For any D-terms t and s, $(t = s)^{\mathfrak{D}} = \mathbf{t}$ iff $t^{\mathfrak{D}} = s^{\mathfrak{D}}$, where only the leftmost occurrence of "=" here is formal (i.e., the one in $\mathcal{V}$); the others are metamathematical.

III. For any D-terms $t_1, \ldots, t_n$ and n-ary predicate symbol ϕ,
$$\left(\phi(t_1, \ldots, t_n)\right)^{\mathfrak{D}} = \mathbf{t} \text{ iff } \phi^{\mathfrak{D}}\left(t_1^{\mathfrak{D}}, \ldots, t_n^{\mathfrak{D}}\right) = \mathbf{t}.$$

Assuming that we have defined M for all D-formulae B that are less complex than A, we next define $M(A)$ using the notation in 1.3.4:

IV. If A is $\neg B$, then $A^{\mathfrak{D}} = F_{\neg}(B^{\mathfrak{D}})$.

V. If A is $B \circ C$—where $\circ$ is any of $\wedge, \vee, \rightarrow, \equiv$—then $A^{\mathfrak{D}} = F_{\circ}(B^{\mathfrak{D}}, C^{\mathfrak{D}})$.

VI. If A is $(\forall \mathbf{x})B$, then $A^{\mathfrak{D}} = \mathbf{t}$ iff *for all* $i \in D$ we have $M(B[\mathbf{x} := i]) = \mathbf{t}$. $\square$

Pause. Why not define the extension of M over all of $L(D)$ by induction on D-formulae A, therefore assuming the definition as given for the *immediate predecessors* of A? Because $B[\mathbf{x} := i]$ is not an i.p. of $(\forall \mathbf{x})B$. This is why we resorted to doing the recursive definition with respect to the complexity of formulae instead.

A.1.7 Exercise. Given a language L and an interpretation $\mathfrak{D} = (D, M)$ for it, let t be a D-term. By induction on the complexity of t show that $t^{\mathfrak{D}} \in D$. $\square$

A.1.8 Exercise. Given a language L and an interpretation $\mathfrak{D} = (D, M)$ for it. Let A be a D-formula. By induction on the complexity of A show that $A^{\mathfrak{D}} \in \{\mathbf{t}, \mathbf{f}\}$. $\square$

A.1.9 Definition. (Truth and Models) Let $\mathfrak{D} = (D, M)$ be a interpretation for L, and A be a D-formula. We say "A is *true (false)* in $\mathfrak{D}$" iff $A^{\mathfrak{D}} = \mathbf{t}$ (**f**). *If in particular A is over L*—i.e., *it contains no constants imported from D*—then we say "$\mathfrak{D}$ is a model of A"—and we write $\models_{\mathfrak{D}} A$—iff $A^{\mathfrak{D}} = \mathbf{t}$.

If $\Sigma \subseteq \mathbf{WFF}$, then we say "$\mathfrak{D}$ is a model of Σ" and write "$\models_{\mathfrak{D}} \Sigma$", iff $\models_{\mathfrak{D}} A$ for each $A \in \Sigma$. We say that "Σ is *satisfiable*" iff it has a model.

When $\Sigma \cup \{A\} \subseteq \mathbf{WFF}$, the notation $\Sigma \models A$ denotes *semantic implication*, also called *logical implication*. It means: "Every model of Σ is a model of A." We write $\models A$ for $\emptyset \models A$. Since every interpretation is, vacuously, a model of $\emptyset$, "$\emptyset \models A$" amounts to saying that every interpretation (appropriate for the language of A) is a model of A. We say then that "A is *logically valid*", *universally valid*, or just *valid*. $\square$

The following lemmata will make the flow of exposition in the proof of the soundness metatheorem smoother.

A.1.10 Lemma. *Let $\mathfrak{D} = (D, M)$ be an interpretation for L and t be a D-term that contains no $\mathbf{x}$. Consider a new interpretation $\mathfrak{D}' = (D', M')$ where $D = D'$ and M' agrees with M everywhere, except possibly at the variable $\mathbf{x}$. Then $M(t) = M'(t)$. This last "$=$" is, of course, on D.*

Proof. See the exercise below. $\square$

A.1.11 Exercise. Prove A.1.10 by induction on the complexity of terms. $\square$

For the benefit of the following lemma we revisit Definition 4.1.21 with an inductive definition of "free (bound) variable".

A.1.12 Definition. (Free and Bound Variables)
  The case of terms: The concept "$\mathbf{x}$ is free in t" for terms coincides with the concept "$\mathbf{x}$ occurs in t". Namely, if t is $\mathbf{x}$, then $\mathbf{x}$ is free in t. If t is a constant or a

$y \neq x$, then x is *not* free in t. If t is $f(t_1, \ldots, t_n)$, then x is free in t iff it is free in *at least one* of the t_i.

The case of formulae:

Atomic case:

(1) x is *not* free in any of $\bot, \top, p$.

(2) x is free in $t = s$ iff it is so in at least one of t or s.

(3) x is free in $\phi(t_1, \ldots, t_n)$ iff it is so in at least one of the t_i.

Nonatomic case:

 (i) x is free in $\neg A$ iff it is so in A.

(ii) x is free in $A \circ B$—where $\circ$ is one of $\wedge, \vee, \rightarrow, \equiv$— iff it is so in at least one of A and B.

(iii) x is free in $(\forall y)A$ iff it is free in A *and* it is *not* the same as y.

A variable that occurs in a formula A, yet is not free, is called *bound* in A. $\square$

Thus, in case (iii) above, x is *not* free in $(\forall y)A$ in precisely two (not mutually exclusive) cases: x is not free in A, or $x = y$.

   Notwithstanding the fact that the Tarski semantics of a formula are the truth values **t** or **f**—*not* a "concrete" metamathematical formula—nevertheless, the above definition allows us to "translate" a first-order formula *with free variables* into a concrete formula *with the same number of free variables* and *vice versa*. We present the process that achieves this as a definition (A.1.14) below, but first we will introduce the notation "$\vec{x}_n$".

A.1.13 Definition. The symbol $\vec{x}_n$ denotes the *ordered sequence* $x_1, x_2, \ldots, x_n$. We will simply write $\vec{x}$ when the length n is either understood or is unimportant to our discussion. We call $\vec{x}_n$ an "n-tuple" or an "n-vector". $\square$

A.1.14 Definition. Let L be a first-order language and $\mathfrak{D} = (D, M)$ an interpretation appropriate for L. A *set* S of n-tuples from D (synonymously, *relation* $S(\vec{x}_n)$) is *first-order definable in $\mathfrak{D}$ over L*—simply put "definable in $\mathfrak{D}$" *if the language is understood*—iff for some formula A *of the language* L that has $x_1, \ldots, x_n$ as its free variables, we have, for all m_j in D ($j = 1, \ldots, n$):

$$S(m_1, \ldots, m_n)^2 \text{ iff } \models_{\mathfrak{D}} A[x_1 := m_1] \cdots [x_n := m_n]$$

It is usual to write "$A[\![t_1, \ldots, t_n]\!]$" for "$A[x_1 := t_1] \cdots [x_n := t_n]$" and any terms t_i as long as it is understood that the substituted variables are the x_i. A

---

[2]"Is true" is always implied in informal mathematics when a relation "$S(m_1, \ldots, m_n)$" is stated.

function f with inputs from D and outputs in D is definable in $\mathfrak{D}$ iff the relation $y = f(x_1, \ldots, x_n)$—known as the *graph* of f—is so definable. Some authors use the term "(first-order) *expressible*" (e.g., [48]) rather than "(first-order) definable" in an interpretation $\mathfrak{D}$. □

The above definition gives precision to statements such as "we code (or express) an informal statement (i.e., relation) $S(x_1, \ldots, x_n)$ into the formal language" or that "the (informal) statement $S(x_1, \ldots, x_n)$ can be *written* (or *made*) in the *formal language*". What *makes* the statement *in the formal language* is a first-order formula A that defines it in the sense of A.1.14.

Conversely, any first-order formula B, of n free variables, over a language L defines (in $\mathfrak{D}$) the *set* of n-tuples

$$\{(k_1, \ldots, k_n) : \models_{\mathfrak{D}} B[\![k_1, \ldots, k_n]\!]\}$$

If we call the above set (relation) R, then we can state that the formula B informally says "$R(x_1, \ldots, x_n)$" in the sense that, for all k_i in D, $R(k_1, \ldots, k_n)$ holds iff $\models_{\mathfrak{D}} B[\![k_1, \ldots, k_n]\!]$.

We next prove a few lemmata that lead to the proof of the soundness metatheorem.

A.1.15 Lemma. *Let* $\mathfrak{D} = (D, M)$ *be an interpretation for* L *and* A *be a* D-*formula that contains* no *free occurrences of* $\mathbf{x}$. *Consider a new interpretation* $\mathfrak{D}' = (D', M')$ *where* $D = D'$ *and* M' *agrees with* M *everywhere, except possibly at the variable* $\mathbf{x}$. *Then* $M(A) = M'(A)$. *This last "=" is, of course, on* $\{\mathbf{t}, \mathbf{f}\}$.

Proof. By induction on the complexity of formulae, mindful of Definitions A.1.6 and A.1.12. We note at the outset that according to the hypothesis, $M(\cdots) = M'(\cdots)$ for every symbol "$\cdots$" of the alphabet $\mathcal{V}(D)$ except, possibly, $\mathbf{x}$.

Atomic case:

(1) The D-formula A is one of $\mathbf{p}, \top, \bot$. Then $M(A) = M'(A)$.

(2) A is $t = s$. Thus

$$M(t = s) = \mathbf{t} \text{ iff } M(t) = M(s)$$
$$\text{iff } M'(t) = M'(s) \text{ by A.1.10}$$
$$\text{iff } M'(t = s) = \mathbf{t} \text{ by A.1.6}$$

(3) A is $\phi(t_1, \ldots, t_n)$. Thus, noting that $M(\phi) = M'(\phi)$,

$$M(\phi(t_1, \ldots, t_n)) = \mathbf{t} \text{ iff } M(\phi)(M(t_1), \ldots, M(t_n)) = \mathbf{t}$$
$$\text{iff } M'(\phi)(M'(t_1), \ldots, M'(t_n)) = \mathbf{t} \text{ by A.1.10}$$
$$\text{iff } M'(\phi(t_1, \ldots, t_n)) = \mathbf{t} \qquad \text{by A.1.6}$$

Nonatomic case:

(i) A is $\neg B$. The I.H. applies to B and yields $M(B) = M'(B)$ by A.1.12(i). We are done by A.1.6(IV).

(ii) A is $B \circ C$. The I.H. applies to B and C and yields $M(B) = M'(B)$ by A.1.12(ii). We are done by A.1.6(V).

(iii) A is $(\forall \mathbf{y})B$. The I.H. applies to $B[\mathbf{y} := i]$, for every $i \in D$, since $\mathbf{x}$ is not free in this D-formula (cf. comment immediately following A.1.12). Thus, we have $M(B[\mathbf{y} := i]) = M'(B[\mathbf{y} := i])$, for all $i \in D$. By (VI) of A.1.6 the above yields $M((\forall \mathbf{y})B) = M'((\forall \mathbf{y})B)$. $\qquad \square$

A.1.16 Lemma. *Let $\mathfrak{D} = (D, M)$ be an interpretation for L, t be a D-term, and $i \in D$. Consider a new interpretation $\mathfrak{D}' = (D', M')$ where $D = D'$ and M' agrees with M everywhere, except possibly at the variable $\mathbf{x}$: M' sets $M'(\mathbf{x}) = i$. Then $M(t[\mathbf{x} := i]) = M'(t)$.*

Proof. By induction on the complexity of the term t.
(1) t is $\mathbf{x}$. Then $M(t[\mathbf{x} := i]) = M(i) = i$ (cf. A.1.5), while $M'(t) = M'(\mathbf{x}) = i$.
(2) t is $\mathbf{y}$ (not $\mathbf{x}$), or is c or is $k \in D$. Then (cf. A.1.5)

$$M(t[\mathbf{x} := i]) = M(t) = \begin{cases} M(\mathbf{y}) & \text{thus, the same as } M'(\mathbf{y}) \\ M(c) & \text{thus, the same as } M'(c) \\ M(k) & \text{i.e., } k, \text{ thus, the same as } M'(k) \end{cases}$$

(3) t is $f(t_1, \ldots, t_n)$. Thus (cf. 4.1.28 and A.1.6)

$$
\begin{aligned}
M(t[\mathbf{x} := i]) &= M(f)\Big(M\big(t_1[\mathbf{x} := i]\big), \ldots, M\big(t_n[\mathbf{x} := i]\big)\Big) \\
&= M(f)\Big(M'(t_1), \ldots, M'(t_n)\Big) && \text{by I.H.} \\
&= M'(f)\Big(M'(t_1), \ldots, M'(t_n)\Big) && M \text{ and } M' \text{ agree on } f \\
&= M'(t) && \square
\end{aligned}
$$

A.1.17 Lemma. *Given a term t, distinct variables $\mathbf{x}$ and $\mathbf{y}$, where $\mathbf{y}$ does not occur in t, and a constant a, then, for any term s and formula A, $s[\mathbf{x} := t][\mathbf{y} := a]$ is the same as $s[\mathbf{y} := a][\mathbf{x} := t]$ and $A[\mathbf{x} := t][\mathbf{y} := a]$ is the same as $A[\mathbf{y} := a][\mathbf{x} := t]$.*

Proof. By induction on the complexity of t (done first) and A. Exercise A.1.18 asks you to fill in the details of the proof. $\qquad \square$

A.1.18 Exercise. Prove Lemma A.1.17. $\qquad \square$

A.1.19 Lemma. *Let $\mathfrak{D} = (D, M)$ be an interpretation for L, A be a D-formula, and $i \in D$. Consider a new interpretation $\mathfrak{D}' = (D', M')$ where $D = D'$ and M' agrees with M everywhere, except possibly at the variable $\mathbf{x}$: M' sets $M'(\mathbf{x}) = i$. Then $M(A[\mathbf{x} := i]) = M'(A)$.*

Proof. By induction on the complexity of formula A.

(1) A is one of $\mathbf{p}, \perp, \top, (\forall \mathbf{x})B$. Then (cf. 4.1.28, A.1.6, and A.1.15)

$$M(A[\mathbf{x} := i]) = M(A) = \begin{cases} M(\mathbf{p}) & \text{this is } M'(\mathbf{p}) \\ M(\perp) & \text{this is } M'(\perp) \\ M(\top) & \text{this is } M'(\top) \\ M((\forall \mathbf{x})B) & \text{this is } M'((\forall \mathbf{x})B) \text{ (by A.1.15)} \end{cases}$$

(2) A is $t = s$. Now $M(A[\mathbf{x} := i]) = \mathbf{t}$ iff $M(t[\mathbf{x} := i]) = M(s[\mathbf{x} := i])$. By A.1.16 the latter is true iff $M'(t) = M'(s)$, i.e., iff $M'(t = s) = \mathbf{t}$.

(3) A is $\phi(t_1, \ldots, t_n)$. $M(\phi(t_1, \ldots, t_n)[\mathbf{x} := i]) = \mathbf{t}$ iff $M(\phi)(M(t_1[\mathbf{x} := i]), \ldots, M(t_n[\mathbf{x} := i])) = \mathbf{t}$. By A.1.16 and $M(\phi) = M'(\phi)$ the latter is true iff $M'(\phi)(M'(t_1), \ldots, M'(t_n)) = \mathbf{t}$, i.e., iff $M'(\phi(t_1, \ldots, t_n)) = \mathbf{t}$.

Nonatomic case:

 (i) A is $\neg B$. By I.H. $M(B[\mathbf{x} := i]) = M'(B)$ and we are done by A.1.6(IV).

 (ii) A is $B \circ C$. By I.H. $M(B[\mathbf{x} := i]) = M'(B)$ and $M(C[\mathbf{x} := i]) = M'(C)$. We are done by A.1.6(V).

(iii) A is $(\forall \mathbf{y})B$ and $((\forall \mathbf{y})B)[\mathbf{x} := i]$ is $(\forall \mathbf{y})B[\mathbf{x} := i]$; the substitution being defined (cf. 4.1.28). Thus $M((\forall \mathbf{y})B[\mathbf{x} := i]) = \mathbf{t}$ iff

$$M(B[\mathbf{x} := i][\mathbf{y} := k]) = \mathbf{t}, \text{ for all } k \in D \qquad (*)$$

By Lemma A.1.17, $(*)$ is equivalent to

$$M(B[\mathbf{y} := k][\mathbf{x} := i]) = \mathbf{t}, \text{ for all } k \in D$$

and hence, by the I.H., to

$$M'(B[\mathbf{y} := k]) = \mathbf{t}, \text{ for all } k \in D \qquad (**)$$

By A.1.6(VI), $(**)$ is equivalent to $M'((\forall \mathbf{y})B) = \mathbf{t}$. $\qquad \square$

We will need one final lemma, to ease the handling of axiom groups **Ax2** and **Ax6** in the proof of A.1.21 below. This lemma embodies a mathematically rigorous formulation, and proof, of the remark made in 8.2.2 on p. 203: "The trickiest part to agree on is that the part $(A[\mathbf{x} := t])^{\mathfrak{D}}$ of (1) is $A_{\mathbf{x}}^{\mathfrak{D}}[\mathbf{x} := t^{\mathfrak{D}}] \ldots$"

In the present notation, we want to show that $M(A[\mathbf{x} := M(t)]) = M(A[\mathbf{x} := t])$. More pleasing intuitively is the notation introduced in Definition A.1.14: $M(A[\![M(t)]\!]) = M(A[\![t]\!])$.

A.1.20 Lemma. *Let $\mathfrak{D} = (D, M)$ be an interpretation for a language L and s, t, and A be D-terms and a D-formula respectively, while $M(t) = i$. Then $M\big(s[\mathbf{x} := t]\big) = M\big(s[\mathbf{x} := i]\big)$ and $M\big(A[\mathbf{x} := t]\big) = M\big(A[\mathbf{x} := i]\big)$.*

Proof. We first do induction on the complexity of s. If s is a constant or $\mathbf{y}$ ($\mathbf{y} \neq \mathbf{x}$), then both $s[\mathbf{x} := t]$ and $s[\mathbf{x} := i]$ are just s. Thus, $M\big(s[\mathbf{x} := t]\big) = M(s) = M\big(s[\mathbf{x} := i]\big)$. If s is $\mathbf{x}$, then $s[\mathbf{x} := t]$ is t while and $s[\mathbf{x} := i]$ is i. Thus, $M\big(s[\mathbf{x} := t]\big) = M(t) = i = M(i) = M\big(s[\mathbf{x} := i]\big)$.

For the induction step let s be $f(t_1, \ldots, t_n)$.

Then $M\big(s[\mathbf{x} := t]\big) = M(f)\big(M\big(t_1[\mathbf{x} := t]\big), \ldots, M\big(t_n[\mathbf{x} := t]\big)\big)$. By the I.H. this is $M(f)\big(M\big(t_1[\mathbf{x} := i]\big), \ldots, M\big(t_n[\mathbf{x} := i]\big)\big)$; that is, $M\big(s[\mathbf{x} := i]\big)$.

We next do an induction on the complexity of A. For the atomic case, the subcases where A is one of $\bot, \top, \mathbf{p}$ are trivial. If on the other hand A is $\phi(t_1, \ldots, t_n)$, then $M\big(A[\mathbf{x} := t]\big) = M(\phi)\big(M\big(t_1[\mathbf{x} := t]\big), \ldots, M\big(t_n[\mathbf{x} := t]\big)\big)$. By the case for terms, this is $M(\phi)\big(M\big(t_1[\mathbf{x} := i]\big), \ldots, M\big(t_n[\mathbf{x} := i]\big)\big)$; that is, $M\big(\phi[\mathbf{x} := i]\big)$. Similarly if A is $t = s$.

The induction step in the case of Boolean connectives being straightforward, let us do the induction step just in the case where A is $(\forall \mathbf{w})B$. If $\mathbf{w}$ is the same as $\mathbf{x}$, then the result is trivial. As usual, we assume that the noted substitutions are defined, otherwise there is nothing to prove. This entails, in particular, that either $\mathbf{w}$ does not occur in t (the interesting case) or that $\mathbf{x}$ is not free in B—this being the trivial case where both substitutions produce the same formula, $(\forall \mathbf{w})B$ (cf. 4.1.28 and 4.1.33). We display the interesting case.

$$
\begin{aligned}
M(A[\mathbf{x} := t]) = \mathbf{t} \text{ iff } & M\Big(\big((\forall \mathbf{w})B\big)[\mathbf{x} := t]\Big) = \mathbf{t} \\
\text{iff } & M\Big(\big((\forall \mathbf{w})B[\mathbf{x} := t]\big)\Big) = \mathbf{t} \\
\text{iff } & M\big(B[\mathbf{x} := t][\mathbf{w} := j]\big) = \mathbf{t} \text{ for all } j \in D, \text{ by A.1.6} \\
\text{iff } & M\big(B[\mathbf{w} := j][\mathbf{x} := t]\big) = \mathbf{t} \text{ for all } j \in D, \text{ by A.1.17} \\
\text{iff } & M\Big(\big(B[\mathbf{w} := j]\big)[\mathbf{x} := t]\Big) = \mathbf{t} \text{ for all } j \in D \\
\text{iff } & M\Big(\big(B[\mathbf{w} := j]\big)[\mathbf{x} := i]\Big) = \mathbf{t} \text{ for all } j \in D, \text{ by I.H.} \\
\text{iff } & M\big(B[\mathbf{w} := j][\mathbf{x} := i]\big) = \mathbf{t} \text{ for all } j \in D \\
\text{iff } & M\big(B[\mathbf{x} := i][\mathbf{w} := j]\big) = \mathbf{t} \text{ for all } j \in D, \text{ by A.1.17} \\
\text{iff } & M\Big(\big((\forall \mathbf{w})B[\mathbf{x} := i]\big)\Big) = \mathbf{t} \text{ by A.1.6} \\
\text{iff } & M\Big(\big((\forall \mathbf{w})B\big)[\mathbf{x} := i]\Big) = \mathbf{t} \\
\text{iff } & M(A[\mathbf{x} := i]) = \mathbf{t} \qquad \square
\end{aligned}
$$

We are ready to revisit soundness within the definition of Tarski semantics. "Our logic" will here be "logic 2" of Chapter 5.

A.1.21 Metatheorem. (Soundness in First-Order Logic) *Let Σ be any theory over a language L and let A be a formula of L. Then $\Sigma \vdash A$ implies $\Sigma \models A$.*

Proof. Assume $\Sigma \vdash A$. Then we have one of the three cases:

  (i) $A \in \Sigma$ (trivial).

  (ii) A is derived using MP.

  (iii) $A \in \Lambda_1$.

Toward case (ii) we show that the rule MP preserves truth. Let then $\mathfrak{D}$ be an arbitrary model of Σ, and B satisfy $B^{\mathfrak{D}} = \mathbf{t}$ and $(B \rightarrow A)^{\mathfrak{D}} = \mathbf{t}$. Thus $A^{\mathfrak{D}} = \mathbf{t}$ by Definition A.1.6, parts (IV) and (V).

Most of the work is for case (iii), and we might as well prove a bit more, namely, that $\models A$.

The main effort here is to show that each instance A of the schemata in **Ax1**–**Ax6** of Definition 4.2.1, *as they appear* in the list—i.e., *prior to prefixing universal quantifiers to form a partial generalization*—satisfies $\models A$.

However, we postpone this task until after we show that prefixing universal quantifiers preserves truth. All this completes the argument.

Generalization preserves truth; that is, if $\models A$, then $\models (\forall \mathbf{x})A$. This translates as "if $M(A) = \mathbf{t}$ in every interpretation $\mathfrak{D} = (D, M)$, then also $M((\forall \mathbf{x})A) = \mathbf{t}$ in every interpretation $\mathfrak{D} = (D, M)$". Well, suppose instead that

$$\models A \qquad\qquad (*)$$

yet, *for some interpretation $\mathfrak{D} = (D, M)$ we have $M((\forall \mathbf{x})A) = \mathbf{f}$.* By A.1.6, this says that *for some $i \in D$, we have*

$$M(A[\mathbf{x} := i]) = \mathbf{f} \qquad\qquad (**)$$

Choose now a new interpretation, $\mathfrak{D}' = (D', M')$ where $D = D'$ and $M = M'$ agree everywhere on the alphabet of L, except possibly at $\mathbf{x}$ where $M'(\mathbf{x}) = i$. By Lemma A.1.19, we have

$$M(A[\mathbf{x} := i]) = M'(A) \qquad\qquad (***)$$

But $M'(A) = \mathbf{t}$ by $(*)$, which contradicts $(**)$.

We return to case (iii).

Case of Ax1: $\models_{\text{taut}} A$. We pick an arbitrary $\mathfrak{D} = (D, M)$ and show that $M(A) = \mathbf{t}$. Let $\mathbf{p}_1, \ldots, \mathbf{p}_n$ be all the propositional variables that occur in A— including under *propositional variables* both those that are from the alphabet $\mathcal{V}$ of L and those that are actually prime subformulae.[3] Define a state v by setting $v(\mathbf{p}_i) = M(\mathbf{p}_i)$, for $i = 1, \ldots, n$. By induction on

---

[3]Prime subformulae were defined in 4.1.25.

the complexity of "P-formulae"—cf. Exercise 4 on p. 149—we show that $v(A) = M(A)$ for each P-formula A (i.e., for each $A \in$ **WFF**). The basis is settled at once by the way v was defined and since, by the definition of state, and A.1.5, v and M agree on $\top, \bot$. Let then A be $\neg B$. By I.H. it is $v(B) = M(B)$. Thus, by 1.3.5 (first "=") and A.1.6(IV) (last "=") we have $v(\neg B) = F_\neg(v(B)) = F_\neg(M(B)) = M(\neg B)$. If finally A is $B \circ C$ (where $\circ$ is as before), then the I.H. yields $v(B) = M(B)$ and $v(C) = M(C)$. Thus, by 1.3.5 (first "=") and A.1.6(V) (last "=") we have $v(B \circ C) = F_\circ(v(B), v(C)) = F_\circ(M(B), M(C)) = M(B \circ C)$.

Now, the assumption yields $v(A) = \mathbf{t}$. By the preceding result, we have $M(A) = \mathbf{t}$.

Case of Ax2: A is $(\forall \mathbf{x})B \to B[\mathbf{x} := t]$. We pick an arbitrary $\mathfrak{D} = (D, M)$ and show that $M(A) = \mathbf{t}$. As $M(A) = F_\to\big(M((\forall \mathbf{x})B), M(B[\mathbf{x} := t])\big)$ we need establish that if $M((\forall \mathbf{x})B) = \mathbf{t}$, then $M(B[\mathbf{x} := t]) = \mathbf{t}$. Well, the hypothesis yields (cf. A.1.6(VI)) that $M(B[\mathbf{x} := i]) = \mathbf{t}$, for all $i \in D$. In particular, $M(B[\mathbf{x} := M(t)]) = \mathbf{t}$. Hence $M(B[\mathbf{x} := t]) = \mathbf{t}$ by A.1.20.

Case of Ax3: A is $(\forall \mathbf{x})(B \to C) \to (\forall \mathbf{x})B \to (\forall \mathbf{x})C$. We pick an arbitrary $\mathfrak{D} = (D, M)$ and show that $M(A) = \mathbf{t}$. Thus, by the truth table for $F_\to$ (1.3.4) I assume $M((\forall \mathbf{x})(B \to C)) = \mathbf{t}$ and $M((\forall \mathbf{x})B) = \mathbf{t}$ and prove $M((\forall \mathbf{x})C) = \mathbf{t}$. The assumptions mean

$$B[\mathbf{x} := i] \to C[\mathbf{x} := i] = \mathbf{t}, \text{ for all } i \in D \qquad (*)$$

$$B[\mathbf{x} := i] = \mathbf{t}, \text{ for all } i \in D \qquad (**)$$

hence $C[\mathbf{x} := i] = \mathbf{t}$, for all $i \in D$, and we are done (A.1.6(VI)).

Case of Ax4: A is $B \to (\forall \mathbf{x})B$, where $\mathbf{x}$ is not free in B. As before, we work with an arbitrary $\mathfrak{D} = (D, M)$ and show that $M(A) = \mathbf{t}$. So assume $M(B) = \mathbf{t}$ and check whether $M((\forall \mathbf{x})B) = \mathbf{t}$. The latter will be so precisely if $M(B[\mathbf{x} := i]) = \mathbf{t}$, for all $i \in D$. But, by the assumption on $\mathbf{x}$, $B[\mathbf{x} := i]$ is just B (cf. 4.1.33).

Case of Ax5: We want $\mathbf{x}^\mathfrak{D} = \mathbf{x}^\mathfrak{D}$ (cf. A.1.6(II)) for any $\mathfrak{D}$. This is immediate as "$i = i$" holds in the metatheory.

Case of Ax6: A is $t = s \to (A[\mathbf{x} := t] \equiv A[\mathbf{x} := s])$. So, pick an arbitrary $\mathfrak{D} = (D, M)$ and assume $M(t = s) = \mathbf{t}$, that is, $M(t) = M(s)$. We want $M(A[\mathbf{x} := t] \equiv A[\mathbf{x} := s]) = \mathbf{t}$, that is (cf. $F_\equiv$ in 1.3.4), $M(A[\mathbf{x} := t]) = M(A[\mathbf{x} := s])$. By A.1.20 that last equality is equivalent to $M(A[\mathbf{x} := M(t)]) = M(A[\mathbf{x} := M(s)])$ and thus holds. □

A.1.22 Exercise. Prove that if a set of formulae Γ has a model (cf. A.1.9), then it is consistent. □

A.2 COMPLETENESS

For ages mathematicians were content with arguing informally as they were building the edifice that is mathematics, until they encountered the various paradoxes that an undisciplined informal approach entails, such as Russell's paradox. This led to a movement, major proponents of which were Whitehead and Russell ([56]), Hilbert ([22]), and Bourbaki ([2]), that advocated the construction, using the techniques of mathematics, of a robust "engine" with the help of which mathematicians could prove their various theorems within rigid, finitary processes (formal proofs), and with absolutely clear rules of reasoning and selection of assumptions. This "engine" is, of course, first-order logic. By "robustness" I refer to the *inherent* inability of this logic to derive contradictions. By "inherent" I mean here that the absolute (nonapplied) first-order logic—having no nonlogical axioms that can "go wrong"— is contradiction-free: Indeed, $\nvdash \bot$ (cf. also 2.6.6); otherwise, by the soundness metatheorem of the previous section (A.1.21), we obtain the absurd $\bot^{\mathfrak{D}} = \mathbf{t}$ in every interpretation $\mathfrak{D}$.

But how many "truths" can we prove within first-order logic? Does this logic tell the whole truth? Gödel proved that it does—the logic is *complete*—in this sense: If a formula is true in *every model* of a chosen set of assumptions (i.e., it is *semantically* implied by these assumptions; cf. A.1.9), then it is also *provable* from these assumptions. In this section we prove Gödel's completeness theorem for so-called *countable languages* L, i.e., languages over countable alphabets (see the discussion under the heading "A brief course on countable sets" on p. 224 below for the definition of *countable*). The proof given here is not Gödel's original ([15]) but is the "modernized" argument due to Leon Henkin ([20]).

The strategy for establishing the completeness of our logic, that is, the implication "if $\Sigma \models A$, then $\Sigma \vdash A$", is to prove the equivalent *contrapositive*: If

$$\Sigma \nvdash A \tag{1}$$

then

$$\Sigma \nvDash A \tag{2}$$

We will recall that on p. 116 we introduced the concept of an *applied first-order logic* or *theory*. It was stated that a theory is a toolbox consisting of

(i) A first-order language that has a hand-picked, specific, set of nonlogical symbols that is appropriate for the intended application (e.g., for set theory we just include the predicate $\in$; nothing else).

(ii) Special axioms that give the basic properties of the nonlogical symbols. Intuitively, these state selected fundamental relative truths that characterize the theory.

(iii) The logical axioms that are common to all theories. Intuitively, these state absolute truths *that are valid in all theories.*
(iv) Rules of inference.

It is standard practice in the literature to take most of these tools for granted and identify a theory with the set of its special axioms Σ, which subsume item (i) above. Thus, a theory is simply any set of formulae, Σ. Any such set, and therefore any theory, is called *consistent* precisely as we defined in the remark following 2.6.6: *if and only if it fails to prove at least one formula; equivalently, if and only if it fails to prove* $\perp$. Thus, the contrapositive formulation of the completeness theorem, (1), immediately yields that *a consistent theory has a model.*

To prove that (1) implies (2) we fix a countable first-order language L and a theory Σ over L that does *not* prove A, and proceed to *construct* a model $\mathfrak{D} = (D, M)$ for Σ that is *not* a model of A (cf. also the approach in the proof of 3.2.1).

To make the construction—which has substantial methodological overlap with the one involved in the proof of Post's theorem in Section 3.2—self-contained, I outline below, with straightforward informal proofs where needed, a number of facts from set theory that we will need.

A brief course in countable sets. A set A[4] is *countable*, if it is empty or (in the opposite case) if there is a way to arrange all its members—*possibly with repetitions*—in an *infinite* linear array, in a "row of locations", utilizing one location for each member of $\mathbb{N}$. Since it is allowed to repeatedly list any element of A, even infinitely many times, all *finite sets are countable*.

We can convert a two-dimensional enumeration

$$(m_{i,j}) \text{ for all } i,j \text{ in } \mathbb{N}$$

into a one-dimensional (one row) enumeration quite easily. The "linearization" or "unfolding" of the infinite matrix of rows is effected by walking along the arrows as follows:

$$
\begin{array}{cccc}
(0,0) & (0,1) & (0,2) & (0,3) \quad \ldots \\
& \nearrow & \nearrow & \nearrow \\
(1,0) & (1,1) & (1,2) \\
& \nearrow & \nearrow \\
(2,0) & (2,1) \\
& \nearrow \\
(3,0) \\
\vdots
\end{array}
$$

Suppose now that A is a countable set. It is clear that every subset of A is countable: If $\emptyset \neq B \subseteq A$, then we enumerate the elements of B as follows:

---

[4]This appendix often refers to sets by name. Much of the time such names are capital Latin letters, A, B, C, unless I refer to a set of formulae (Σ, Δ, Γ). The context will protect us from confusing these A, B, C for formulae.

Fix a member $b \in B$ that will be used as explained below. We now form two arrays, an auxiliary *array and a* target *array. The latter will contain an enumeration of B. Enumerate A by steps. In each step we produce the next member of A, say a, which we put in the first unused location of the auxiliary array. If $a \in B$, then we also place it in the first unused location of the target array. Otherwise we put b in that location at this step.*

Another fact that we will need is that if $A \neq \emptyset$ is countable, then it also has an enumeration where *every element appears infinitely often*. Indeed, fix an enumeration $a_0, a_1, a_2, \ldots$ of A. Form the infinite matrix whose every row is $a_0, a_1, a_2, \ldots$ and linearize the matrix in the manner we linearized $(m_{i,j})$for all i,j in $\mathbb{N}$ above. That is, enumerate A as follows:

$$a_0,$$
$$a_0, a_1,$$
$$a_0, a_1, a_2,$$
$$a_0, a_1, a_2, a_3,$$
$$\ldots$$

Examples of nonempty countable sets are: any finite set; $\mathbb{N}$; the set of all even integers; the set of all integers;[5] the set of all nonnegative rational numbers.[6]

Let now A be some countable non empty set (possibly infinite), and fix an enumeration of it, $a_0, a_1, \ldots$ For example, A might be the alphabet $\mathcal{V}$ of a countable first-order language. The set of all strings of length two over A is the set of all $a_i a_j$ for $i \geq 0, j \geq 0$, and is linearizable in the manner of $(m_{i,j})$for all i,j in $\mathbb{N}$. That is, the set of all such strings is countable. This extends to strings of any length, as we can see via simple induction. Fix an $n \geq 0$ and assume that the set of strings of length n is countable, with an enumeration $d_0, d_1, \ldots$ But then so is the set of strings of length $n + 1$, since these are all the strings of the form $d_i a_j$, for $i \geq 0, j \geq 0$.

But how about the set of *all* strings over A? This is countable too! Indeed, let $a_0^n, a_1^n, a_4^n, \ldots$ be an infinite enumeration of all strings of length $n \geq 0$. Then we can enumerate all strings by linearizing the infinite matrix a_j^i, for $i \geq 0, j \geq 0$.[7] Note that the first row, a_j^0, for $j \geq 0$, consists of the empty string, ϵ, everywhere.

---

[5]Think of an integer as a pair (n, m), where $m \in \mathbb{N}$ and $n = 0$ or $n = 1$. The intention is to have $(0, m)$ "code" (stand for) m while $(1, m)$ code $-m$. We know that the set of *all* (n, m) for $n \geq 0, m \geq 0$ can be linearized just as the matrix above was. The set of all integers is (coded as) a subset of this matrix.

[6]Think of such a rational as a pair (n, m), where $n \in \mathbb{N}$ and $m \in \mathbb{N}$ with $m \neq 0$ so that (n, m) stands for n/m. The set of all (n, m) for $n \geq 0, m \geq 0$ can be linearized just as the matrix above was. The set of the nonnegative rationals is (coded as) a subset of this matrix.

[7]There is some esoteric small print here: In the proof of the countability of the set of *all strings* we tacitly used a set-theoretical principle known as the *axiom of choice*. It enables us to make mathematical constructions that involve an infinite set of selections from a set in the absence of a "precise rule" that lets us *specify* these selections. In our case, out of many possible enumerations for each string length, we chose one, *for each* $n \geq 0$. Even more esoteric is a result of Feferman and Levy that shows the *necessity* of the axiom of choice principle if one wants, in general, to show that the *union* of a countable set—$A_0, A_1, A_2, \ldots$—of countable sets is itself countable ([14]). The said union is denoted by $\bigcup_{n \geq 0} A_n$ and means the set S with an "entrance condition" $x \in S$ that is equivalent to "for some $i, x \in A_i$". That is, S contains all the objects found in all the A_i and nothing else.

We now turn to the construction of the model we announced prior to our preceding digression into the properties of countable sets. At first we pick an infinite countable set N—for example, we can take N to be $\mathbb{N}$—and fix an enumeration $m_0, m_1, m_2, \ldots$ of N. As usual, $\mathbf{WFF}(N)$ is the set of all N-formulae over L. We note that $\Sigma \subseteq \mathbf{WFF}(N)$. We next define an *extension* Γ of Σ—which simply means a superset of Σ, i.e., $\Sigma \subseteq \Gamma \subseteq \mathbf{WFF}(N)$). Γ will have a number of key properties that will lead to the *Main Lemma* (A.2.4).

To this end, we fix attention on an enumeration $G_0, G_1, G_2, \ldots$ of all formulae of $\mathbf{WFF}(N)$, and on an enumeration $E_0, E_1, \ldots$ of all "existential" formulae among the G_i, that is, those that have the form $(\exists \mathbf{x})B$. We assume without loss of generality that each formula E_i *occurs infinitely often in the list*.

Pause. Can we do this? For sure: $\mathbf{WFF}(N)$ is a subset of the countable set of all strings over $\mathcal{V}(N)$, thus it is countable itself. In turn, the subset of $\mathbf{WFF}(N)$ that contains all the "existential" formulae $(\exists \mathbf{x})B$, but just those, is countable, and by the results in our "course in countable sets" can be enumerated so that every member is repeated infinitely often.

We can now define a sequence $\Gamma_0, \Gamma_1, \ldots$ by recursion, in two steps, using the intermediate Δ_n-sequence that we define in parallel: Let $\Gamma_0 = \Sigma$, and for $n = 0, 1, \ldots$ let

$$\Delta_n = \begin{cases} \Gamma_n \cup \{G_n\} & \text{if } \Gamma_n \cup \{G_n\} \not\vdash A \\ \Gamma_n \cup \{\neg G_n\} & \text{otherwise} \end{cases} \tag{3}$$

This is almost identical to the construction in the proof of 3.2.1 (see also Claim Four on p. 96, and (i) below). We next let

$$\Gamma_{n+1} = \begin{cases} \Delta_n \cup \{B[\mathbf{x} := c]\} & \text{if } \Delta_n \vdash E_n \text{ where } E_n \text{ is } (\exists \mathbf{x})B \\ \Delta_n & \text{otherwise} \end{cases} \tag{4}$$

In (4) we choose the so-called *Henkin constant* $c \in N$ so that $c = m_i$ where i is *the smallest* such that m_i *does not occur* in any of $A, G_0, \ldots, G_n, E_0, \ldots, E_n$.

We now note a set of properties of the Γ_n sequence of formulae:

(i) If $\Gamma_n \not\vdash A$, then $\Delta_n \not\vdash A$. This is clear if Δ_n is given by the top case. Otherwise, we have $\Gamma_n \cup \{G_n\} \vdash A$, which precludes $\Gamma_n \cup \{\neg G_n\} \vdash A$ (see the analogous argument in Claim Four on p. 96).

(ii) If $\Delta_n \not\vdash A$, then $\Gamma_{n+1} \not\vdash A$. Let $\Gamma_{n+1} \vdash A$ instead. Thus, $\Delta_n \cup \{B[\mathbf{x} := c]\} \vdash A$. As c cannot occur in Σ (why?), and $\Delta_n \vdash (\exists \mathbf{x})B$, the conditions of 6.5.19 apply and we have $\Delta_n \vdash A$; a contradiction.

(iii) For every $n \geq 0$, $\Gamma_n \not\vdash A$. By trivial induction on n, via (i) and (ii), given that $\Gamma_0 = \Sigma \not\vdash A$.

We now define Γ by

$$\Gamma = \bigcup_{n \geq 0} \Gamma_n \tag{5}$$

(iv) $\Gamma \not\vdash A$. If not, let $B_1, \ldots, B_n$ be all the formulae of Γ used in a proof that certifies $\Gamma \vdash A$. Let Γ_m, for appropriate m, contain all the B_i. Then $\Gamma_m \vdash A$, contradicting (iii).

(v) For every $B \in \mathbf{WFF}(N)$, either $B \in \Gamma$ or $(\neg B) \in \Gamma$, *but not both*. Indeed, each B is some G_i. By (3) and (4), at stage i of the construction, one of B_i or $\neg B_i$ is placed in Γ_{i+1}. If at different stages (why "different"?) both some B and $\neg B$ enter in Γ, then, by 2.5.7, $\Gamma \vdash \bot$, and hence (cf. 2.6.6) $\Gamma \vdash A$, contradicting (iv).

(vi) Γ is a *maximal consistent theory* in that whenever $B \notin \Gamma$, then $\Gamma \cup \{B\}$ is inconsistent. Indeed, if $B \notin \Gamma$, then $(\neg B) \in \Gamma$, and hence $\Gamma \cup \{B\} \vdash \neg B$. But $\Gamma \cup \{B\} \vdash B$, too.

(vii) Γ is *deductively closed*; that is, $\Gamma \vdash B$ implies $B \in \Gamma$. Otherwise, by maximal consistency (cf. 2.6.6), we get $\Gamma \cup \{B\} \vdash A$; hence $\Gamma \vdash A$ (2.1.6), contradicting (iv).

(viii) $\Lambda_1 \subseteq \Gamma$. Indeed, by deductive closure and 2.1.1, $\vdash B$ entails $\Gamma \vdash B$ for any B.

(ix) Γ is an *N-Henkin theory*, meaning that, $\Gamma \vdash (\exists \mathbf{x})B$ implies, for some $k \in N$, $\Gamma \vdash B[\mathbf{x} := k]$. Indeed, let $\Gamma \vdash (\exists \mathbf{x})B$, where $(\exists \mathbf{x})B$ is E_n for some n. By (vii), $E_n \in \Gamma_m$ and hence, by (3), $E_n \in \Delta_m$ for some m. *Without loss of generality we may assume $m = n$.* Indeed,

Case 1: If $m < n$ is our original situation, then note that $\Delta_m \subseteq \Delta_n$

Case 2: If $m > n$ is our original situation, then, as $(\exists \mathbf{x})B$ is enumerated as an E_i infinitely often, take an $i > m$ such that E_i is still $(\exists \mathbf{x})B$. But now we are back to **Case 1**.

Thus, by (4), $B[\mathbf{x} := k]$ is placed in Γ_{m+1} for some $k \in N$.

So far, the construction presented and its properties track fairly closely the one given in Section 3.2, and its properties. The primary design aim in either construction was to construct a maximal consistent theory Γ that contains Σ—but fails to prove some given formula A—and use the theory's properties to construct a model for Σ. But what is (ix) good for? In informal mathematics the truth of an existential statement such as $(\exists \mathbf{x})B$, where B has at most only $\mathbf{x}$ free, entails the existence of an object (constant) c that makes $B[\mathbf{x} := c]$ true—this is *the semantics of* $\exists$. *This phenomenon is not replicated in formal logic in general*; that is, $(\exists \mathbf{x})B \to B[\mathbf{x} := c]$ is not a theorem schema. See Exercise A.2.13. Henkin's construction in (4), which has as a consequence (ix) above, is the additional work we have to do to make the huge (cf. (v) and (vi)) theory Γ behave, with respect to the quantifier $\exists$, according to what the quantifier's semantics dictate. Such behavior in turn supports our task to build a model of Σ whose informal logical properties we can mimic formally within the theory Γ.

The final concern is to allow Γ to handle constants in the domain of the (not-yet-constructed) model of Σ with some of the effectiveness informal mathematics can exhibit. In particular, we will need one more property in order to finally define our model: that Γ can *distinguish constants*; that is, if $m \neq n$ in N, metamathematically speaking, then Γ *certifies* this by proving $\neg m = n$, which by (vii) means that the "certificate" is just $(\neg m = n) \in \Gamma$. But this is not necessarily true for the arbitrarily chosen N! However, we can make it happen for a "smaller" set D obtained by judiciously discarding members of N. And this involves a few more technical steps!

We start by defining the "equality class of n" for each $n \in N$:

$$e(n) = \{m \in N : \Gamma \vdash m = n\} \tag{6}$$

The $e(n)$ sets have some interesting properties:

e1. $n \in e(n)$: Indeed, $\Gamma \vdash n = n$ by **Ax5** and 6.1.19.

e2. If $e(n)$ and $e(m)$ both contain the same i from N, then $e(n) = e(m)$: First, note that the hypothesis translates to $\Gamma \vdash i = n$ and $\Gamma \vdash i = m$. By 7.0.1 via 6.1.19 we get

$$\Gamma \vdash n = m \tag{7}$$

and

$$\Gamma \vdash m = n \tag{7'}$$

If now $k \in e(n)$, then $\Gamma \vdash k = n$; hence $\Gamma \vdash k = m$ by (7) and 7.0.1 via 6.1.19. That is, $k \in e(m)$. The converse follows using $(7')$ instead.

e3. If $k \in e(n)$, then $e(k) = e(n)$: by **e1** and **e2**.

e4. If $e(k) = e(n)$, then $\Gamma \vdash k = n$: by assumption and **e1**, $k \in e(n)$. We now invoke (6).

We next define the set

$$D = \{\min e(n) : n \in N\} \tag{8}$$

where "$\min S$" for any $\emptyset \neq S \subseteq N$ denotes the element m_i of S that has *the smallest index i* in the fixed (in our discussion) enumeration $m_0, m_1, \dots$ of N (cf. p. 226).

By **e1**, the selection "$\min e(n)$" is always possible. Let next $k \neq n$ in D. Thus, $k = \min e(i)$ and $n = \min e(j)$ for some i, j in N. It follows that $e(k) = e(i)$ and $e(n) = e(j)$ by **e3**. Can we have $\Gamma \vdash k = n$? If yes, then $k \in e(n)$ by (6); hence $e(k) = e(n)$ by **e3**. We conclude that $e(i) = e(j)$ and therefore this set has *two distinct elements*, k and n, of *smallest index* in the $(m_l)_{l \geq 0}$ enumeration of N, which is absurd![8] We have:

A.2.1 Lemma. Γ *distinguishes the members of D, that is, $m \neq n$ in D implies $\Gamma \vdash \neg m \neq n$.*

---

[8]The reader who is conversant with the concepts of *equivalence relations* and *equivalence classes* will recognize the relation between n and m in N given by $\Gamma \vdash n = m$ as an equivalence relation, i.e., one that is reflexive, symmetric, and transitive. He will also recognize the set $e(n)$ as the equivalence class with representative n. We avoided this terminology in the interest of those readers who have not seen these concepts before.

A.2.2 Lemma. Γ *is an D-Henkin theory (cf. (ix)).*

Proof. So let $\Gamma \vdash (\exists\mathbf{x})B$. We know from (ix) that for some $k \in N$, we have $\Gamma \vdash B[\mathbf{x} := k]$. Let $n = \min e(k)$. Then $n \in D$ and $\Gamma \vdash n = k$ (by (6) and (8)). By **Ax6**, $\Gamma \vdash B[\mathbf{x} := n] \equiv B[\mathbf{x} := k]$; hence $\Gamma \vdash B[\mathbf{x} := n]$. □

A.2.3 Exercise. $D \neq \emptyset$. □

In summary, we have the statement below:

A.2.4 Lemma. (Main Semantic Lemma) *If Σ is a theory over a first-order language L that cannot prove A (also over L), and N is any infinite countable set, then there is a nonempty subset $D \subseteq N$ and a D-Henkin theory $\Gamma \subseteq \mathbf{WFF}(N)$ that extends Σ—that is, $\Sigma \subseteq \Gamma$—which distinguishes the constants of D (in the sense of A.2.1) and satisfies (iv)–(viii) as well.*

A.2.5 Remark. We continue working with L, Σ, N, D, Γ as in A.2.4. By **Ax5** and 6.1.19, $\vdash t = t$ for any N-term t; hence $\Gamma \vdash t = t$. By 6.5.2 we have $\Gamma \vdash (\exists\mathbf{x})\mathbf{x} = t$. Thus, $\Gamma \vdash m = t$ for some $m \in D$, since Γ is D-Henkin. This m is *unique* in D for, if not, then we also have $\Gamma \vdash n = t$, for some $n \in D$ where $n \neq m$. By 7.0.1, $\Gamma \vdash m = n$. But A.2.1 yields $\Gamma \vdash \neg m = n$, contradicting consistency of Γ (iv). □

We are ready to define an interpretation $\mathfrak{D} = (D, M)$ of L:

A.2.6 Definition. We start with a consistent theory Σ over a first-order language L, and an arbitrary countable infinite set N. We will actually define the interpretation $\mathfrak{D} = (D, M)$ for the augmented language L', which has as alphabet $\mathcal{V}' = \mathcal{V} \cup (N - D)$.[9] Trivially, this will induce an interpretation for L itself: All we have to do is to forget that we have interpreted also the constants from $N - D$. In what follows we faithfully track Definition A.1.5. We thus give meaning to all elements of $\mathcal{V}'(D)$:

(1) For each object variable $\mathbf{y}$ from L' we have $\Gamma \vdash m = \mathbf{y}$ for a unique $m \in D$ (cf. A.2.5). We set $M(\mathbf{y}) = m$.

(2) For each Boolean variable $\mathbf{q}$ from L', we set $M(\mathbf{q}) = \mathbf{t}$ iff $\mathbf{q} \in \Gamma$.

(3) We set $M(\bot) = \mathbf{f}$ and $M(\top) = \mathbf{t}$.

(4) For each constant symbol c from L' we have $\Gamma \vdash m = c$ for a unique $m \in D$ by Remark A.2.5. We set $M(c) = m$.

 Note that if $c \in D$, then—m being also in D—$c = m$ by an argument based on the concluding remarks (uniqueness) in A.2.5. This is as it should be! (Cf. A.1.5(5).)

---

[9]Recall that $N - D$, set difference, denotes all the members of N that are not in D.

(5) Let f be a function symbol from L' of arity $k > 0$. We want to specify a "concrete" (metamathematical) $M(f)$. We do so by specifying what inputs (from D) generate what outputs (in D) under this $M(f)$. Let then $m_1, \ldots, m_k \in D$. By A.2.5, $\Gamma \vdash m = f(m_1, \ldots, m_k)$ for a unique $m \in D$. Thus we *define* $M(f)(m_1, \ldots, m_k)$—the output—to be this unique m: $M(f)(m_1, \ldots, m_k) = m$.

(6) For each k-ary predicate symbol ϕ from L', we let $M(\phi)$ be the metamathematical relation that has the following input/output behavior: $M(\phi)(n_1, \ldots, n_k) = \mathbf{t}$ iff $\phi(n_1, \ldots, n_k) \in \Gamma$. $\square$

We will prove that $\mathfrak{D}$ is a model of Σ with the help of two lemmata.

A.2.7 Lemma. *For every D-term t over L' (same as N-term over $L!$) $\Gamma \vdash t = m$ if $m = M(t)$.*[10]

Proof. By induction on the complexity of t: If t is a variable or constant we are done by (1) and (4) of A.2.6 (being mindful of 7.0.1). Suppose that t is $f(t_1, \ldots, t_n)$. By the I.H. we have

$$\Gamma \vdash t_i = k_i, \text{ where } k_i = M(t_i), \text{ for } i = 1, \ldots, n$$

Thus, by 7.0.7, and since k_i are also formal names of constants,

$$\Gamma \vdash f(t_1, \ldots, t_n) = f(k_1, \ldots, k_n) \tag{$*$}$$

Now, $M(t) = M(f)\big(M(t_1), \ldots, M(t_n)\big)$ by A.1.6. In other words, $M(t) = M(f)(k_1, \ldots, k_n)$, say, $= j$. By A.2.6(5), $\Gamma \vdash j = f(k_1, \ldots, k_n)$, which by 7.0.1 and $(*)$ yields $\Gamma \vdash f(t_1, \ldots, t_n) = j$. $\square$

A.2.8 Lemma. *For every D-formula B over L', $M(B) = \mathbf{t}$ iff $B \in \Gamma$.*

Proof. By deductive closure of Γ (cf. (vii), p. 227), we can use $B \in \Gamma$ and $\Gamma \vdash B$ interchangeably. We use induction on the complexity of B (cf. A.1.6).

The atomic cases:

(i) B is $\mathbf{p}$. We are done by A.2.6(2).

(ii) B is $\top$. As $M(\top) = \mathbf{t}$ (A.2.6(3)), we need to show $\top \in \Gamma$. This is so by (viii) on p. 227.

(iii) B is $\bot$. As $M(\top) = \mathbf{f}$ (A.2.6(3)), we need to show $\bot \notin \Gamma$. This is so by (iv) on p. 227 (cf. also 2.6.6).

---

[10]This "$\Gamma \vdash t = m$ if $m = M(t)$" may appear a bit roundabout. Why not just say "$\Gamma \vdash t = M(t)$"? Well, we allowed letters such as i, j, k, m, n to have a dual role, as *formal names* of constants imported into L' from D and as *informal names* of members of D. However, we have made no agreements to adopt notation such as "$M(\ldots)$" formally, nor will we. M remains a symbol outside our formal alphabet $\mathcal{V}'$.

(iv) B is $t = s$. In one direction, let $M(t = s) = \mathbf{t}$, that is (A.1.6), $M(t) = M(s)$. Let us call i this member of D. By A.2.7 we have $\Gamma \vdash t = i$ and $\Gamma \vdash s = i$, hence (7.0.1) $\Gamma \vdash t = s$.

In the other direction we start with what we have just concluded. Then A.2.7 yields $\Gamma \vdash t = i$ and $\Gamma \vdash s = j$, where $i = M(t)$ and $j = M(s)$. By 7.0.1, $\Gamma \vdash i = j$. This entails $i = j$ metamathematically (in D)—and hence $M(t = s) = \mathbf{t}$—since otherwise $\Gamma \vdash \neg i = j$ (A.2.1) contradicting Γ's consistency.

(v) B is $\phi(t_1, \ldots, t_n)$. Let $k_i = M(t_i)$ $(i = 1, \ldots, n)$.

By A.2.6(6), $M(\phi)(k_1, \ldots, k_n) = \mathbf{t}$ iff $\phi(k_1, \ldots, k_n) \in \Gamma$ iff

$$\Gamma \vdash \phi(k_1, \ldots, k_n) \tag{$**$}$$

By repeated application of **Ax6** and A.2.7—the latter yielding $\Gamma \vdash t_i = k_i$ $(i = 1, \ldots, n)$—we see that $(**)$ is equivalent to $\Gamma \vdash \phi(t_1, \ldots, t_n)$.

The nonatomic cases:

If B is any of $\neg C$ or $C \circ D$ for $\circ \in \{\wedge, \vee, \rightarrow, \equiv\}$, then the argument is precisely the same as the one given in the proof of 3.2.1, under "Main Claim" (p. 97). Thus we will consider here only the case where B is $(\forall \mathbf{x})C$. Let then $M\big((\forall \mathbf{x})C\big) = \mathbf{t}$. Thus (A.1.6), $M\big(C[\mathbf{x} := i]\big) = \mathbf{t}$, for all $i \in D$. By the I.H. we have

$$C[\mathbf{x} := i] \in \Gamma, \text{ for all } i \in D \tag{$***$}$$

We want to conclude that $(\forall \mathbf{x})C \in \Gamma$. If not, then $\neg(\forall \mathbf{x})C$ is in Γ ((v), p. 227), which via 3.2.1 and the definition of $\exists$, is equivalent to $(\exists \mathbf{x})\neg C \in \Gamma$. By the D-Henkin property of Γ, for some $k \in D$ we have $\neg C[\mathbf{x} := k]$ in Γ, which along with $(***)$ contradicts the consistency of Γ ((iv), p. 227).

Conversely, let $(\forall \mathbf{x})C \in \Gamma$. Then (6.1.5 and deductive closure of Γ), $C[\mathbf{x} := i] \in \Gamma$, for all $i \in D$. By the I.H. $M\big(C[\mathbf{x} := i]\big) = \mathbf{t}$, for all $i \in D$; hence $M\big((\forall \mathbf{x})C\big) = \mathbf{t}$ by A.1.6. $\qquad\square$

Thus, by $\Sigma \subseteq \Gamma$ and the just proved A.2.8, $M(B) = \mathbf{t}$ for all $B \in \Sigma$, while $M(A) = \mathbf{f}$ by (iv) of p. 227. In summary,

A.2.9 Lemma. *With Σ and A as given ((1) on p. 223) and $\mathfrak{D}$ as constructed, $\mathfrak{D}$ is a model of Σ, but not of A. Thus $\Sigma \not\models A$.*

A.2.10 Corollary. (The Consistency Theorem for First-Order Logic) *Every consistent theory has a model.*

Proof. Start with a consistent Σ. Thus for some A in its language, $\Sigma \not\vdash A$. $\qquad\square$

A.2.11 Metatheorem. (Gödel's Completeness Theorem) *Given a theory Σ over a first-order language L. If A is a formula over L, then $\Sigma \models A$ implies $\Sigma \vdash A$.*

Proof. We have already shown that if $\Sigma \not\vdash A$, then $\Sigma \not\models A$. $\qquad\square$

A.2.12 Exercise. (Compactness of First-Order Logic) Let Σ be a *finitely satisfiable* theory over some language L, that is, every *finite* subset of Σ is satisfiable (A.1.9). Prove that the entire Σ is also satisfiable. □

A.2.13 Exercise. Let L be a first-order language with only one constant, c. Show that $\not\models (\exists \mathbf{x})B \to B[\mathbf{x} := c]$. □

A.3   A BRIEF THEORY OF COMPUTABILITY

Computability is the part of logic that gives a mathematically precise formulation to the concepts *algorithm*, *mechanical procedure*, and *calculable function* (or relation). Its advent was strongly motivated, in the 1930s, by Hilbert's program, in particular by his belief that the *Entscheidungsproblem*, or *decision problem*, for axiomatic theories, that is, the problem "Is this formula a theorem of that theory?" was solvable by a mechanical procedure that was yet to be discovered.

Now, since antiquity, mathematicians have invented "mechanical procedures", e.g., Euclid's algorithm for the "greatest common divisor",[11] and had no problem recognizing such procedures when they encountered them. But how do you mathematically *prove* the *nonexistence* of such a mechanical procedure for a particular problem? You need a *mathematical formulation* of what *is* a "mechanical procedure" in order to do that!

Intensive activity by many (Post [37, 38], Kleene [26], Church [4], Turing [55], Markov [34]) led in the 1930s to several alternative formulations, each purporting to mathematically characterize the concepts *algorithm*, *mechanical procedure*, and *calculable function*. All these formulations were quickly proved to be equivalent; that is, the calculable functions admitted by any one of them were the same as those that were admitted by any other. This led Alonzo Church to formulate his conjecture, famously known as "Church's Thesis", that any *intuitively* calculable function is also calculable within any of these mathematical frameworks of calculability or computability.[12]

By the way, Church proved ([3, 4]) that Hilbert's *Entscheidungsproblem* admits no solution by functions that are calculable within any of the known mathematical frameworks of computability. Thus, if we accept his "thesis", the Entscheidungsproblem admits no algorithmic solution, period!

The eventual introduction of computers further fueled the study of and research on the various mathematical frameworks of computation, "models of computation" as we often say, and "computability" is nowadays a vibrant and very extensive field. The model of computation that I will present here, due to Shepherdson and Sturgis

---

[11] That is, the largest positive integer that is a common divisor of two given integers.

[12] I stress that even if this sounds like a "completeness *theorem*" in the realm of computability, it is not. It is just an empirical belief, rather than a provable result. For example, Péter [36] and Kalmár [25], have argued that it is conceivable that the intuitive concept of calculability may in the future be extended so much as to transcend the power of the various mathematical models of computation that we currently know.

[44], is a later model that has been informed by developments in computer science, in particular by the advent of so-called *high-level*[13] programming languages.

A.3.1   A Programming Framework for Computable Functions

So, what *is* a computable function, mathematically speaking? There are two main ways to approach this question. One is to define a programming formalism—that is, a programming language—and say "a function is computable precisely if it can be 'programmed' in the programming language". Such programming languages are the *Turing Machines* (or TMs) of Turing and the *unbounded register machines* (or URMs) of Shepherdson and Sturgis. Note that the term *machine* in each case is a misnomer, as both the TM and the URM formulations are really programming languages, the first being very much like assembly language of "real" computers, the latter reminding us more of (subsets of) Algol (or Pascal).

The other main way is to define a set of computable functions inductively, starting with some initial functions, and allowing the iteration of function-building operations to build all the remaining functions of the set. This approach (originally due to Dedekind [8] for what we nowadays call *primitive recursive functions*, and later due to Kleene [26] for what we nowadays call *partial recursive functions*) is very elegant, but is less intuitively immediate, whereas the programming approach has the attraction of being natural to those who have done some programming.

We now embark on defining the high-level programming language *URM*. The alphabet of the language is

$$\leftarrow, +, \dot{-}, :, X, 0, 1, 2, 3, 4, 5, 6, 7, 8, 9, \textbf{if}, \textbf{else}, \textbf{goto}, \textbf{stop} \tag{1}$$

Just like any other high level programming language, URM manipulates the contents of *variables*. However, these are restricted to be of *natural number type*—i.e., the only type of data such variables can denote (or "hold", or "contain", in programming jargon) are members of $\mathbb{N}$. Since this programming language is for theoretical considerations only—rather than practical implementation—every variable is allowed to hold any natural number whatsoever, hence the "UR" in the language name ("unbounded register", used synonymously with *variable of unbounded capacity*).

The syntax of the variables is simple: A variable (name) is a string that starts with X and continues with one or more 1:

$$\text{URM variable set:} \quad X1, X11, X111, X1111, \ldots \tag{2}$$

Nevertheless, as we have been doing in the case of first-order languages, we will more conveniently utilize the bold face lower case letters $\textbf{x}, \textbf{y}, \textbf{z}, \textbf{u}, \textbf{v}, \textbf{w}$, with or without subscripts or primes as metavariables in our discussions of the URM, and in examples of programs.

Rather than employing "BNF" notation to define the language (cf. p. 17)—that is, the syntax of URM programs—I will simply say that a URM program is a finite

---

[13]The level is "higher" the more the programming language is distanced from machine-dependent details.

(ordered) sequence of instructions (or commands) of the following five types:

$$L : \mathbf{x} \leftarrow a$$
$$L : \mathbf{x} \leftarrow \mathbf{x} + 1$$
$$L : \mathbf{x} \leftarrow \mathbf{x} \div 1 \tag{3}$$
$$L : \mathbf{stop}$$
$$L : \text{if } \mathbf{x} = 0 \text{ goto } M \text{ else goto } R$$

where L, M, R, a, written in decimal notation, are in $\mathbb{N}$, and $\mathbf{x}$ is some variable. We call instructions of the last type *if-statements*.

Each instruction in a URM program must be numbered by its *position number*, L, in the program—":" separating the position number from the instruction. We call these numbers *labels*. Thus, the label of the first instruction is always "1". The instruction **stop** must occur only once in a program, as the last instruction.

The semantics of each command is given in the context of a URM *computation*. The latter we will let have its intuitive meaning in this subsection, and we will defer a mathematical definition until Subsection A.3.3, where such a definition will be needed.

Thus, *for now*, a computation is the process that cycles along the instructions of a program, during which process each instruction that is visited upon—the *current instruction*—causes an *action* that we usually term "the result of the execution" of the instruction. I said "cycles along" because instructions of the last two types (may) cause the computation to loop back or cycle, revisiting an instruction that was already visited by the computation.

Every computation begins with the instruction labeled "1" as the *current* instruction. The semantic action of instructions of each type is defined if and only if they are current, and is as follows:

(i) $L : \mathbf{x} \leftarrow a$. Action: The value of $\mathbf{x}$ becomes the (natural) number a. Instruction $L + 1$ will be the next current instruction.

(ii) $L : \mathbf{x} \leftarrow \mathbf{x} + 1$. Action: This causes the value of $\mathbf{x}$ to increase by 1. The instruction labeled $L + 1$ will be the next current instruction.

(iii) $L : \mathbf{x} \leftarrow \mathbf{x} \div 1$. Action: This causes the value of $\mathbf{x}$ to decrease by 1, *if* it was originally nonzero. Otherwise it remains 0. The instruction labeled $L + 1$ will be the next current instruction.

(iv) $L : \mathbf{stop}$. Action: No variable (referenced in the program) changes value. The next current instruction is still the one labeled L.

(v) $L : \text{if } \mathbf{x} = 0 \text{ goto } M \text{ else goto } R$. Action: No variable (referenced in the program) changes value. The next current instruction is numbered M if $\mathbf{x} = 0$; otherwise it is numbered R.

 This command is syntactically illegal (meaningless) if any of M or R exceed the label of the program's **stop** instruction.

We say that a computation *terminates,* or *halts,* iff it ever makes (as we say "reaches") the instruction **stop** current. Note that the semantics of "L : **stop**" *appear* to require the computation to continue *ad infinitum,* but it does so in a trivial manner where no variable changes value, and the current instruction remains the same: Practically, the computation is over.

One usually gives names to URM programs, or as we just say, "to URMs", such as M, N, P, Q, R, F, H, G.

A.3.1 Definition. (Computing a Function) We say that a URM, M, *computes* a function f of n arguments *provided*—for some choice of variables $x_1, \ldots, x_n$ of M that we designate as *input variables* and a choice of a variable y that we designate as *the output variable*—the following precise conditions hold for *every choice* of input sequence (or "n-tuple"), $a_1, \ldots, a_n$ from $\mathbb{N}$:

(1) We *initialize the computation,* by doing two things:

    (a) We *initialize* the input variables with the input values $a_1, \ldots, a_n$. We *initialize all other variables* of M to be 0.

    (b) We next make the instruction labeled "1" current, and thus start the computation.

(2) The computation terminates iff $f(a_1, \ldots, a_n)$ is defined, or, symbolically, iff "$f(a_1, \ldots, a_n) \downarrow$".

(3) If the computation terminates, that is, if at some point the instruction **stop** becomes current, then the value of y at that point (and hence at any future point, by (iv) above), is $f(a_1, \ldots, a_n)$. $\qquad\square$

(1) The notation "$f(a_1, \ldots, a_n) \uparrow$" means that $f(a_1, \ldots, a_n)$ is undefined.

(2) The function computed by a URM, M, with inputs and output designated as above, can also be denoted with the symbol $M_y^{x_1, \ldots, x_n}$. This symbol, with no need for comment, makes it clear as to which are the input variables (superscript) of M, and which is the output variable (subscript). The variables $x_1, \ldots, x_n$ in $M_y^{x_1, \ldots, x_n}$ are "apparent", or not free for substitution; since $M_y^{x_1, \ldots, x_n}$ is not a term (in the predicate logic sense of the word), it does not denote a value. Note also that any attempt to effect such substitutions, for example, $M_y^{3, x_2, \ldots, x_n}$, would lead, in general, to nonsensical situations like "$L : 3 \leftarrow 3 + 1$", a command that wants to change the (standard) value of the symbol "3" (from 3 to 4)!

Thus, we may write $f = M_y^{x_1, \ldots, x_n}$, but *not* $f(a_1, \ldots, a_n) = M_{f(a_1, \ldots, a_n)}^{a_1, \ldots, a_n}$.

Note that f denotes, by name, a function, that is, a *potentially infinite table* of input/output pairs, where the input is always an n-tuple. On the other hand, $M_y^{x_1, \ldots, x_n}$ goes a step further: It *finitely represents the table* f, being able to do so because it is a finite set of instructions that can be used to compute the output for each input where f is defined.

A.3.2 Definition. (Computable Functions) A function f of n variables $x_1, \ldots, x_n$ is called *partial computable* iff for some URM, M, we have $f = M_{\mathbf{y}}^{\mathbf{x}_1, \ldots, \mathbf{x}_n}$. The set of all partial computable functions is denoted by $\mathcal{P}$. The set of all the *total* functions in $\mathcal{P}$—that is, those that are defined on *all inputs* from $\mathbb{N}$—is the set of *computable* functions and is denoted by $\mathcal{R}$. The term *recursive* is used in the literature synonymously with the term *computable*. □

Note that since a URM is a *theoretical*, rather than practical, model of computation we do *not* include *human-computer-interface* considerations in the computation. Thus, the "input" and "output" phases just happen during initialization—they are *not* part of the computation. That is why we have dispensed with both **read** and **write** instructions and speak instead of *initialization* in (1) of A.3.1. This approach to input/output is entirely analogous with the input/output convention for the other well-known model of computation, the Turing machine (cf. [6, 24, 31, 46, 49]).

A.3.3 Example. Let M be the program

$$1 : \mathbf{x} \leftarrow \mathbf{x} + 1$$
$$2 : \mathbf{stop}$$

Then $M_{\mathbf{x}}^{\mathbf{x}}$ is the function f given for all $x \in \mathbb{N}$ by $f(x) = x + 1$, the *successor* function. □

A.3.4 Remark. (λ Notation) To avoid saying verbose things such as "$M_{\mathbf{x}}^{\mathbf{x}}$ is the function f given for all $x \in \mathbb{N}$ by $f(x) = x + 1$", we will often use Church's λ-notation and write instead "$M_{\mathbf{x}}^{\mathbf{x}} = \lambda x.x + 1$".

In general, the notation "$\lambda \cdots .$" marks the beginning "λ" and the end "." of a sequence of input variables "$\cdots$". What comes after the period "." is the "rule" that indicates how the output relates to the input. The template for λ-notation thus is

$$\lambda\text{"input"}.\text{"output-rule"}$$

Relating to the above example, we note that $f = \lambda x.x + 1 = \lambda y.y + 1 = \lambda z.f(z)$ is correct. To the left and right of each "=" we have the table for a function, and we are saying all these tables are the same. Note that x, y, z are "apparent" variables ("dummy", bound) and are not free (for substitution). In particular, $f = f(x)$ is incorrect as we have distinct types to the left and right of "=": a table and a number (albeit unspecified number).

Pause. Why bother with these notational acrobatics? Because well-chosen notation protects against meaningless statements, such as

$$M_{\mathbf{x}}^{\mathbf{x}} = \mathbf{x} + 1 \tag{1}$$

that one might make in the context of the above example. As remarked before, "$M_{\mathbf{x}}^{\mathbf{x}}$" is not a term, nor are the occurrences of $\mathbf{x}$ in it free (for substitution). For example, "$M_3^3 = 4$" (substituting 3 for $\mathbf{x}$ throughout in (1)) is totally meaningless, as it says

$$\begin{matrix} 1 : 3 \leftarrow 3 + 1 \\ 2 : \mathbf{stop} \end{matrix} = 4$$

However, $M_\mathbf{x}^\mathbf{x} = \lambda\mathbf{x}.\mathbf{x}+1$ *does* make syntactic *and* semantic sense; indeed it is true, as two tables are compared and are found to be equal! Since $\lambda\mathbf{x}.\mathbf{x}+1 = \lambda y.y+1$ the following three tables are identical:[14]

$\mathbf{x} \to \boxed{M}$	$\boxed{M} \to \mathbf{x}$
0	1
1	2
2	3
3	4
$\vdots$	$\vdots$

Input $\mathbf{x}$	Output $\mathbf{x}+1$
0	1
1	2
2	3
3	4
$\vdots$	$\vdots$

Input y	Output $y+1$
0	1
1	2
2	3
3	4
$\vdots$	$\vdots$

In programming circles, the distinction between *function definition* or *declaration*, $\lambda\vec{x}.f(\vec{x})$, and *function invocation* (or *call*, or *application*, or "use")—what we call a *term*, $f(\vec{x})$, in first-order language parlance—is well known. The definition part, in programming, uses various notations depending on the programming language and corresponds to writing a program that implements the function, just as we did with M here.

There is a double standard in notation, when it comes to relations. A relation R, in the metatheory, is a table (i.e., set) of n-tuples. Its counterpart in formal logic is a formula. But where in the formal theory we almost never write a formula A as $A(x)$ in order to draw attention to our interest in its (free) variable x, in the metatheory most frequently we write a relation R as $R(\vec{x}_n)$—without λ notation—thus drawing attention to its "input slots", which here are $x_1, \ldots, x_n$ (i.e., its "free variables").

Since stating "$R(\vec{a}_n)$", by convention, is short for "$\vec{a}_n \in R$", we have two notations for a relation: *Relational*, i.e., $R(\vec{x}_n)$, and *set-theoretic*, i.e., $\vec{x}_n \in R$, both without the benefit of λ notation. There are exceptions to this practice, for example, when we define one relation from another one via the process of "freezing" some of the original relation's inputs. For example, writing $x < y$ (the standard "less than" on $\mathbb{N}$) means that *both* x and y are meant to be inputs; we have a table of ordered pairs. However, we will write $\lambda x.x < y$ to convey that y is fixed and that the input is just x. Clearly, a different relation arises for each y; we have an infinite family of tables: For $y = 0$ we have the empty table; for $y = 1$ one that contains just 0; for $y = 2$ one that contains just 0, 1; etc. $\square$

A.3.5 Example. Let M be the program

$$1 : \mathbf{x} \leftarrow \mathbf{x} \dot- 1$$
$$2 : \mathbf{stop}$$

[14] $\mathbf{x} \to \boxed{M}$ means "$\mathbf{x}$ is input to M" and $\boxed{M} \to \mathbf{x}$ indicates "$\mathbf{x}$ is output from M".

Then M_x^x is the function $\lambda x.x \dotdiv 1$, the *predecessor* function. The operation $\dotdiv$ is called "proper subtraction" and is in general defined by

$$x \dotdiv y = \begin{cases} x - y & \text{if } x \geq y \\ 0 & \text{otherwise} \end{cases}$$

It ensures that subtraction (as modified) does not take us out of the set of the so-called *number-theoretic* functions, which are those with inputs from $\mathbb{N}$ and outputs in $\mathbb{N}$.

□

Pause. Why are we restricting computability theory to number-theoretic functions? Surely, in practice we can compute with negative numbers, rational numbers, and with nonnumerical entities, such as graphs, trees, etc. Theory ought to reflect, and explain, our practices, no? It does. Negative numbers and rational numbers can be coded by natural number pairs. Computability of number-theoretic functions can handle such pairing (and unpairing; decoding). Moreover, finite objects such as graphs, trees, and the like that we manipulate via computers can be also coded (and decoded) by natural numbers. After all, the internal representation of data in computers is, at the lowest level, via natural numbers represented in binary notation. Computers cannot handle infinite objects such as (irrational) real numbers. But there is an extensive computability theory (which originated with the work of Kleene, [27]) that can handle such numbers as inputs and also compute with them. But this is beyond our scope.

A.3.6 Example. Let M be the program

$$1 : \mathbf{x} \leftarrow 0$$
$$2 : \mathbf{stop}$$

Then M_x^x is the function $\lambda x.0$, the *zero function*. □

In Definition A.3.2 we spoke of partial computable and total computable functions. We retain the qualifiers *partial* and *total* for all number-theoretic functions, even for those that may not be computable. Thus a function is *total* iff it is everywhere defined and is *nontotal* (no hyphen) otherwise. The set union of all total and nontotal number-theoretic functions is the set of all *partial functions*. Thus *partial* is *not* synonymous with *nontotal*. Compare with the so-called *partial order*[15] of discrete mathematics (and set theory). A partial order *may* be also a total (or *linear*) order.

---

[15] A two-place (binary) relation R on a set D is a *partial order* iff it is *irreflexive*, that is, for no x can we have $R(x, x)$, and *transitive*, that is, $R(x, y)$ and $R(y, z)$ imply $R(x, z)$. It is a total or *linear* order if, moreover, we have *trichotomy*: For any two elements x and y of D, $R(x, y) \lor x = y \lor R(y, x)$ is true.

If D is $\mathbb{N}$, then the "less than" relation, $<$, is a total order on D. If D is the set of all subsets of $\mathbb{N}$, then $\subset$ (proper subset relation) is a partial but not total order on D.

A.3.7 Example. The *unconditional* goto instruction, namely, "L : goto L''", can be simulated by L : if $\mathbf{x} = 0$ goto L' else goto L'. $\square$

A.3.8 Example. Let M be the program segment

$$
\begin{aligned}
&k-1 : \mathbf{x} \leftarrow 0 \\
&k : \quad\; \mathbf{x} \leftarrow \mathbf{x} + 1 \\
&k+1 : \mathbf{z} \leftarrow \mathbf{z} \dot{-} 1 \\
&k+2 : \text{if } \mathbf{z} = 0 \text{ goto } k+3 \text{ else goto } k \\
&k+3 : \ldots
\end{aligned}
$$

What it does, by the time the computation reaches instruction $k + 3$, is to have set the value of $\mathbf{z}$ to 0, and to make the value of $\mathbf{x}$ equal to the value that $\mathbf{z}$ had when instruction $k - 1$ was current. In short, the above sequence of instructions simulates the following sequence

$$
\begin{aligned}
&L : \quad \mathbf{x} \leftarrow \mathbf{z} \\
&L+1 : \mathbf{z} \leftarrow 0 \\
&L+2 : \ldots
\end{aligned}
$$

where the semantics of L : $\mathbf{x} \leftarrow \mathbf{z}$ are standard in programming: They require that upon execution of the instruction the value of $\mathbf{z}$ is copied into $\mathbf{x}$, but the value of $\mathbf{z}$ remains unchanged. $\square$

A.3.9 Exercise. Write a program segment that simulates precisely L : $\mathbf{x} \leftarrow \mathbf{z}$; that is, copy the value of $\mathbf{z}$ into $\mathbf{x}$ without causing $\mathbf{z}$ to change as a side effect. $\square$

Because of the above, without loss of generality, one may assume of any input variable, $\mathbf{x}$, of a program M that it is *read-only*. This means that its value remains invariant throughout any computation of the program. Indeed, if $\mathbf{x}$ is not so, a new input variable, $\mathbf{x}'$, can be introduced as follows to relieve $\mathbf{x}$ from its input role: Add at the very beginning of M the (derived) instruction $1 : \mathbf{x} \leftarrow \mathbf{x}'$, where $\mathbf{x}'$ is a variable that does not occur in M. Adjust all the following labels consistently, including, of course, the ones referenced by if-statements—a tedious but straightforward task. Call M' the so-obtained URM. Clearly, $M'\,{}^{\mathbf{x}',\mathbf{y}_1,\ldots,\mathbf{y}_n}_{\mathbf{z}} = M^{\mathbf{x},\mathbf{y}_1,\ldots,\mathbf{y}_n}_{\mathbf{z}}$.

A.3.10 Example. (Composing Computable Functions) Suppose that $\lambda x \vec{y}.f(x, \vec{y})$ and $\lambda \vec{z}.g(\vec{z})$ are partial computable, and say $f = F^{\mathbf{x},\vec{\mathbf{y}}}_{\mathbf{u}}$ while $g = G^{\vec{\mathbf{z}}}_{\mathbf{x}}$.

Since we can rewrite any program renaming its variables at will, we assume without loss of generality that $\mathbf{x}$ is the only variable common to F and G. Thus, if we concatenate the programs G and F in that order, and (1) remove the last instruction of G (k : **stop**, for some k)—call the program segment that results from this G', and (2) renumber the instructions of F as $k, k + 1, \ldots$ (and, as a result, the references that if-statements of F make) in order to give $(G'F)$ the correct program structure, then, $\lambda \vec{y} \vec{z}.f(g(\vec{z}), \vec{y}) = (G'F)^{\vec{\mathbf{y}},\vec{\mathbf{z}}}_{\mathbf{u}}$. Note that all non-input variables of F will hold 0 as soon as the execution of $(G'F)$ makes the first instruction of F current *for the first time*. This is because none of these can be changed by G' under our assumption, thus ensuring that F works as designed. $\square$

Thus, we have, by repeating the above a finite number of times:

A.3.11 Proposition. *If* $\lambda \vec{y}_n.f(\vec{y}_n)$ *and* $\lambda \vec{z}.g_i(\vec{z})$, *for* $i = 1, \dots, n$, *are partial computable, then so is* $\lambda \vec{z}.f(g_1(\vec{z}), \dots, g_n(\vec{z}))$.

We can rephrase A.3.11, saying simply that $\mathcal{P}$ *is closed under composition*. For the record, we will define *composition* to mean the somewhat rigidly defined operation used in A.3.11, that is:

A.3.12 Definition. Given any partial functions (computable or not) $\lambda \vec{y}_n.f(\vec{y}_n)$ and $\lambda \vec{z}.g_i(\vec{z})$, for $i = 1, \dots, n$, we say that $\lambda \vec{z}.f(g_1(\vec{z}), \dots, g_n(\vec{z}))$ is the result of their *composition*. □

We characterized the definition as "rigid". Indeed note that it requires that *all* the arguments of f be substituted by a $g_i(\vec{z})$—unlike Example A.3.10, where we *substituted* a function invocation (cf. terminology in A.3.4) in *one* variable of f there, and did nothing with the variables $\vec{y}$—and for each application $g_i(\dots)$ the argument list, "$\dots$", *must be the same*, for example $\vec{z}$. This rigidity is only apparent, as we show in examples in the next subsection (A.3.2).

Composing a number of times that *depends on the value of an input variable* is *iteration*. The general case of iteration is called *primitive recursion*.

A.3.13 Definition. (Primitive Recursion) A number-theoretic function f is defined by *primitive recursion* from given functions $\lambda \vec{y}.h(\vec{y})$ and $\lambda x \vec{y} z.g(x, \vec{y}, z)$ provided, *for all* $x, \vec{y}$, its values are given by the two equations below:

$$\begin{aligned} f(0, \vec{y}) &= h(\vec{y}) \\ f(x+1, \vec{y}) &= g(x, \vec{y}, f(x, \vec{y})) \end{aligned}$$

h is the *basis function*, while g is the *iterator*.
It will be useful to use the notation $f = prim(h, g)$ to indicate in shorthand that f is defined as above from h and g (note the order). □

Note that $f(1, \vec{y}) = g(0, \vec{y}, h(\vec{y}))$, $f(2, \vec{y}) = g(1, \vec{y}, g(0, \vec{y}, h(\vec{y})))$, $f(3, \vec{y}) = g(2, \vec{y}, g(1, \vec{y}, g(0, \vec{y}, h(\vec{y}))))$, etc. Thus the "$x$-value", 0, 1, 2, 3, etc., equals the number of times we compose g with itself. Hence "iteration", i.e., composition as many times as an input value dictates.

A.3.14 Example. (Iterating Computable Functions) Suppose that $\lambda x \vec{y} z.g(x, \vec{y}, z)$ and $\lambda \vec{y}.h(\vec{z})$ are partial computable, and say $g = G_{\mathbf{z}}^{\mathbf{i}, \vec{\mathbf{y}}, \mathbf{z}}$ while $h = H_{\mathbf{z}}^{\vec{\mathbf{y}}}$.
By earlier remarks we may assume:
(i) The only variables that H and G have in common are $\mathbf{z}, \vec{\mathbf{y}}$.
(ii) $\vec{\mathbf{y}}$ are read-only in both H and G.
(iii) $\mathbf{i}$ is read-only in G.
(iv) $\mathbf{x}$ does not occur in any of H or G.

We can now argue that the following program, let us call it F, computes f defined as in A.3.13 from h and g, where $\boxed{H'}$ is program H with the **stop** instruction removed, $\boxed{G'}$ is program G with the **stop** instruction removed, and instructions have been renumbered (and if-statements adjusted) as needed:

$$\boxed{H'}$$

$$
\begin{array}{ll}
r: & \mathbf{i} \leftarrow 0 \\
r+1: & \text{if } \mathbf{x} = 0 \text{ goto } k+m+2 \text{ else goto } r+2 \\
r+2: & \mathbf{x} \leftarrow \mathbf{x} \doteq 1
\end{array}
$$

$$\boxed{G'}$$

$$
\begin{array}{ll}
k: & \mathbf{i} \leftarrow \mathbf{i} + 1 \\
k+1: & \mathbf{w}_1 \leftarrow 0 \\
\vdots & \\
k+m: & \mathbf{w}_m \leftarrow 0 \\
k+m+1: & \text{goto } r+1 \\
k+m+2: & \text{stop}
\end{array}
$$

The instructions $\mathbf{w}_i \leftarrow 0$ set explicitly to zero all the variables of G' other than $\mathbf{i}, \mathbf{z}, \vec{\mathbf{y}}$ to ensure correct behavior of G'. Note that the $\mathbf{w}_i$ are *implicitly* initialized to zero *only* the first time G' is executed. Clearly, $f = F_{\mathbf{z}}^{\mathbf{x}, \vec{\mathbf{y}}}$. $\square$

We have at once:

A.3.15 Proposition. *If f, g, h relate as in Definition A.3.13 and h and g are in $\mathcal{P}$, then so is f. We say that $\mathcal{P}$ is* closed under primitive recursion.

A.3.16 Example. (Unbounded Search) Suppose that $\lambda x\vec{y}.g(x,\vec{y})$ is partial computable, and say $g = G_{\mathbf{z}}^{\mathbf{x}, \vec{\mathbf{y}}}$. By earlier remarks we may assume that $\vec{\mathbf{y}}$ and $\mathbf{x}$ are read-only in G and that $\mathbf{z}$ is *not* one of them.

Consider the following program F, where $\boxed{G'}$ is program G with the **stop** instruction removed, and instructions have been renumbered (and if-statements adjusted) as needed so that its first command has label 2.

$$
\begin{array}{ll}
1: & \mathbf{x} \leftarrow 0
\end{array}
$$

$$\boxed{G'}$$

$$
\begin{array}{ll}
k: & \text{if } \mathbf{z} = 0 \text{ goto } k+l+3 \text{ else goto } k+1 \\
k+1: & \mathbf{w}_1 \leftarrow 0 \ \{\textbf{Comment. } \text{Setting all non-input variables to 0; cf. A.3.14.}\} \\
\vdots & \\
k+l: & \mathbf{w}_l \leftarrow 0 \ \{\textbf{Comment. } \text{Setting all non-input variables to 0; cf. A.3.14.}\} \\
k+l+1: & \mathbf{x} \leftarrow \mathbf{x} + 1 \\
k+l+2: & \text{goto } 2 \\
k+l+3: & \text{stop}
\end{array}
$$

Let us set $f = F_{\mathbf{x}}^{\vec{\mathbf{y}}}$. Note that, for any $\vec{a}$, $f(\vec{a}) \downarrow$ precisely if the URM F, initialized with $\vec{a}$ as the input values in $\vec{\mathbf{y}}$, ever reaches **stop**. This condition becomes true as long as the two conditions, (1) and (2), are fulfilled:

(1) Instruction k just found that $\mathbf{z}$ holds 0. This value of $\mathbf{z}$ is the result of an execution of G (i.e., G' with the **stop** instruction added) with input values $\vec{a}$ in $\vec{\mathbf{y}}$ and, say, b in $\mathbf{x}$, the latter being the *iteration counter*—$0, 1, 2, \ldots$—that indicates how many times instruction 2 becomes current,

(2) In none of the previous iterations (with $\mathbf{x}$-value $< b$) did G' (essentially, G) get into a nonending computation (*infinite loop*).

Correspondingly, the computation of F will never halt for an input $\vec{a}$ if either G loops for ever at some step, or, if it halts in every iteration b, but nevertheless it never exits with a $\mathbf{z}$-value of 0.

Thus, for all $\vec{a}$,

$$f(\vec{a}) = \min\{x : g(x, \vec{a}) = 0 \wedge (\forall y)(y < x \rightarrow g(y, \vec{a}) \downarrow)\} \qquad \square$$

A.3.17 Definition. The operation on partial functions g given for all $\vec{a}$ by

$$\min\{x : g(x, \vec{a}) = 0 \wedge (\forall y)(y < x \rightarrow g(y, \vec{a}) \downarrow)\}$$

is called *unbounded search* (along the variable x) and is denoted by the symbol $(\mu x)g(x, \vec{a})$. The function $\lambda \vec{y}.(\mu x)g(x, \vec{y})$ is defined precisely when the minimum exists. $\qquad \square$

The result of Example A.3.16 yields at once:

A.3.18 Proposition. $\mathcal{P}$ *is closed under unbounded search; that is, if* $\lambda x \vec{y}.g(x, \vec{y})$ *is in* $\mathcal{P}$, *then so is* $\lambda \vec{y}.(\mu x)g(x, \vec{y})$.

A.3.19 Example. Is the function $\lambda \vec{x}_n.x_i$, where $1 \leq i \leq n$, in $\mathcal{P}$? Yes, and here is a program, M, for it:

$$
\begin{aligned}
&1: && \mathbf{w}_1 \leftarrow 0 \\
&\;\vdots \\
&i: && \mathbf{z} \leftarrow \mathbf{w}_i \; \{\textbf{Comment.} \text{ Cf. Exercise A.3.9}\} \\
&\;\vdots \\
&n: && \mathbf{w}_n \leftarrow 0 \\
&n+1: \textbf{stop}
\end{aligned}
$$

$\lambda \vec{x}_n.x_i = M_{\mathbf{z}}^{\vec{\mathbf{w}}_n}$. To ensure that M indeed *has* the $\mathbf{w}_i$ as variables we reference them in instructions at least once, in any manner whatsoever. $\qquad \square$

A.3.2 Primitive Recursive Functions

Exercises A.3.3, A.3.6, and A.3.19 show that the successor, the zero, and the *generalized identity* functions respectively—which we will often name S, Z and U_i^n respectively—are in $\mathcal{P}$; thus, not only are they "intuitively computable", but they are so in a precise mathematical sense. We have also shown that "computability" of functions is preserved by the operations of composition, primitive recursion, and unbounded search. In this subsection we will explore the properties of the important set of functions known as *primitive recursive*. We introduce them by derivations just as we introduced the theorems of logic.

A.3.20 Definition. ($\mathcal{PR}$**-derivations;** $\mathcal{PR}$**-functions**) A $\mathcal{PR}$-derivation is a finite sequence of number-theoretic functions that obeys, in its step-by-step construction, the following requirements. At each step we may write:

(1) Any one of Z, S, U_i^n (for any $n > 0$ and any $0 < i \leq n$).

(2) $\lambda \vec{z}.f(g_1(\vec{z}), \ldots, g_n(\vec{z}))$, *provided* each of $f, g_1, \ldots, g_n$ has already been written.

(3) $prim(h, g)$, provided appropriate h and g have already been written. Note that h and g are "appropriate" (cf. A.3.13) as long as g has two more arguments than h.

A function f is *primitive recursive*, or a $\mathcal{PR}$-function, iff it occurs in some $\mathcal{PR}$-derivation. The set of functions allowed in step (1) are called *initial functions*. We will denote this set by $\mathcal{I}$. The set of *all* $\mathcal{PR}$-functions will be denoted by $\mathcal{PR}$. $\square$

A.3.21 Remark. The above definition defines essentially Dedekind's ([8]) "recursive" functions. Subsequently they have been renamed *primitive recursive* allowing the unqualified term *recursive* to be synonymous with *computable* and apply to the functions of $\mathcal{R}$ (cf. A.3.2).

The concept of a $\mathcal{PR}$-derivation is entirely analogous with those of proof, formula-calculation, and term-calculation. Vis-à-vis proofs, derivations have the following analogous elements: initial functions (vs. axioms) and the operations composition and primitive recursion (vs. the rules of inference, Leib and Eqn).

As was the case with proofs (1.4.8 and 4.2.9), we can cut the tail off a derivation and still have a derivation. Thus, a $\mathcal{PR}$-function is one that appears at the end of a $\mathcal{PR}$-derivation.

Properties of primitive recursive functions can be proved by induction on derivation length, just as properties of theorems can be (and have been in this volume) proved by induction on the length of proofs.

That a certain function is primitive recursive can be proved by exhibiting a derivation for it, just as is done for the certification of a theorem: We exhibit a proof. However, in proving theorems we accept the use of known theorems in proofs (cf. 2.1.6). Similarly, if we know that certain functions are primitive recursive, then we immediately infer that so is one obtained from them by an allowed operation (composition, primitive recursion, or yet-to-be-introduced derived operations). For example, if h and g are in $\mathcal{PR}$ and $prim(h, g)$ makes sense according to A.3.13, then the latter is in $\mathcal{PR}$, too, since we can concatenate derivations of h and g and add $prim(h, g)$ to the right end.

In analogy to the case of theorem proving, where we benefit from powerful derived *rules*, in the same way certifying functions as primitive recursive is greatly facilitated by the introduction of derived *operations* on functions beyond the two we assumed as given outright (*primary operations*) in Definition A.3.20. $\qquad\square$

A.3.22 Theorem. $\mathcal{PR}$ *contains* $\mathcal{I}$ *and is closed under primitive recursion and composition. Indeed, of all possible sets that include* $\mathcal{I}$ *and are closed under these two operations,* $\mathcal{PR}$ *is the smallest with respect to inclusion.*

Proof. That $\mathcal{PR}$ contains $\mathcal{I}$ is immediate from A.3.20. Why it is closed under primitive recursion was outlined in Remark A.3.21 and the case of composition is analogous (see Exercise A.3.23). Let now S be a set that includes $\mathcal{I}$ and is closed under the two operations. By induction on the length of derivations, we can prove that if $f \in \mathcal{PR}$, then $f \in S$.

For the basis, let f occur in a derivation of length 1. Then it is in $\mathcal{I}$ and we are done. Assume the claim for all f that appear in derivations of length $\leq n$ and consider an f that appears in one of length $n + 1$. If it appears before the last step, then we are done by the I.H. Let it then appear only at the last step. If it is in $\mathcal{I}$, we are done by assumption on S. Let then $f = prim(h, g)$ where h and g show up earlier in the derivation under consideration. By I.H. both h and g are in S. As this set is closed under primitive recursion, it contains f as well. The case of composition causing the presence of f in the derivation is similar (see Exercise A.3.23). $\qquad\square$

A.3.23 Exercise. Provide the missing details in the proof of A.3.22. $\qquad\square$

 A.3.24 Remark. (Induction on $\mathcal{PR}$) Just as we do "induction on formulae" we can do *induction on* $\mathcal{PR}$ toward proving a property $\mathscr{P}(f)$ for all $f \in \mathcal{PR}$. We prove:

(1) (*Basis*) Any one of Z, S, U_i^n (for any $n > 0$ and any $0, i \leq n$) has the property $\mathscr{P}$.

(2) $\lambda\vec{z}.f(g_1(\vec{z}), \ldots, g_n(\vec{z}))$ has the property, *provided* each of $f, g_1, \ldots, g_n$ do.

(3) $prim(h, g)$ has the property, provided h and g do.

The above procedure is more elegant (and more widely used) than induction on the length of $\mathcal{PR}$-derivation and is immediately based on the latter. Alternatively, it directly follows from A.3.22: Let $S = \{f : f$ is number-theoretic and $\mathscr{P}(f)$ holds$\}$ where $\mathscr{P}$ satisfies (1)–(3). But then S contains the initial functions and is closed under composition and primitive recursion (verify!). Thus $\mathcal{PR} \subseteq S$, or $f \in \mathcal{PR}$ implies that $\mathscr{P}(f)$ holds. $\qquad\square$

A.3.25 Example. If $\lambda xyw.f(x, y, w)$ and $\lambda z.g(z)$ are in $\mathcal{PR}$, how about $\lambda xzw.f(x, g(z), w)$? It is in $\mathcal{PR}$ since

$$\lambda xzw.f(x, g(z), w) = \lambda xzw.f(U_1^3(x, z, w), g(U_2^3(x, z, w)), U_3^3(x, z, w))$$

and the U_i^n are primitive recursive. The reader will see at once that to the right of "=" we have correctly formed compositions as expected by A.3.12.

Similarly, for the same functions above,

(1) $\lambda yw.f(2, y, w)$ is in $\mathcal{PR}$. Indeed, this function can be obtained by composition, since
$$\lambda yw.f(2, y, w) = \lambda yw.f\Big(SSZ\big(U_1^2(y, w)\big), y, w\Big)$$
where I wrote "$SSZ(\ldots)$" as short for $S(S(Z(\ldots)))$ for visual clarity. Clearly, using $SSZ\big(U_2^2(y, w)\big)$ above works as well.

(2) $\lambda xyw.f(y, x, w)$ is in $\mathcal{PR}$. Indeed, this function can be obtained by composition, since
$$\lambda xyw.f(y, x, w) = \lambda xyw.f\Big(U_2^3(x, y, w), U_1^3(x, y, w), U_3^3(x, y, w)\Big)$$

 In this connection, note that while $\lambda xy.g(x, y) = \lambda yx.g(y, x)$, yet $\lambda xy.g(x, y) \neq \lambda xy.g(y, x)$ in general. For example, $\lambda xy.x \dotminus y$ asks that we subtract the second input (y) from the first (x), but $\lambda xy.y \dotminus x$ asks that we subtract the first input (x) from the second (y).

(3) $\lambda xy.f(x, y, x)$ is in $\mathcal{PR}$. Indeed, this function can be obtained by composition, since
$$\lambda xy.f(x, y, x) = \lambda xy.f\big(U_1^2(x, y), U_2^2(x, y), U_1^2(x, y)\big)$$

(4) $\lambda xyzwu.f(x, y, w)$ is in $\mathcal{PR}$. Indeed, this function can be obtained by composition, since
$$\lambda xyzwu.f(x, y, w) =$$
$$\lambda xyzwu.f(U_1^5(x, y, z, w, u), U_2^5(x, y, z, w, u), U_4^5(x, y, z, w, u))$$

$\square$

The above are summarized, named, and generalized in the following straightforward exercise:

A.3.26 Exercise. (Grzegorczyk Substitution Operations [18]) $\mathcal{PR}$ is closed under the following operations:

(i) *Substitution of a function invocation for a variable*:
From $\lambda\vec{x}y\vec{z}.f(\vec{x}, y, \vec{z})$ and $\lambda\vec{w}.g(\vec{w})$ obtain $\lambda\vec{x}\vec{w}\vec{z}.f(\vec{x}, g(\vec{w}), \vec{z})$.

(ii) *Substitution of a constant for a variable*:
From $\lambda\vec{x}y\vec{z}.f(\vec{x}, y, \vec{z})$ obtain $\lambda\vec{x}\vec{z}.f(\vec{x}, k, \vec{z})$.

(iii) *Interchange of two variables*:
From $\lambda\vec{x}y\vec{z}w\vec{u}.f(\vec{x}, y, \vec{z}, w, \vec{u})$ obtain $\lambda\vec{x}y\vec{z}w\vec{u}.f(\vec{x}, w, \vec{z}, y, \vec{u})$.

(iv) *Identification of two variables*:

From $\lambda \vec{x} y \vec{z} w \vec{u}. f(\vec{x}, y, \vec{z}, w, \vec{u})$ obtain $\lambda \vec{x} y \vec{z} \vec{u}. f(\vec{x}, y, \vec{z}, y, \vec{u})$.

(v) *Introduction of "don't care" variables*:

From $\lambda \vec{x}. f(\vec{x})$ obtain $\lambda \vec{x} \vec{z}. f(\vec{x})$. □

By A.3.26 composition can simulate the Grzegorczyk operations if the initial functions $\mathcal{I}$ are present. Of course, (i) alone can in turn simulate composition. With these comments out of the way, we see that the "rigidity" of Definition A.3.12 is gone.

A.3.27 Example. The definition of primitive recursion is also rigid, but this rigidity is removable as well. For example, natural and simple recursions such as $p(0) = 0$ and $p(x + 1) = x$—this one defining $p = \lambda x.x \doteq 1$—do not fit the schema of Definition A.3.13, which requires that the defined function has one more variable than the basis, so it cannot have only one variable! We can get around this. Define first $\widetilde{p} = \lambda xy.x \doteq 1$ as follows: $\widetilde{p}(0, y) = 0$ and $\widetilde{p}(x + 1, y) = x$. Now this can be dressed up according to the syntax of the schema in A.3.13,

$$
\begin{aligned}
\widetilde{p}(0, y) &= Z(y) \\
\widetilde{p}(x + 1, y) &= U_1^3(x, y, \widetilde{p}(x, y))
\end{aligned}
$$

that is, $\widetilde{p} = prim(Z, U_1^3)$. Then we can get p by (Grzegorczyk) substitution: $p = \lambda x. \widetilde{p}(x, 0)$. Incidentally, this shows that both p and $\widetilde{p}$ are in $\mathcal{PR}$.

Another rigidity in the definition of primitive recursion is that, apparently, one can use only the first variable as the iterating variable. Consider, for example, $sub = \lambda xy.x \doteq y$. Clearly, $sub(x, 0) = x$ and $sub(x, y + 1) = p(sub(x, y))$ is correct semantically, but the format is wrong: We are not supposed to iterate along the second variable! Well, define instead $\widetilde{sub} = \lambda xy.y \doteq x$:

$$
\begin{aligned}
\widetilde{sub}(0, y) &= U_1^1(y) \\
\widetilde{sub}(x + 1, y) &= p\big(U_3^3(x, y, \widetilde{sub}(x, y))\big)
\end{aligned}
$$

Then, using variable swapping (Grzegorczyk operation (iii)), we can get sub: $sub = \lambda xy.\widetilde{sub}(y, x)$. Clearly, both $\widetilde{sub}$ and sub are in $\mathcal{PR}$. With practice, one gets used to accepting at once simplified recursions like the one for p and sub. One needs to make them conform to the format of A.3.13 only if the instructor insists! □

A.3.28 Exercise. Prove that $\lambda xy.x + y$ and $\lambda xy.x \times y$ are primitive recursive. Of course, we will usually write multiplication $x \times y$ in "implied notation", xy. □

A.3.29 Example. The very important "*switch*" (or "if-then-else") function $sw = \lambda xyz.$if $x = 0$ then y else z is primitive recursive. It is directly obtained by primitive recursion on initial functions: $sw(0, y, z) = y$ and $sw(x + 1, y, z) = z$. □

A.3.30 Exercise. Dress up the recursion $sw(0, y, z) = y$ and $sw(x + 1, y, z) = z$ to bring it into the format required by Definition A.3.13. □

A.3.31 Exercise. Prove by induction on derivation lengths that all functions in $\mathcal{PR}$ are total.   □

A.3.32 Proposition. $\mathcal{PR} \subseteq \mathcal{R}$.

Proof. By A.3.11, A.3.15, and A.3.22, $\mathcal{PR} \subseteq \mathcal{P}$. But all the functions in $\mathcal{PR}$ are total (cf. A.3.31 and Definition A.3.2).   □

Indeed, the above inclusion is proper, but the proof is beyond our scope (cf. for example, [49]). We also state for the record:

A.3.33 Proposition. $\mathcal{R}$ *is closed under both composition and primitive recursion.*

Proof. Because $\mathcal{P}$ is, and both operations conserve totalness.   □

A.3.34 Example. Consider the function ex given by

$$
\begin{aligned}
ex(x,0) &= 1 \\
ex(x,y+1) &= ex(x,y)x
\end{aligned}
$$

Thus, if $x = 0$, then $ex(x,0) = 1$, but $ex(x,y) = 0$ for all $y > 0$. On the other hand, if $x > 0$, then $ex(x,y) = x^y$ for all y.

Note that x^y is "mathematically" undefined when $x = y = 0$.[16] Thus, by Exercise A.3.31 the exponential cannot be a primitive recursive function!

This is rather silly, since the computational process for the exponential is so straightforward; thus it is a shame to declare the function non-$\mathcal{PR}$. After all, we know *exactly where and how it is undefined* and we can remove this undefinability by *redefining "x^y" to mean $ex(x,y)$ for all inputs.*

Clearly $ex \in \mathcal{PR}$. We do this kind of redefinition a lot in computability in order to remove easily recognizable points of "nondefinition" of calculable functions. We will see further examples, such as the remainder, quotient, and logarithm functions.

Caution! We cannot always remove points of nondefinition of a calculable function and still obtain a computable function.   □

A.3.35 Definition. A relation $R(\vec{x})$ is *(primitive) recursive* iff its *characteristic function,*

$$
\chi_R = \lambda\vec{x}. \begin{cases} 0 & \text{if } R(\vec{x}) \\ 1 & \text{if } \neg R(\vec{x}) \end{cases}
$$

is (primitive) recursive. The set of all primitive recursive (respectively, recursive) *relations* is denoted by $\mathcal{PR}_*$ (respectively, $\mathcal{R}_*$).   □

Computability theory practitioners often call relations *predicates*. It is clear that one can go from relation to characteristic function and back in a unique way, since $R(\vec{x}) \equiv \chi_R(\vec{x}) = 0$. Thus, we may think of relations as "0-1 valued" functions. The concept of relation simplifies further development of the theory of primitive recursive functions.

---

[16]In first-year university calculus we learn that "0^0" is an "indeterminate form".

The following is useful:

A.3.36 Proposition. $R(\vec{x}) \in \mathcal{PR}_*$ iff some $f \in \mathcal{PR}$ exists such that, for all $\vec{x}$, $R(\vec{x}) \equiv f(\vec{x}) = 0$.

Proof. For the *if*-part, I want $\chi_R \in \mathcal{PR}$. This is so since $\chi_R = \lambda\vec{x}.1 \doteq (1 \doteq f(\vec{x}))$ (using Grzegorczyk substitution and $\lambda xy.x \doteq y$). For the *only if*-part, $f = \chi_R$ will do. $\square$

A.3.37 Corollary. $R(\vec{x}) \in \mathcal{R}_*$ iff some $f \in \mathcal{R}$ exists such that, for all $\vec{x}$, $R(\vec{x}) \equiv f(\vec{x}) = 0$.

Proof. By the above proof, A.3.32, and A.3.33. $\square$

A.3.38 Corollary. $\mathcal{PR}_* \subseteq \mathcal{R}_*$.

Proof. By the above corollary and A.3.32. $\square$

A.3.39 Theorem. $\mathcal{PR}_*$ *is closed under the Boolean operations.*

Proof. It suffices to look at the cases of $\neg$ and $\vee$.

($\neg$) Say, $R(\vec{x}) \in \mathcal{PR}_*$. Thus (A.3.35), $\chi_R \in \mathcal{PR}$. But then $\chi_{\neg R} \in \mathcal{PR}$, since $\chi_{\neg R} = \lambda\vec{x}.1 \doteq \chi_R(\vec{x})$, by Grzegorczyk substitution and $\lambda xy.x \doteq y \in \mathcal{PR}$.

($\vee$) Let $R(\vec{x}) \in \mathcal{PR}_*$ and $Q(\vec{y}) \in \mathcal{PR}_*$. Then $\lambda\vec{x}\vec{y}.\chi_{R\vee Q}(\vec{x},\vec{y})$ is given by

$$\chi_{R\vee Q}(\vec{x},\vec{y}) = \text{if } R(\vec{x}) \text{ then } 0 \text{ else } \chi_Q(\vec{y})$$

and therefore is in $\mathcal{PR}$. $\square$

It is common practice to use $R(\vec{x})$ and $\chi_R(\vec{x})$ (almost) interchangeably. For example, "if $R(\vec{x})$ then ..." is the same as "if $\chi_R(\vec{x}) = 0$ then ...". The latter more directly shows that a (Grzegorczyk) substitution was effected into an argument of the if-then-else (A.3.29) function:

$$\begin{array}{c} \chi_R(\vec{x}) \\ \downarrow \\ \text{if } \quad x \quad = 0 \text{ then } \dots \end{array}$$

thus establishing the primitive recursiveness of the resulting function.

A.3.40 Remark. Alternatively, note that $\chi_{R\vee Q}(\vec{x},\vec{y}) = \chi_R(\vec{x}) \times \chi_Q(\vec{y})$. $\square$

A.3.41 Corollary. $\mathcal{R}_*$ *is closed under the Boolean operations.*

Proof. As above, mindful of A.3.32, and A.3.33. $\square$

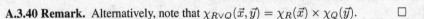

A.3.42 Example. The relations $x \leq y$, $x < y$, $x = y$ are in $\mathcal{PR}_*$. See A.3.4 for a refresher on our conventions regarding lambda notation and relations.

With this out of the way, note that $x \leq y \equiv x \dotminus y = 0$ and invoke A.3.36. Finally invoke Boolean closure and note that $x < y \equiv \neg y \leq x$ while $x = y$ is equivalent to $x \leq y \wedge y \leq x$. □

A.3.43 Proposition. *If $R(\vec{x}, y, \vec{z}) \in \mathcal{PR}_*$ and $\lambda \vec{w}.f(\vec{w}) \in \mathcal{PR}$, then $R(\vec{x}, f(\vec{w}), \vec{z})$ is in $\mathcal{PR}_*$.*

Proof. If $Q(\vec{x}, \vec{w}, \vec{z})$ denotes $R(\vec{x}, f(\vec{w}), \vec{z})$, then $\chi_Q(\vec{x}, \vec{w}, \vec{z}) = \chi_R(\vec{x}, f(\vec{w}), \vec{z})$. □

A.3.44 Proposition. *If $R(\vec{x}, y, \vec{z}) \in \mathcal{R}_*$ and $\lambda \vec{w}.f(\vec{w}) \in \mathcal{R}$, then $R(\vec{x}, f(\vec{w}), \vec{z})$ is in $\mathcal{R}_*$.*

Proof. Similar to that of A.3.43. □

A.3.45 Corollary. *If $f \in \mathcal{PR}$ (respectively, in $\mathcal{R}$), then its graph, $z = f(\vec{x})$ is in $\mathcal{PR}_*$ (respectively, in $\mathcal{R}_*$).*

Proof. Using the relation $z = y$ and A.3.43. □

The following converse of A.3.45, "if $z = f(\vec{x})$ is in $\mathcal{PR}_*$ and f is total, then $f \in \mathcal{PR}$" is *not* true. A counterexample is provided by the Ackermann function.[17] However, "if $z = f(\vec{x})$ is in $\mathcal{R}_*$ and f is total, then $f \in \mathcal{R}$" *is* true. Cf. [49] and the exercise below.

A.3.46 Exercise. Using unbounded search, prove that if $z = f(\vec{x})$ is in $\mathcal{R}_*$ and f is total, then $f \in \mathcal{R}$. □

A.3.47 Definition. (Bounded Quantifiers) The shorthand notations $(\forall y)_{<z} R(z, \vec{x})$ and $(\exists y)_{<z} R(z, \vec{x})$ stand for $(\forall y)\big(y < z \rightarrow R(z, \vec{x})\big)$ and $(\exists y)\big(y < z \wedge R(z, \vec{x})\big)$ respectively, and similarly for the nonstrict inequality "$\leq$" (cf. the general case on p. 175). □

A.3.48 Theorem. $\mathcal{PR}_*$ *is closed under bounded quantification.*

Proof. By A.3.39 it suffices to look at the case of $(\exists y)_{<z}$ since $(\forall y)_{<z} R(y, \vec{x}) \equiv \neg(\exists y)_{<z} \neg R(y, \vec{x})$.

Let then $R(y, \vec{x}) \in \mathcal{PR}_*$ and let us give the name $Q(z, \vec{x})$ to $(\exists y)_{<z} R(y, \vec{x})$. We note that $Q(0, \vec{x})$ is false (why?) and $Q(z + 1, \vec{x}) \equiv Q(z, \vec{x}) \vee R(z, \vec{x})$. Thus,

$$\chi_Q(0, \vec{x}) = 1$$
$$\chi_Q(z + 1, \vec{x}) = \chi_Q(z, \vec{x})\chi_R(z, \vec{x})$$ □

---

[17]The so-called Ritchie version of the Ackermann function $\lambda n x.A_n(x)$ is given by a "double recursion" for all n, x: $A_0(x) = x + 2$, $A_{n+1}(0) = 2$, $A_{n+1}(x + 1) = A_n(A_{n+1}(x))$.

A.3.49 Corollary. $\mathcal{R}_*$ *is closed under bounded quantification.*

A.3.50 Exercise. The operations of *bounded summation* and *bounded multiplication* are quite handy.

These are applied on a function f and yield the functions $\lambda z \vec{x}. \sum_{i<z} f(i, \vec{x})$ and $\lambda z \vec{x}. \prod_{i<z} f(i, \vec{x})$ respectively, where $\sum_{i<0} f(i, \vec{x}) = 0$ and $\prod_{i<0} f(i, \vec{x}) = 1$ by definition. Prove that $\mathcal{PR}$ and $\mathcal{R}$ are closed under both operations; i.e., if f is in $\mathcal{PR}$ (respectively, in $\mathcal{R}$), then so are $\lambda z \vec{x}. \sum_{i<z} f(i, \vec{x})$ and $\lambda z \vec{x}. \prod_{i<z} f(i, \vec{x})$. $\qquad\square$

A.3.51 Definition. (Bounded Search) Let f be a total number-theoretic function of $n + 1$ variables. The symbol $(\mu y)_{<z} f(y, \vec{x})$, for all $z, \vec{x}$, stands for

$$\begin{cases} \min\{y : y < z \land f(y, \vec{x}) = 0\} & \text{if } (\exists y)_{<z} f(y, \vec{x}) = 0 \\ z & \text{otherwise} \end{cases}$$

We define "$(\mu y)_{\leq z}$" to mean "$(\mu y)_{<z+1}$". $\qquad\square$

A.3.52 Theorem. $\mathcal{PR}$ *is closed under the bounded search operation* $(\mu y)_{<z}$. *That is, if* $\lambda y \vec{x}. f(y, \vec{x}) \in \mathcal{PR}$, *then* $\lambda z \vec{x}. (\mu y)_{<z} f(y, \vec{x}) \in \mathcal{PR}$.

Proof. Set $g = \lambda z \vec{x}. (\mu y)_{<z} f(y, \vec{x})$. Then the following primitive recursion settles it:

$$g(0, \vec{x}) = 1$$
$$g(z + 1, \vec{x}) = \text{if } g(z, \vec{x}) < z \text{ then } g(z, \vec{x})$$
$$\text{else if } f(z, \vec{x}) = 0 \text{ then } z$$
$$\text{else } z + 1 \qquad\square$$

A.3.53 Exercise. Mindful of the comment following A.3.39, dress up the primitive recursion that defined g above so that it conforms to the Definition A.3.13. $\qquad\square$

A.3.54 Corollary. $\mathcal{PR}$ *is closed under the bounded search operation* $(\mu y)_{\leq z}$.

A.3.55 Exercise. Prove the corollary. $\qquad\square$

A.3.56 Corollary. $\mathcal{R}$ *is closed under the bounded search operations* $(\mu y)_{<z}$ *and* $(\mu y)_{\leq z}$.

Consider now a set of *mutually exclusive* relations $R_i(\vec{x})$, $i = 1, \ldots, n$, that is, $R_i(\vec{x}) \land R_j(\vec{x})$ is false for each $\vec{x}$ as long as $i \neq j$.

Then we can define a function f *by cases R_i* from given functions f_j by the requirement (for all $\vec{x}$) given below:

$$f(\vec{x}) = \begin{cases} f_1(\vec{x}) & \text{if } R_1(\vec{x}) \\ f_2(\vec{x}) & \text{if } R_2(\vec{x}) \\ \dots & \dots \\ f_n(\vec{x}) & \text{if } R_n(\vec{x}) \\ f_{n+1}(\vec{x}) & \text{otherwise} \end{cases}$$

where, as is usual in mathematics, "if $R_j(\vec{x})$" is short for "if $R_j(\vec{x})$ is true" and "otherwise" is the condition $\neg(R_1(\vec{x}) \vee \cdots \vee R_n(\vec{x}))$. We have the following result:

A.3.57 Theorem. (Definition by Cases) *If the functions f_i, $i = 1, \ldots, n + 1$ and the relations $R_i(\vec{x})$, $i = 1, \ldots, n$ are in $\mathcal{PR}$ and $\mathcal{PR}_*$ respectively, then so is f above.*

Proof. Either by repeated use (composition) of if-then-else or by noting—mindful of A.3.39—that

$$f(\vec{x}) = f_1(\vec{x})(1 \dotminus \chi_{R_1}(\vec{x})) + \cdots + f_n(\vec{x})(1 \dotminus \chi_{R_n}(\vec{x})) + \\ f_{n+1}(\vec{x})(1 \dotminus \chi_{\neg(R_1 \vee \cdots \vee R_n)}(\vec{x})) \qquad \square$$

A.3.58 Corollary. *Same statement as above, replacing $\mathcal{PR}$ and $\mathcal{PR}_*$ by $\mathcal{R}$ and $\mathcal{R}_*$ respectively.*

The tools we now have at our disposal allow easy certification of the primitive recursiveness of some very useful functions and relations. But first a definition:

A.3.59 Definition. $(\mu y)_{<z} R(y, \vec{x})$ means $(\mu y)_{<z} \chi_R(y, \vec{x})$. $\qquad \square$

Thus, if $R(y, \vec{x}) \in \mathcal{PR}_*$ (resp. $\in \mathcal{R}_*$), then $\lambda z \vec{x}.(\mu y)_{<z} R(y, \vec{x}) \in \mathcal{PR}$ (resp. $\in \mathcal{R}$), since $\chi_R \in \mathcal{PR}$ (resp. $\in \mathcal{R}$).

A.3.60 Example. The following are in $\mathcal{PR}$ or $\mathcal{PR}_*$ as appropriate:

(1) $\lambda xy.\left\lfloor \dfrac{x}{y} \right\rfloor$ [18] (the quotient of the division x/y). This is another instance of a nontotal function with an "obvious" way to remove the points where it is undefined. Thus the symbol is extended to *mean* $(\mu z)_{\leq x}\big((z + 1)y > x\big)$ for all x, y. It follows that, for $y > 0$, $\lfloor x/y \rfloor$ is as expected in normal math, while $\lfloor x/0 \rfloor = x + 1$.

(2) $\lambda xy.rem(x, y)$ (the remainder of the division x/y). $rem(x, y) = x \dotminus y\lfloor x/y \rfloor$.

---

[18] The symbol "$\lfloor x \rfloor$" is called the *floor* of x. It succeeds in the literature (with the same definition) the so-called "greatest integer function, $[x]$", i.e., the *integer part* of the real number x.

(3) $\lambda xy.x|y$ (x divides y). $x|y \equiv rem(y,x) = 0$. Note that if $y > 0$, we cannot have $0|y$—a good thing!—since $rem(y,0) = y$. Our redefinition of $\lfloor x/y \rfloor$ yields, however, $0|0$, but we can live with this in practice.

(4) $Pr(x)$ (x is a prime). $Pr(x) \equiv x > 1 \wedge (\forall y)_{\leq x}(y|x \rightarrow y = 1 \vee y = x)$.

(5) $\pi(x)$ (the number of primes $\leq x$).[19] The following primitive recursion certifies the claim: $\pi(0) = 0$, and $\pi(x+1) =$ if $Pr(x+1)$ then $\pi(x) + 1$ else $\pi(x)$.

(6) $\lambda n.p_n$ (the nth prime). First note that the graph $y = p_n$ is primitive recursive: $y = p_n \equiv Pr(y) \wedge \pi(y) = n + 1$. Next note that, for all n, $p_n \leq 2^{2^n}$ (see Exercise A.3.61 below), thus $p_n = (\mu y)_{\leq 2^{2^n}}(y = p_n)$, which settles the claim.

(7) $\lambda nx.\exp(n,x)$ (the exponent of p_n in the prime factorization of x). $\exp(n,x) = (\mu y)_{\leq x}\neg(p_n^{y+1}|x)$.

(8) $Seq(x)$ (x's prime number factorization contains at least one prime, but no gaps) $Seq(x) \equiv x > 1 \wedge (\forall y)_{\leq x}(\forall z)_{\leq x}(Pr(y) \wedge Pr(z) \wedge y < z \wedge z|x \rightarrow y|x)$.   □

A.3.61 Exercise. Prove by induction on n, that for all n we have $p_n \leq 2^{2^n}$.

Hint. Consider, as Euclid did,[20] $p_0 p_1 \cdots p_n + 1$. If this number is prime, then it is greater than or equal to p_{n+1} (why?). If it is composite, then none of the primes up to p_n divide it. So any prime factor of it is greater than or equal to p_{n+1} (why?).   □

A.3.62 Exercise. Prove that $\lambda x.\lfloor \log_2 x \rfloor \in \mathcal{PR}$. Remove the undefinedness at $x = 0$ in some convenient manner.   □

A.3.63 Definition. (Coding Sequences) Any sequence of numbers, $a_0, \ldots, a_n, n \geq 0$, is *coded* by the number $\langle a_0, \ldots, a_n \rangle$ defined as

$$\prod_{i \leq n} p_i^{a_i+1}$$ □

For coding to be useful, we need a simple decoding scheme. Define then the expressions:

(i) $(z)_i$ as shorthand for $\exp(i,z) \dotminus 1$

(ii) $lh(z)$ (pronounced "length of z") as shorthand for $(\mu y)_{\leq z}\neg(p_y|z)$

Note that

(a) $\lambda iz.(z)_i$ and $\lambda z.lh(z)$ are in $\mathcal{PR}$.

---

[19]The π-function plays a central role in number theory figuring in the so-called *prime number theorem*. See for example, [30].

[20]In his proof that there are infinitely many primes.

(b) If $Seq(z)$, then $z = \langle a_0, \ldots, a_n \rangle$ for some $a_0, \ldots, a_n$. In this case, $lh(z)$ equals the number of distinct primes in the decomposition of z, that is, the length $n + 1$ of the coded sequence. Then $(z)_i$, for $i < lh(z)$, equals a_i. For larger i, $(z)_i = 0$. Note that if $\neg Seq(z)$ then $lh(z)$ need not equal the number of distinct primes in the decomposition of z. For example, 10 has 2 primes, but $lh(10) = 1$.

The tools lh, $Seq(z)$, and $\lambda iz.(z)_i$ are sufficient to perform *decoding*, primitive recursively, once the truth of $Seq(z)$ is established. This coding/decoding is essentially that of Gödel's ([16]) and we will use it in the following subsection.

We conclude this subsection with a flexible (seeming) extension of primitive recursion. Simple primitive recursion defines a function "at $n + 1$" in terms of its value "at n". However we also have examples of "recursions" (or "recurrences"), one of the best known perhaps being the *Fibonacci sequence*, $0, 1, 1, 2, 3, 5, 8, \ldots$, that is given by $F_0 = 0$, $F_1 = 1$ and (for $n > 1$) $F_{n+1} = F_n + F_{n-1}$, where the value at $n+1$ depends on both the values at n and $n - 1$. This generalizes to recursions where the value at $n+1$ depends on the entire *history*, or *course-of-values*, of the function values at $n, n - 1, n - 2, \ldots, 1, 0$. Compare with the contrast between simple induction and "strong" induction (cf. "A crash course on induction", p. 17). The easiest way to utilize the course of values of a function f: $f(0, \vec{y}), f(1, \vec{y}), \ldots, f(x, \vec{y})$ prior to $x + 1$, or "at x", is to code it by a single number!

A.3.64 Definition. (Course-of-Values Recursion) We say that f, of $n + 1$ arguments, is defined by a *basis* function $\lambda \vec{y}_n.b(\vec{y}_n)$ and an *iterator* $\lambda x \vec{y}_n z.g(x, \vec{y}_n, z)$ by *course-of-values recursion* if for all $x, \vec{y}_n$ the following equations hold

$$f(0, \vec{y}_n) = b(\vec{y}_n)$$
$$f(x + 1, \vec{y}_n) = g(x, \vec{y}_n, H(x, \vec{y}_n))$$

where $\lambda x \vec{y}_n.H(x, \vec{y}_n)$ is the *history function*, which "at x" is given (for all $\vec{y}_n$) by

$$\langle f(0, \vec{y}), f(1, \vec{y}), \ldots, f(x, \vec{y}) \rangle \qquad \square$$

The major result here is:

A.3.65 Theorem. $\mathcal{PR}$ *is closed under course-of-values recursion.*

Proof. So, let b and g be in $\mathcal{PR}$. We will show that $f \in \mathcal{PR}$. It suffices to prove that the history function H is primitive recursive, for then $f = \lambda x \vec{y}_n.\big(H(x, \vec{y}_n)\big)_x$ and we are done by Grzegorczyk substitution. To this end, the following equations—true for all $x, \vec{y}_n$—settle the case:

$$H(0, \vec{y}_n) = \langle b(\vec{y}_n) \rangle$$
$$H(x + 1, \vec{y}_n) = H(x, \vec{y}_n) p_{x+1}^{g(x, \vec{y}_n, H(x, \vec{y}_n))} \qquad \square$$

A.3.66 Example. The Fibonacci sequence, $(F_n)_{n \geq 0}$, can be viewed as the function $\lambda n.F_n$. As such it is in $\mathcal{PR}$. Indeed, letting H_n be the history of the sequence at n—that is, $\langle F_0, \ldots, F_n \rangle$—we have the following course-of-values recursion for $\lambda n.F_n$ in terms of functions known to be in $\mathcal{PR}$.

$$F_0 = 0$$
$$F_{n+1} = \text{if } n = 0 \text{ then } 1$$
$$\text{else } (H_n)_n + (H_n)_{n \dot{-} 1} \qquad \square$$

A.3.3 URM Computations

As an "agent" executes some URM's, M, instructions, it generates at each step *instantaneous descriptions* (*IDs*) of a computation. The information each such description includes is simply the values of each variable of M, and the label (instruction number) of *the instruction that is about to be executed next*—the *current* instruction.

In this subsection we will *arithmetize* URMs and their computations—just as Gödel did in the case of formal arithmetic and its proofs ([16])—and prove a cornerstone result of computability, the "normal form theorem" of Kleene that, essentially, says that the URM programming language is rich enough to allow us write a *universal program for functions*. Such a program, U, receives two inputs: One is a URM description, M, and the other is "data", x. U then simulates M on the data, behaving exactly as M would on input x. Programmers may call such a program an *interpreter* or *compiler*.

Toward this end, we will first define a relation $T(z, x, y)$—known in the literature as the *Kleene T-predicate*—that is true iff the URM coded by z, when it receives input x in its variable $X1$, will have a terminating computation that is coded by y. Thus we turn to coding URMs and URM computations.

 Normalizing Input/Output: There is clearly no loss of generality in assuming that any URM that computes a unary function does so using X1 as input *and* output variable. Such a URM will have at least two instructions, since the **stop** instruction does not reference any variables.

We arithmetize or code a URM M as a sequence of numbers—coded by a single number as in A.3.63—where each number of the sequence is the code of some instruction.

A.3.67 Definition. (Codes for Instructions) The instructions are coded as follows, where $X1^i$ is short for

$$X\overbrace{1 \cdots 1}^{i \text{ ones}}$$

(1) $L: X1^i \leftarrow a$ has code $\langle 1, L, i, a \rangle$.

(2) $L: X1^i \leftarrow X1^i + 1$ has code $\langle 2, L, i \rangle$.

(3) $L: X1^i \leftarrow X1^i \dot{-} 1$ has code $\langle 3, L, i \rangle$.

(4) $L: \textbf{if } X1^i = 0 \textbf{ goto } P \textbf{ else goto } R$ has code $\langle 4, L, i, P, R \rangle$.

(5) $L :$ **stop** has code $\langle 5, L \rangle$. $\square$

The first component of each instruction code z, $(z)_0$ denotes the *instruction type*, the second—$(z)_1$—the label, and the remaining components give enough information for us to uniquely know what precise instruction we are talking about. For example, in $z = \langle 3, L, i \rangle$ we read that we are talking about the "decrement by one" instruction $((z)_0 = 3)$ applied to $X1^i$ $((z)_2 = i)$, which is found at label L $((z)_1 = L)$.

In turn, we code a URM M as an ordered sequence of numbers, each being a code for an instruction. Thus given a code z (i.e., z codes something: $Seq(z)$ holds) we can determine algorithmically whether z codes some URM. More precisely:

A.3.68 Theorem. *The relation $URM(z)$ that holds precisely if z codes a URM is in* $\mathcal{PR}_*$.

Proof. In what follows we employ shorthand such as $(\exists z, w)_{<u}$ for $(\exists z)_{<u}(\exists w)_{<u}$, and similarly for longer quantifier groupings and for $\forall$.

$$
\begin{aligned}
URM(z) \equiv\ & Seq(z) \wedge (z)_{lh(z) \dot- 1} = \langle 5, lh(z) \rangle^{21} \wedge \\
& (\forall i)_{<lh(z)}(i \neq lh(z) \dot- 1 \rightarrow ((z)_i)_0 \neq 5) \wedge \\
& (\forall L)_{<lh(z)}\Big(Seq((z)_L) \wedge \\
& \Big[(\exists i, a)_{\leq z}(z)_L = \langle 1, L+1, i+1, a \rangle \vee \\
& (\exists i)_{\leq z}\Big\{ (z)_L = \langle 2, L+1, i+1 \rangle \vee \\
& (z)_L = \langle 3, L+1, i+1 \rangle \vee \\
& (\exists M, R)_{<lh(z)}(z)_L \\
& = \langle 4, L+1, i+1, M+1, R+1 \rangle \Big\} \Big] \Big)
\end{aligned}
$$
$\square$

A.3.69 Definition. An ID of a computation of a URM M is an ordered sequence $L; a_1, \dots, a_r$, where all of M's variables are among the $X1, X11, \dots X1^r$ that we will denote by $\mathbf{x}_1, \dots, \mathbf{x}_r$, and a_i is the current value of $\mathbf{x}_i$ immediately before instruction L is executed. L points precisely to the *current instruction*, meaning the next to be executed.

All IDs have the same length, and we say that ID $I_1 = L; a_1, \dots, a_r$ *yields* ID $I_2 = P; b_1, \dots, b_r$, in symbols $I_1 \vdash I_2$, exactly when

(i) L labels "$\mathbf{x}_i \leftarrow c$", and I_1 and I_2 are identical, except that $b_i = c$ and $P = L+1$.

---

[21] Note that $z = \langle (z)_0, \dots, (z)_{lh(z) \dot- 1} \rangle$, but we have used positive labels; thus the last label is $lh(z)$. Similar comment about "$(\exists i, a)_{\leq z}(z)_L = \langle 1, L+1, i+1, a \rangle$", etc. Why $i+1$? Because the variables are $X1, X11, X111, \dots$

(ii) L labels "$x_i \leftarrow x_i + 1$", and I_1 and I_2 are identical, except that $b_i = a_i + 1$ and $P = L + 1$.

(iii) L labels "$x_i \leftarrow x_i \dot- 1$", and I_1 and I_2 are identical, except that $b_i = a_i \dot- 1$ and $P = L + 1$.

(iv) L labels "if $x_i = 0$ goto R else goto Q", and I_1 and I_2 are identical, except that $P = R$ if $a_i = 0$, while $P = Q$ otherwise.

(v) L labels "**stop**", and I_1 and I_2 are identical.

A *terminating computation* of M with input $a_1, \ldots, a_k$ is a sequence $I_1, \ldots, I_n$ such that for all $i < n$ we have $I_i \vdash I_{i+1}$ and for some $j \leq n$, I_j has as 0th member the label of **stop**. Moreover, I_1 is *initial*; that is, if the input variables of M are, without loss of generality, the variables $x_1, \ldots, x_k$, then

$$I_1 = 1; a_1, \ldots, a_k, \underbrace{0, \ldots, 0}_{r-k \text{ 0s}} \qquad \square$$

We code an ID $I = L; a_1, \ldots, a_r$ as $code(I) = \langle L, a_1, \ldots, a_r \rangle$ and a terminating computation $I_1, \ldots, I_n$ by $\langle code(I_1), \ldots, code(I_n) \rangle$.

A.3.70 Theorem. *The relation $Comp(z, y)$ that is true iff y codes a terminating computation of the URM coded by z, which has "$X1$" as its only input variable (cf. preambular remarks of subsection A.3.3, p. 254), is primitive recursive.*

Proof. By the remark on p. 254 that normalizes the input/output convention, it must be that $lh(y) \geq 2$. Definition A.3.69 allows us the technical convenience to include into an ID enough room for more variables than may actually be present in the URM (that is coded by) z. Clearly (cf. A.3.67), a generous allowance for the length of an ID, as this is determined by the largest j such that $X1^j$ occurs in z, is $\max\{(z)_i : i < lh(z)\}$. Even simpler (and more generous) is just z, which is the one we will work with. Thus, our IDs will each have length $z + 1$, allowing for the label information. Observe next that

$$Comp(z, y) \equiv URM(z) \wedge Seq(y) \wedge (\forall i)_{<lh(y)} \big[Seq((y)_i) \wedge lh((y)_i) = z + 1 \big] \wedge$$
$$lh(y) > 1 \wedge (\forall j)_{<lh(y)\dot-1} yield(z, (y)_j, (y)_{j+1}) \wedge$$

{**Comment.** The last ID surely has the label of z's **stop**.} $((y)_{lh(y)\dot-1})_0 = lh(z) \wedge$

{**Comment.** The initial ID.} $((y)_0)_0 = 1 \wedge (\forall i)_{\leq z} (1 < i \to ((y)_0)_i = 0)$

The relation "$yield(z, (y)_j, (y)_{j+1})$" above says "URM z causes $(y)_j \vdash (y)_{j+1}$". The notation "$yield(z, u, v)$" is thus shorthand that expands as follows (cf. A.3.69 and A.3.67):

$$yield(z, u, v) \equiv (\exists k)_{\leq z}(\exists L)_{<lh(z)}\Big(L+1 = (u)_0 \wedge k > 0 \wedge \Big\{$$

$$(\exists a)_{\leq z}\big((z)_L = \langle 1, L+1, k, a \rangle \wedge v = 2p_k^{a+1}\lfloor u/p_k^{\exp(k,u)} \rfloor 2^2\big) \vee$$

$$((z)_L = \langle 2, L+1, k \rangle \wedge v = 2p_k u) \vee$$

$$((z)_L = \langle 3, L+1, k \rangle \wedge v = 2(\text{if } (u)_k = 0 \text{ then } u \text{ else } \lfloor u/p_k \rfloor)) \vee$$

$$(\exists P, R)_{\leq lh(z)}\big((z)_L = \langle 4, L+1, k, P, R \rangle \wedge P > 0 \wedge R > 0 \wedge$$

$$v = \text{if } (u)_k = 0 \text{ then } \lfloor u/2^{L+2}\rfloor 2^{P+1}$$

$$\text{else } \lfloor u/2^{L+2} \rfloor 2^{R+1}\big) \vee$$

$$((z)_L = \langle 5, L+1 \rangle \wedge v = u)\Big\}\Big) \qquad \qquad \square$$

A.3.71 Corollary. (The Kleene T-predicate) *The Kleene predicate $T(z, x, y)$ that is true precisely when the URM z with input x has a terminating computation y, is primitive recursive.*

Proof. By earlier remarks, $T(z, x, y) \equiv Comp(z, y) \wedge ((y)_0)_1 = x.$ $\qquad \square$

Noting that for any predicate $R(y, \vec{x})$, $(\mu y)R(y, \vec{x})$ is alternative notation for $(\mu y)\chi_R(y, \vec{x})$, we have:

A.3.72 Corollary. (The Kleene Normal Form Theorem)

(1) *For any URM M, if z is its code (A.3.67), then we have M_{X1}^{X1} is defined on input x iff $(\exists y)T(z, x, y)$.*

(2) *There is a primitive recursive function d such that for any $\lambda x.f(x) \in \mathcal{P}$ there is a number z and we have for all x:*

$$f(x) = d\big((\mu y)T(z, x, y)\big)$$

Proof. Statement (1) is immediate as "$(\exists y)T(z, x, y)$" says that there is a terminating computation of M (coded as z) on input x.

For (2), we first remark that "=" means that the two sides are both undefined, or both defined and numerically equal. Now, the role of d is to extract from a terminating computation's *last* ID its 1st component—recalling our definition, A.3.2, *and* our convention that unary functions are computed, without loss of generality (cf. p. 254), as M_{X1}^{X1}, for various M. Thus, for all y, $d(y) = \big((y)_{lh(y) \dot- 1}\big)_1$ will do. $\qquad \square$

---

[22] The effect of "$L+1 : X1^k \leftarrow a$" on ID $u = \langle L+1, \ldots \rangle$ is to change $L+1$ to $L+2$ (effected by the factor 2) and change the current value of $X1^k$, $(u)_k$—stored in the ID as a factor $p_k^{\exp(k,u)}$, a factor that we remove by dividing u by it—to a, this being stored in v as a factor p_k^{a+1}.

A.3.73 Remark. The *normal form theorem* says that every computable unary function, in the technical sense of A.3.2, can be expressed as an unbounded search followed by a composition, using a toolbox of just *two* primitive recursive functions(!): d and $\lambda zxy.\chi_T(z,x,y)$. This representation, or "normal form", is parametrized by z, which denotes a URM M that computes the function in a normalized manner: as M_{X1}^{X1}. Thus what we set out to do at the beginning of this section is done: The two-input URM U that computes $\lambda zx.d\big((\mu y)T(z,x,y)\big)$—clearly a computable function this, by closure properties of $\mathcal{P}$ (A.3.11 and A.3.18) and A.3.32—*is* universal, just as compilers are in computing. U accepts as inputs a program M coded as a number z, and data for said program, x. It then acts exactly as program z would on x, i.e., as M_{X1}^{X1}. □

A.3.74 Definition. (Rogers's ϕ-Notation ([41]) We denote by ϕ_z the zth partial recursive function, in the sense that, for all z, $\phi_z = \lambda x.d\big((\mu y)T(z,x,y)\big)$. □

A.3.75 Remark. (1) From Definition A.3.67 it is clear that *not* every $z \in \mathbb{N}$ represents a URM. Nevertheless, Definition A.3.74 indexes *all* partial computable functions using *all* numbers from $\mathbb{N}$ as indices, not just those z that represent URMs. This is so because the term "$d\big((\mu y)T(z,x,y)\big)$" is meaningful for *any* z. Thus, if z is *not* a URM code, then $T(z,x,y)$ will simply be false for all x and all y; thus $\phi_z(x) \uparrow$ for all x. This is fine! Indeed it is consistent with the phenomenon where a real-life computer program that is not syntactically correct (like our z here) will not be translated by the compiler and thus will not run. Therefore, for any input it will decline to offer an output; the corresponding function will be totally undefined.

(2) Definition A.3.2, for the unary case, can now be rephrased as "$\lambda x.f(x) \in \mathcal{P}$ iff, for some $z \in \mathbb{N}$, $f = \phi_z$". We say that z is a "ϕ-index" of f. □

A.3.76 Exercise. Prove that every function of $\mathcal{P}$ has infinitely many ϕ-indices.

Hint. There are infinitely many ways to modify a program and yet have all programs so obtained compute the same function. □

A.3.77 Example. The nowhere-defined function can also be obtained from a program that compiles all right. Setting $\widetilde{S} = \lambda yx.x + 1$ we note:

(1) $\lambda x.(\mu y)\widetilde{S}(y,x) \in \mathcal{P}$ by A.3.32 and A.3.18.

(2) By the techniques of A.3.16 we can write a program for $\lambda x.(\mu y)\widetilde{S}(y,x)$.

As a side-effect we have that $\mathcal{PR} \neq \mathcal{P}$ and $\mathcal{R} \neq \mathcal{P}$. □

A.3.78 Exercise. (URM-independent Characterization of $\mathcal{P}$) Define the concept of $\mathcal{P}$-*derivations* as in A.3.20, however, adding a 4th case of what we may write at each step: We may also write $\lambda\vec{x}.(\mu y)f(y,\vec{x})$, if $\lambda y\vec{x}.f(y,\vec{x})$ is already written.

Prove:

(1) $f \in \mathcal{P}$ (as in Definition A.3.2) iff f appears in some $\mathcal{P}$-derivation.

(2) Of all possible sets of functions that include $\mathcal{I}$ and are closed under primitive recursion, composition and unbounded search, $\mathcal{P}$ is the smallest with respect to inclusion. □

A.3.79 Remark. (Case of Many Inputs) It has been convenient to present the normal form theorem, in particular the Kleene predicate, in terms of unary (1-ary) functions. It is good to know that in doing so, in the presence of coding, we did not restrict generality at all. Indeed, if $\lambda \vec{x}_n . f(\vec{x}_n) \in \mathcal{P}$, then so is $g = \lambda z . f((z)_0, \ldots, (z)_n)$ by composition. Thus, every n-input computable function is expressible via coding through a unary computable function: For all $\vec{x}_n$, we have $f(\vec{x}_n) = g(\langle \vec{x}_n \rangle)$.

In particular, if $g = \phi_i$, then for all $\vec{x}_n$, we have $f(\vec{x}_n) = \phi_i(\langle \vec{x}_n \rangle)$ and hence (by A.3.72) $f(\vec{x}_n) = d((\mu y) T(i, \langle \vec{x}_n \rangle, y))$. We call i a ϕ-index of f.

It is customary in the literature to introduce the notations (for $n > 1$) $T^{(n)}(z, \vec{x}_n, y)$, the n-input Kleene predicate, as shorthand for $T(z, \langle \vec{x}_n \rangle, y)$, and $\phi_a^{(n)}$ as shorthand for $\lambda \vec{x}_n . \phi_a(\langle \vec{x}_n \rangle)$. Thus the n variables normal form theorem can be expressed as: For all $\vec{x}_n$, $\phi^{(n)}(\vec{x}_n) = d((\mu y) T^{(n)}(z, \vec{x}_n, y))$. $\qquad\square$

A.3.4 Semi-computable Relations; Unsolvability

We next define a $\mathcal{P}$-counterpart of $\mathcal{R}_*$ and $\mathcal{PR}_*$ and look into some of its closure properties.

A.3.80 Definition. (Semi-computable Relations) A relation $P(\vec{x})$ is called *semi-computable* iff for some $f \in \mathcal{P}$, we have, for all $\vec{x}_n$,

$$P(\vec{x}_n) \equiv f(\vec{x}_n) \downarrow \qquad (1)$$

The set of all semi-computable relations is denoted by $\mathcal{P}_*$.[23]

If $f = \phi_a^{(n)}$ in (1) above, then we say that "a is *a semi-computable index* or just a *semi-index* of $P(\vec{x}_n)$". If $n = 1$ (thus $P \subseteq \mathbb{N}$) and a is one of the semi-indices of P, then we write $P = W_a$ ([41]). $\qquad\square$

We have at once:

A.3.81 Theorem. (Normal Form Theorem for Semi-Computable Relations) $P(\vec{x}_n) \in \mathcal{P}_*$ iff, for some $a \in \mathbb{N}$, we have (for all $\vec{x}_n$) $P(\vec{x}_n) \equiv (\exists z) T^{(n)}(a, \vec{x}_n, z)$.

Proof. Only if-part. Let $P(\vec{x}_n) \equiv f(\vec{x}_n) \downarrow$, with $f \in \mathcal{P}$. By A.3.79 $f = \phi_a^{(n)}$ for some $a \in \mathbb{N}$.

If-part: By A.3.80 and A.3.79, $P(\vec{x}_n) \equiv \phi_a^{(n)}(\vec{x}_n) \downarrow$. But $\phi_a \in \mathcal{P}$. $\qquad\square$

Rephrasing the above (hiding the "a" and remembering that $\mathcal{PR}_* \subseteq \mathcal{R}_*$), we have:

A.3.82 Corollary. (Strong Projection Theorem) $P(\vec{x}_n) \in \mathcal{P}_*$ iff, for some recursive predicate $Q(\vec{x}_n, z)$, we have (for all $\vec{x}_n$) $P(\vec{x}_n) \equiv (\exists z) Q(\vec{x}_n, z)$.

---

[23]We are making this symbol up. It is not standard in the literature.

Proof. For the *only if*, take $Q(\vec{x}_n, z)$ to be $\lambda \vec{x}_n z . T^{(n)}(a, \vec{x}_n, z)$ for appropriate $a \in \mathbb{N}$. For the *if,*, take $f = \lambda \vec{x}_n . (\mu z) Q(\vec{x}_n, z)$. Then $f \in \mathcal{P}$ and $P(\vec{x}_n) \equiv f(\vec{x}_n) \downarrow$. □

A.3.83 Remark. (Deciders and Verifiers) A computable relation $P(\vec{x}_n)$ is, by definition, one for which $\chi_P \in \mathcal{R}$; thus it has an associated URM M that *decides membership* of any $\vec{a}_n$ in P:[24] "yes" (output 0) if it is in, "no" (output 1) if it is not. Thus this M is a *decider* for $P(\vec{x}_n)$.

A semi-computable relation $Q(\vec{x}_m)$, on the other hand, comes equipped only with a *verifier*, i.e., a URM N that verifies $\vec{a}_m \in Q$, *if true*, by virtue of halting on input $\vec{a}_m$.

While *mathematically* speaking $\vec{a}_m \notin Q$ is also "verified" by virtue of *looping forever* on input $\vec{a}_m$, *algorithmically speaking* this is *no verification at all* as we do *not* have a way of knowing whether N is looping forever as opposed to being awfully sluggish and being about to halt in a couple of trillion years (cf. halting problem A.3.88).

In the algorithmic sense, a verifier (of a semi-computable set of m-tuples) verifies *only* the "yes" instances of questions such as "Is $\vec{a}_m \in Q$?" □

Clearly, though, if we have a verifier for a relation $Q(\vec{x}_n)$ *and* also have a verifier for its *complement* $\neg Q(\vec{x}_n)$, then we can build a decider for $Q(\vec{x}_n)$: On input $\vec{a}_n$ we simply run both verifiers simultaneously. If the one for Q halts, we print 0 and stop the computation; if the one for $\neg Q$ halts, we print 1 and stop. This computes $\chi_Q(\vec{a}_n)$. Put more mathematically,

A.3.84 Proposition. *If $Q(\vec{x}_n)$ and $\neg Q(\vec{x}_n)$ are in $\mathcal{P}_*$, then both are in $\mathcal{R}_*$.*

Proof. Let i and j be semi-indices of Q and $\neg Q$ respectively, that is (A.3.81),

$$Q(\vec{x}_n) \equiv (\exists z) T^{(n)}(i, \vec{x}_n, z)$$
$$\neg Q(\vec{x}_n) \equiv (\exists z) T^{(n)}(j, \vec{x}_n, z)$$

Define

$$g = \lambda \vec{x}_n . (\mu z) \big(T^{(n)}(i, \vec{x}_n, z) \vee T^{(n)}(j, \vec{x}_n, z) \big)$$

Trivially, $g \in \mathcal{P}$. Hence, $g \in \mathcal{R}$, since it is total (why?). We are done by noticing that $Q(\vec{x}_n) \equiv T^{(n)}(i, \vec{x}_n, g(\vec{x}_n))$. By closure properties of $\mathcal{R}_*$ (A.3.41), $\neg Q(\vec{x}_n)$ is in $\mathcal{R}_*$, too. □

A.3.85 Proposition. $\mathcal{R}_* \subseteq \mathcal{P}_*$.

Proof. Let $Q(\vec{x}) \in \mathcal{R}_*$ and y be a new variable (other than any of the $\vec{x}$). By "$\vdash A \equiv (\exists \mathbf{x}) A$ if $\mathbf{x}$ is not free in A" (cf. p. 188, Exercise 11) and soundness, we have $Q(\vec{x}) \equiv (\exists y) Q(\vec{x})$ is true in the metatheory (where we are developing this section on computability). By A.3.82, $Q(\vec{x}) \in \mathcal{P}_*$. □

---

[24]"$\vec{a}_n \in P$" (set notation) is synonymous with "$P(\vec{a}_n)$ holds" or just "$P(\vec{a}_n)$" (relational notation).

A.3.86 Definition. (Unsolvable or Undecidable Problems) A *problem* is a question "$\vec{x}_n \in R$?" for some set of n-tuples R. "The problem $\vec{x}_n \in R$ is *recursively unsolvable*", or just *unsolvable*, or *undecidable*, means that the set R—equivalently, the relation $\vec{x}_n \in R$ or $R(\vec{x}_n)$—is *not* in $\mathcal{R}_*$. Put colloquially, there is *no* URM-programmable solution for the problem; there is no decider for the question "Is $\vec{x}_n \in R$?".

The *halting problem* has central significance in computability. It is the question whether "program x will ever halt if it starts computing on input x". That is, if we set $K = \{x : \phi_x(x) \downarrow\}$, then the halting problem is $x \in K$. We denote the complement of K by $\overline{K}$. $\qquad\square$

A.3.87 Exercise. The halting problem $x \in K$ is semi-recursive.

Hint. The problem is "$\phi_x(x) \downarrow$". Now invoke the normal form theorem (A.3.72(1)). $\qquad\square$

A.3.88 Theorem. (Unsolvability of the Halting Problem) *The halting problem is unsolvable.*

Proof. In view of the preceding exercise (and A.3.84), it suffices to show that $\overline{K}$ is not semi-computable. Suppose instead that i is a semi-index of the set. Thus, $x \in \overline{K} \equiv (\exists z)T(i, x, z)$, or, making the part $x \in \overline{K}$—that is, $\phi_x(x) \uparrow$—explicit:

$$\neg(\exists z)T(x, x, z) \equiv (\exists z)T(i, x, z) \tag{1}$$

Substituting i into x in (1) we get a contradiction. $\qquad\square$

A.3.89 Remark. (1) Since $K \in \mathcal{P}_*$, we conclude that the inclusion $\mathcal{R}_* \subseteq \mathcal{P}_*$ (A.3.85) is proper, i.e., $\mathcal{R}_* \subset \mathcal{P}_*$.

(2) The characteristic function of K provides an example of a *total* uncomputable function.

(3) In A.3.34 we saw an example of how to remove "points of nondefinition" from a function so that it remains computable but has been now extended to a total function. Can we always do that? No; for example, the function $f = \lambda x.\phi_x(x) + 1$ *cannot* be extended to a *total* computable function. Of course, by A.3.72, $f \in \mathcal{P}$. Here is why: Suppose that $g \in \mathcal{R}$ extends f. Thus, $g = \phi_i$ for some i. Let us look at $g(i)$: We have

$$g(i) \underset{\text{by } g = \phi_i}{=} \phi_i(i) \underset{\text{both sides defined}}{\neq} \phi_i(i) + 1 \underset{\text{def. of } f}{=} f(i)$$

But since $f(i) \downarrow$, we also have $g(i) = f(i)$ as g extends f; a contradiction. $\qquad\square$

A.3.90 Theorem. (Closure Properties of $\mathcal{P}_*$) $\mathcal{P}_*$ *is closed under* $\vee$, $\wedge$, $(\exists y)_{<z}$, $(\exists y)$, *and* $(\forall y)_{<z}$. *It is* not *closed under either* $\neg$ *or* $(\forall y)$.

Proof. Given semi-computable relations $P(\vec{x}_n)$, $Q(\vec{y}_m)$ and $R(y, \vec{u}_k)$ of semi-indices p, q, r respectively. In each case we will express the relation we want to prove semi-computable as a strong projection (A.3.82):

$\vee$

$$P(\vec{x}_n) \vee Q(\vec{y}_m) \equiv (\exists z)T^{(n)}(p, \vec{x}_n, z) \vee (\exists z)T^{(m)}(q, \vec{y}_m, z)$$
$$\equiv (\exists z)\big(T^{(n)}(p, \vec{x}_n, z) \vee T^{(m)}(q, \vec{y}_m, z)\big)$$

$\wedge$

$$P(\vec{x}_n) \wedge Q(\vec{y}_m) \equiv (\exists z)T^{(n)}(p, \vec{x}_n, z) \wedge (\exists z)T^{(m)}(q, \vec{y}_m, z)$$
$$\equiv (\exists w)\big((\exists z)_{<w}T^{(n)}(p, \vec{x}_n, z) \wedge (\exists z)_{<w}T^{(m)}(q, \vec{y}_m, z)\big)$$

$(\exists y)_{<z}$

$$(\exists y)_{<z}R(y, \vec{u}_k) \equiv (\exists y)_{<z}(\exists w)T^{(k+1)}(r, y, \vec{u}_k, w)$$
$$\equiv (\exists w)(\exists y)_{<z}T^{(k+1)}(r, y, \vec{u}_k, w)$$

$(\exists y)$

$$(\exists y)R(y, \vec{u}_k) \equiv (\exists y)(\exists w)T^{(k+1)}(r, y, \vec{u}_k, w)$$
$$\equiv (\exists z)(\exists y)_{<z}(\exists w)_{<z}T^{(k+1)}(r, y, \vec{u}_k, w)$$

$(\forall y)_{<z}$

$$(\forall y)_{<z}R(y, \vec{u}_k) \equiv (\forall y)_{<z}(\exists w)T^{(k+1)}(r, y, \vec{u}_k, w)$$
$$\equiv (\exists v)(\forall y)_{<z}(\exists w)_{<v}T^{(k+1)}(r, y, \vec{u}_k, w)$$

As for possible closure under $\neg$ and $\forall y$, K provides a counterexample to $\neg$, and $\neg T(x, x, y)$ provides a counterexample to $\forall y$. $\qquad\square$

A.3.91 Remark. (Computably Enumerable Sets) There is an interesting characterization of *nonempty* semi-computable sets that is found in all introductions to the theory of computation. These sets are precisely those that can be "enumerated effectively" or "computably", that is,

A nonempty set $S \subseteq \mathbb{N}$ is semi-computable iff some $f \in \mathcal{PR}$ has S as its set of outputs, *or* range *as we say technically.*

Indeed, assume first that, for some semi-index i, $x \in S \equiv (\exists y)T(i, x, y)$ for all x. *Intuitively* now, we can enumerate all pairs x, y of numbers, coded as "$\langle x, y \rangle$", and for every pair that satisfies $T(i, x, y)$ output x.

Rigorously, this is accomplished by f given for all z as follows:

$$f(z) = \begin{cases} (z)_0 & \text{if } T(i, (z)_0, (z)_1) \\ a & \text{otherwise} \end{cases}$$

where "a" is some fixed member of S that we keep outputting every time the condition "$T(i, x, y)$" fails,[25] ensuring that f is total. I wrote x for $(z)_0$ and y for $(z)_1$ to connect with the preceding intuitive construction. Of course f is primitive recursive.

---

[25] Because either we did not let the computation $\phi_i(x)$ to go on long enough, or no terminating computation exists.

Conversely, if $\mathcal{S}$ is the range of some primitive recursive g, that is, $x \in \mathcal{S} \equiv (\exists y)g(y) = x$ holds for all x, we immediately get that $\mathcal{S}$ is semi-recursive by A.3.82, since the graph of g, $g(y) = x$, is in $\mathcal{PR}_*$ (cf. A.3.45).

This result justifies the nomenclature *computably enumerable* (c.e.) and also *recursively enumerable* (r.e.) for all semi-computable sets (the nomenclature applies to the empty set as well on the understanding that its members can trivially be enumerated by doing nothing). There is no loss of generality in presenting the characterization for subsets of $\mathbb{N}$ since via coding $\langle \ldots \rangle$ it can be trivially and naturally extended to sets of n-tuples for $n > 1$.   □

A.3.92 Exercise. Prove that if $\lambda\vec{x}.f(\vec{x}) \in \mathcal{P}$, then its graph $y = f(\vec{x})$ is in $\mathcal{P}_*$.   □

A.3.93 Exercise. Prove that if $y = f(\vec{x})$ is in $\mathcal{P}_*$, then $\lambda\vec{x}.f(\vec{x}) \in \mathcal{P}$.   □

A.3.94 Exercise. (Definition by Positive Cases) Consider a set of *mutually exclusive* relations $R_i(\vec{x})$, $i = 1, \ldots, n$, that is, $R_i(\vec{x}) \wedge R_j(\vec{x})$ is false for each $\vec{x}$ as long as $i \neq j$.

Then we can define a function f *by positive cases* R_i from given functions f_j by the requirement (for all $\vec{x}$) given below:

$$f(\vec{x}) = \begin{cases} f_1(\vec{x}) & \text{if } R_1(\vec{x}) \\ f_2(\vec{x}) & \text{if } R_2(\vec{x}) \\ \ldots & \ldots \\ f_n(\vec{x}) & \text{if } R_n(\vec{x}) \\ \uparrow & \text{otherwise} \end{cases}$$

Prove that if each f_i is in $\mathcal{P}$ and each of the $R_i(\vec{x})$ is in $\mathcal{P}_*$, then $f \in \mathcal{P}$.

Hint. Use A.3.92 and A.3.93 along with closure properties of $\mathcal{P}_*$ relations.   □

A.4 GÖDEL'S FIRST INCOMPLETENESS THEOREM

We prove here a *semantic version* of Gödel's first incompleteness theorem that relies on computability techniques and the semantic notion of *correctness* of an axiomatic system. In this form the theorem states that any "reasonable" axiomatic system that attempts to have as theorems *precisely* all the "true" (first-order) formulae of arithmetic will fail: There will be infinitely many true formulae that are not theorems. The qualifier *reasonable* could well be replaced by *practical*: One must be able to tell, algorithmically, whether a formula is an axiom—how else can one check a proof, let alone write one? "True" means true in the *standard interpretation* $\mathfrak{N} = (\mathbb{N}, M)$ (given below).

To set the stage more precisely, we will need some definitions and notation. In order to do arithmetic, we first need a first-order (logical) language that we use to write down formulae and proofs. The alphabet of arithmetic has as nonlogical symbols the following:

$$0, S, +, \times, <$$

These nonlogical symbols we can, of course, interpret in any way we please. However, the *standard* interpretation is given by the table below:

Abstract (language) symbol	Concrete interpretation via M
0	0 (zero)
S	$\lambda x.x + 1$
+	$\lambda xy.x + y$
×	$\lambda xy.x \times y$
<	$\lambda xy.x < y$

The alphabet has only one constant symbol; however, an arbitrary $n \in \mathbb{N}$ can be captured formally in the language by the string

$$\underbrace{SS \cdots S}_{n \text{ times}} 0$$

which we will denote by $\tilde{n}$ in order to distinguish it from the informal (metamathematical) name n. Thus, any axiomatic system (or theory) that we use to formally prove theorems of arithmetic will contain:

(1) The first-order alphabet described above[26]

(2) The language, that is, the sets of well-formed formulae (**WFF**) and of terms, **Term**

(3) A distinguished *recursive* subset of **WFF**: the special (or nonlogical) axioms for arithmetic[27]

(4) Another distinguished subset of **WFF**: the logical axioms (4.2.1)

(5) The rule of inference: modus ponens

A.4.1 Remark. Several observations will be helpful:

(i) We required that any axiomatic system we devise for arithmetic is "practical" (or "reasonable"). As we noted above, we understand this "reasonableness" as a promise to have a *decider* for axioms. This is why in (3) above we ask for a recursive set of (special) axioms.

This immediately begs the question, "But is not a 'decider' a URM that expects *numerical* inputs, as opposed to *string* inputs?"

---

[26] Of course, this language has the standard logical part that any first-order language has (cf. 4.1.2).

[27] A particularly famous choice of axioms is due to Peano—the so-called *Peano arithmetic* (PA). It has axioms that give the behavior of every nonlogical symbol, plus the induction axiom schema:

$$A[\mathbf{x} := 0] \wedge (\forall \mathbf{x})\big(A \to A[\mathbf{x} := S\mathbf{x}]\big) \to (\forall \mathbf{x})A$$

This schema gives one axiom for each choice of the formula A.

There is no difference in principle, since a number (e.g., if written in standard decimal notation) is a string over $\{0, 1, 2, 3, 4, 5, 6, 7, 8, 9\}$ and, conversely, any string over $\{0, 1, 2, 3, 4, 5, 6, 7, 8, 9\}$ naturally represents a number.

(ii) Applying this observation more generally to strings over *any finite alphabet*, not just over $\{0, 1, 2, 3, 4, 5, 6, 7, 8, 9\}$, we will take a more careful approach that disallows "0" as a digit, because its presence has undesirable side-effects.

For this reason we act as follows: Given an alphabet of $b > 1$ symbols, we first fix an order of its members and assign to every symbol, as its value, its *position number in the order*, that is, $1, 2, 3, \ldots, b$. Then any string $a_0 a_1 a_2 \cdots a_n$ of symbols a_i over this *alphabet* can be thought of as a number expressed in so-called "*b-adic*" notation (*b* being the "base" of the notation), where the symbols a_i *are b-adic digits*:

$$a_0 + a_1 b^1 + a_2 b^2 + \cdots + a_n b^n$$

Conversely, any *positive* integer can be expressed in a unique way in *b*-adic notation (i.e., unique *b*-adic digits a_i can be found) as above (cf. [39, 1, 47, 49] and Remark A.4.2 below).

In our present context, let us fix an *order* for the (finite version of the) alphabet of arithmetic displayed as (1) below. This alphabet has 35 members (separated by commas) that we view as 35-adic digits of value, each, equal to its position in (1). For example, 1 is the value of digit "x", 11 of digit "$\prime$", 12 of digit "0", 35 of digit "$<$".[28]

$$x, y, z, u, v, w, =, p, q, r, {}', 0, 1, 2, 3, 4, 5, 6, 7, 8, 9,$$
$$\top, \bot, (,), \neg, \wedge, \vee, \rightarrow, \equiv, \forall, S, +, \times, < \qquad (1)$$

Boolean variables are generated as in (4), p. 94, while object variables as in the footnote 87, p. 115. Thus *any* string over the alphabet (1) *denotes a number in* "*b-adic*" *notation*,[29] where $b = 35$. For example, the formulae $(\forall x')x' = x'$ and $0 < 1$ have numerical values

$$11 + 1 \cdot 35^1 + 7 \cdot 35^2 + 11 \cdot 35^3 + 1 \cdot 35^4 + 25 \cdot 35^5 + 11 \cdot 35^6 + 1 \cdot 35^7 + 31 \cdot 35^8 + 24 \cdot 35^9$$

and

$$13 + 35 \cdot 35^1 + 12 \cdot 35^2$$

respectively, that is, in *standard decimal notation*, 1961469340480871 and 15938 respectively.

---

[28] Programmers are aware of "hexadecimal" notation, that is, notation base-16, where the digits that are allowed are $0, 1, 2, 3, 4, 5, 6, 7, 8, 9, a, b, c, d, e, f$. Rather than using as digits "10", "15" etc., one uses "a" and "f" respectively to avoid the ambiguities of string notation that does *not* use separators between the symbols of a string.

[29] Sometimes we say a "*b*-adic number", just as we say "binary number", "hexadecimal number".

Pause. So what are the issues with the digit "0"? Why not number the symbols in (1) by 0 through 34 and work in ordinary base 35, named usually b-ary,[30] instead? Because we will have trouble with strings like $x1 < 0$. The "digit" in the most significant position is (of value) 0 and we lose information as we pass to the string's numerical value. That is, both $x1 < 0$ and $1 < 0$ denote the same number. Correspondingly, associativity of concatenation will fail. Concatenating the digits "y" and "x" and "y" of the alphabet, first as $(yx)y$ and then as $y(xy)$, yields different (numerical) results, the first $b^2 + 1$, the second $b + 1$.

At the end of this discussion we see that to speak of a set of strings over the alphabet (1) is the same as talking about a subset of $\mathbb{N}$, while terminology such as "the set **WFF** is *decidable* (or *primitive recursive*, or *semi-computable*, as the case may be)" is now technically meaningful.

(iii) Now that speaking about "recursive sets of *strings*" makes sense, we note further that the sets **WFF**, **Term**, and Λ_1 (logical axioms, cf. 4.2.1) are all recursive. Indeed, at the intuitive level we see that we can parse *algorithmically*, that is, we can write a URM that will do it for us, a string **t** to decide the question "Is $\mathbf{t} \in$ **Term**?" Rather than specifying what such a program might look like, I will rather outline how the characteristic function of the set **Term**, χ_{Term}, will be defined by course-of-values recursion from functions and relations known to be primitive recursive (in which case we just invoke A.3.65).

To this end we just follow the definition (4.1.7) and define as follows, (a)–(d), being careful to use boldface type for variable names, $\mathbf{t}, \mathbf{x}, \mathbf{y}, \mathbf{z}$ etc., so that there will be no clash with the 35-adic *digits* x, y, z, u, v, w—which are just *names for the numbers* $1, 2, 3, 4, 5, 6$—of the alphabet (1) (p. 265):

(a) If **t** is a number whose 35-adic representation is a variable or a constant, then we let $\chi_{\text{Term}}(\mathbf{t}) = 0$.

(b) If **t** is a number whose 35-adic representation is $S\mathbf{x}$, for some string **x** (over alphabet (1)), then we let $\chi_{\text{Term}}(\mathbf{t}) = 0$ precisely when $\chi_{\text{Term}}(\mathbf{x}) = 0$.

(c) If **t** is a number whose 35-adic representation is $+\mathbf{z}$, for some string **z**, then we let $\chi_{\text{Term}}(\mathbf{t}) = 0$ precisely when

$$(\exists \mathbf{x}, \mathbf{y})_{<\mathbf{t}} \big(\chi_{\text{Term}}(\mathbf{x}) = 0 \wedge \chi_{\text{Term}}(\mathbf{y}) = 0 \wedge$$
$$\mathbf{t} \text{ (expressed in 35-adic notation) is the string } + \mathbf{xy} \big)$$

(d) If **t** is a number whose 35-adic representation is $\times \mathbf{z}$, for some string **z**, then we let $\chi_{\text{Term}}(\mathbf{t}) = 0$ precisely when

$$(\exists \mathbf{x}, \mathbf{y})_{<\mathbf{t}} \big(\chi_{\text{Term}}(\mathbf{x}) = 0 \wedge \chi_{\text{Term}}(\mathbf{y}) = 0 \wedge$$

---

[30]"b-ary" signifies base-b and digit range 0 to $b - 1$. Cf. the term *binary*. "b-adic", signifies base-b but digit range 1 to b. Using digits 1 and 2, base-2, we have, *dyadic* notation, not *binary*. The term "b-adic" is due to Smullyan [47]. Interestingly, the suffix *adic* is Greek for *ary*: Cf. *dyadic* vs. *binary* (*dyo* or "$\delta\upsilon o$" is Greek for *two*).

$\mathbf{t}$ (expressed in 35-adic notation) is the string $\times \mathbf{xy})^{31}$

(e) We set $\chi_{\text{Term}}(\mathbf{t}) = 1$ in all other cases.

Similarly, one can outline a course-of-values recursion for the characteristic function of **WFF**, thus showing the primitive recursiveness of the set (equivalently, of the membership relation). Finally, the logical axioms, i.e., the partial generalizations of formulae in the groups **Ax1–Ax6**, can be algorithmically recognized from their *shape* (**Ax2–Ax6**) or via truth tables (**Ax1**) once the partial generalization prefix is removed.

It is also worth noting that the only rule of inference (thinking of logic 2, cf. Chapter 5) is algorithmically applicable. Indeed, the configuration

$$A \to B, A \vdash B$$

is recognized by its *form*. Thus the relation $MP(\mathbf{x}, \mathbf{y}, \mathbf{z})$ on numbers—that is true precisely when the numbers $\mathbf{x}, \mathbf{y}, \mathbf{z}$ expressed in 35-adic notation have the forms $A \to B, A$ and B respectively, for some formulae A and B—is recursive (decidable).

$\square$

A.4.2 Remark. (Digression) (1) One theorem of Euclid states that having fixed a $b > 1$ (from $\mathbb{N}$), any $n \in \mathbb{N}$ has a unique quotient and remainder, that is, unique q and r exist—where $0 \leq r < b$—such that $n = bq + r$.

One immediately obtains, for the same $b > 1$, and any $n > 0$ a unique representation

$$n = b\pi + \upsilon, \text{ where } 0 < \upsilon \leq b, \text{ and both } \pi \text{ and } \upsilon \text{ are in } \mathbb{N} \qquad (*)$$

The existence part in $(*)$ follows from Euclid's theorem: Given $n > 0$, we have q and r, where $0 \leq r < b$, such that $n = bq + r$. If $r \neq 0$, take $\pi = q$ and $\upsilon = r$. Otherwise, since $q \neq 0$ (why?), take $\pi = q - 1$ and $\upsilon = b$. Uniqueness is settled by Euclid's old argument: If $b\pi + \upsilon = b\pi' + \upsilon'$, for $0 < \upsilon \leq b$ and $0 < \upsilon' \leq b$, then b is a factor of the absolute difference $|\upsilon - \upsilon'|$. As $|\upsilon - \upsilon'| < b$ (why?), this forces $\upsilon = \upsilon'$. Trivially, then, $\pi = \pi'$ as well.

Statement $(*)$ leads to the existence and uniqueness of b-adic representations for positive integers exactly in the same manner Euclid's theorem induces the existence and uniqueness of b-ary representations for nonnegative integers. Proceeding by strong induction we note that $n = 1$ has a b-adic representation: "1". Admitting the claim for all $1 \leq m < n$, and using $(*)$, we have (I.H. applied to π) $n = b(a_0 + a_1 b^1 + \cdots + a_k b^k) + \upsilon$ for some appropriate b-adic digits a_i. This settles existence. For uniqueness let

$$a_0 + a_1 b^1 + \cdots + a_k b^k = c_0 + c_1 b^1 + \cdots + c_m b^m \qquad (**)$$

where $0 < a_i, c_j \leq b$ for all i, j. By $(*)$ we have $a_0 = c_0$. Thus

$$a_1 + a_2 b^1 + \cdots + a_k b^{k-1} = c_2 + c_2 b^1 + \cdots + c_m b^{m-1}$$

---

[31]This, "$\mathbf{x}, \mathbf{y} < \mathbf{t}$" is what makes the recursion "course-of-values". In "$+\mathbf{xy}$" we are using the formal prefix notation for terms rather than the friendly (but informal) "$\mathbf{x} + \mathbf{y}$".

and, as above, $a_1 = c_1$. And so on (or use induction).

(2) Now we can go back and prove $\chi_{\text{Term}} \in \mathcal{PR}$ in detail. We will first develop some tools for manipulating b-adic numbers.[32] As before, we will typeset all the variables that we employ in boldface to distinguish them from the digits x, y, z, u, v, w of alphabet (1) p. 265.

We fix the base b throughout our discussion and show that $\lambda \mathbf{x}.|\mathbf{x}|$, the length of the number $\mathbf{x}$ expressed in b-adic notation, is in $\mathcal{PR}$:

$$|0| = 0$$
$$|\mathbf{x} + 1| = \begin{cases} |\mathbf{x}| + 1 & \text{if } b^{|\mathbf{x}|+1} = \mathbf{x}(b-1) + b \\ |\mathbf{x}| & \text{otherwise} \end{cases}$$

We next show that concatenation is a primitive recursive operation. We will denote by $\mathbf{x} * \mathbf{y}$ the (numerical) result of concatenating the b-adic representations of $\mathbf{x}$ and $\mathbf{y}$ in that order. Since $\mathbf{x} * \mathbf{y} = \mathbf{x}b^{|\mathbf{y}|} + \mathbf{y}$, we have $\lambda \mathbf{x} \mathbf{y}.\mathbf{x} * \mathbf{y} \in \mathcal{PR}$. The absence of 0 digits makes clear that $(\mathbf{x} * \mathbf{y}) * \mathbf{z} = \mathbf{x} * (\mathbf{y} * \mathbf{z})$, thus writing "$\mathbf{x} * \mathbf{y} * \mathbf{z}$" is unambiguous.

Note that the number 0 behaves like the empty string, in terms both of its length and its behavior with respect to concatenation: $0 * \mathbf{y} = 0b^{|\mathbf{y}|} + \mathbf{y} = \mathbf{y}$ and $\mathbf{x} * 0 = \mathbf{x}b^{|0|} + 0 = \mathbf{x}$.

Let next $\mathbf{x}B\mathbf{y}$ be the relation "the b-adic representation of $\mathbf{x}$ is a prefix of that of $\mathbf{y}$". This relation is primitive recursive (cf. A.3.45) since $\mathbf{x}B\mathbf{y} \equiv (\exists \mathbf{z})_{\leq \mathbf{y}} \mathbf{y} = \mathbf{x} * \mathbf{z}$. Similarly, for the "postfix relation" $\mathbf{x}E\mathbf{y}$ that means "the b-adic representation of $\mathbf{x}$ is a postfix of that of $\mathbf{y}$" and the "part of" relation, $\mathbf{x}P\mathbf{y}$, meaning "the b-adic representation of $\mathbf{x}$ is a substring of that of $\mathbf{y}$": $\mathbf{x}E\mathbf{y} \equiv (\exists \mathbf{z})_{\leq \mathbf{y}} \mathbf{y} = \mathbf{z} * \mathbf{x}$ and $\mathbf{x}P\mathbf{y} \equiv (\exists \mathbf{z})_{\leq \mathbf{y}}(\mathbf{z}B\mathbf{y} \wedge \mathbf{x}E\mathbf{z})$.

Let now d be any digit among the ones allowed in b-adic notation: $1, 2, \ldots, b$. The relation $\lambda \mathbf{x}.tally_d(\mathbf{x})$ says that $\mathbf{x}$ is nonzero and all its digits (in b-adic notation) are the same as d. This, too, is primitive recursive: $tally_d(\mathbf{x}) \equiv \mathbf{x} > 0 \wedge (\forall \mathbf{z})_{\leq \mathbf{x}}(\mathbf{z}P\mathbf{x} \wedge |\mathbf{z}| = 1 \rightarrow \mathbf{z} = d)$.

Important Notational Convention: *In expressing the various formulae and functions we are consistently using the* value *of an alphabet symbol rather than its* name. *Thus, rather than writing "$0B\mathbf{x}$" to indicate that the alphabet symbol "0" begins the (b-adic representation of*) $\mathbf{x}$, *we write instead "$12B\mathbf{x}$"; rather than* $\mathbf{x} = x$, *we write* $\mathbf{x} = 1$, *etc.*

Most of the above machinery based on b-adic concatenation was developed by Smullyan ([47]) and Bennett ([1]) and is retold in Tourlakis ([49]), the ideas having originated in Quine ([39]).

To conclude our task, we will show first that the relation $Var(\mathbf{x})$ that is true precisely when the b-adic notation of $\mathbf{x}$ has the syntax of an object variable (cf. footnote 87, p. 115) is in $\mathcal{PR}$. We split our task into two subtasks:

---

[32]That is, positive integers written in b-adic notation for some $b > 1$.

(i) The relation $Num(\mathbf{x})$ says that the b-adic notation of the number $\mathbf{x}$ is a string over the subalphabet $\{0, 1, 2, 3, 4, 5, 6, 7, 8, 9\}$ of (1) that does *not* begin with 0. We now have $Num(\mathbf{x}) \equiv \mathbf{x} > 0 \wedge \neg 12^{33} B\mathbf{x} \wedge (\forall \mathbf{y})_{\leq \mathbf{x}} (\mathbf{y} P \mathbf{x} \wedge |\mathbf{y}| = 1 \to \mathbf{y} = 12^{34} \vee \mathbf{y} = 13 \vee \mathbf{y} = 14 \vee \mathbf{y} = 15 \vee \mathbf{y} = 16 \vee \mathbf{y} = 17 \vee \mathbf{y} = 18 \vee \mathbf{y} = 19 \vee \mathbf{y} = 20 \vee \mathbf{y} = 21)$. Clearly, $Num(\mathbf{x}) \in \mathcal{PR}_*$.

(ii) $Var(\mathbf{x}) \equiv (\exists \mathbf{y}, \mathbf{z})_{<\mathbf{x}} ((\mathbf{y} = 0 \vee tally_{11}{}^{35}(\mathbf{y})) \wedge (\mathbf{z} = 0 \vee Num(\mathbf{z})) \wedge (\mathbf{x} = 1 * \mathbf{y} * \mathbf{z} \vee \mathbf{x} = 2 * \mathbf{y} * \mathbf{z} \vee \mathbf{x} = 3 * \mathbf{y} * \mathbf{z} \vee \mathbf{x} = 4 * \mathbf{y} * \mathbf{z} \vee \mathbf{x} = 5 * \mathbf{y} * \mathbf{z} \vee \mathbf{x} = 6 * \mathbf{y} * \mathbf{z})$.

Finally, revisiting χ_{Term} we see at once:

$$\chi_{\text{Term}}(\mathbf{t}) =$$

$$\begin{cases} \text{if } Var(\mathbf{t}) \vee \mathbf{t} = 12 & \text{then } 0 \\ \text{else if } (\exists \mathbf{x})_{<\mathbf{t}} (\mathbf{t} = 32^{36} * \mathbf{x} \wedge \chi_{\text{Term}}(\mathbf{x}) = 0) & \text{then } 0 \\ \text{else if } (\exists \mathbf{x}, \mathbf{y})_{<\mathbf{t}} (\mathbf{t} = 33 * \mathbf{x} * \mathbf{y} \wedge \chi_{\text{Term}}(\mathbf{x}) = 0 \wedge \chi_{\text{Term}}(\mathbf{y}) = 0) & \text{then } 0 \\ \text{else if } (\exists \mathbf{x}, \mathbf{y})_{<\mathbf{t}} (\mathbf{t} = 34 * \mathbf{x} * \mathbf{y} \wedge \chi_{\text{Term}}(\mathbf{x}) = 0 \wedge \chi_{\text{Term}}(\mathbf{y}) = 0) & \text{then } 0 \\ \text{else} & 1 \end{cases}$$

□

A.4.3 Exercise. The reader is invited to similarly prove, in detail, that the relations

(1) $\mathbf{x} \in \mathbf{WFF}$, which holds exactly when the number $\mathbf{x}$, expressed in b-adic notation, is a formula, and

(2) $MP(\mathbf{x}, \mathbf{y}, \mathbf{z})$, which holds precisely when the numbers $\mathbf{x}, \mathbf{y}, \mathbf{z}$, expressed in b-adic notation, are formulae of the forms A, $(A \to B)$ and B respectively, are primitive recursive.   □

A.4.4 Exercise. Elegant as the primitive recursive definition of the b-adic length $|\mathbf{x}|$ may be, it is, in practical terms, computationally very inefficient. For one thing, the recursion has as many levels as the value of $\mathbf{x}$. With more thought the depth of recursion can be reduced to $|\mathbf{x}|$, which is approximately $\log_b(\mathbf{x})$. Effect such a definition by first getting two primitive recursive functions $quot$ and res such that for any $\mathbf{x} > 0$ and $\mathbf{y} > 1$ we have $\mathbf{x} = quot(\mathbf{x}, \mathbf{y})\mathbf{y} + res(\mathbf{x}, \mathbf{y})$ with $0 < res(\mathbf{x}, \mathbf{y}) \leq \mathbf{y}$. Then observe that for any $\mathbf{x} > 0$, the b-adic length of $\mathbf{x}$ is one longer than that of $quot(\mathbf{x}, b)$.   □

Let us now agree on a coding of proofs. Since a proof is a sequence of formulae, $A_0, \ldots, A_n$, and as each formula can be naturally identified with a number, we will code such sequences, and hence proofs, as $\langle A_0, \ldots, A_n \rangle$.

---

[33] Digit "0".
[34] Digit "0".
[35] Digit "/" has value 11.
[36] 32 is the value of digit "S", 33 is that of "+" and 34 that of "×".

A.4.5 Lemma. *Given the stipulations (1)–(5) above (p. 264), the relation*

$$inProof(\mathbf{y}, \mathbf{x})$$

which holds precisely when the number $\mathbf{y}$ *codes a proof of the formula coded by* $\mathbf{x}$, *is decidable.*

Proof. (**Outline**) We have enough confidence by now to accept that the relation $\mathbf{x} \in \Lambda_1$, which holds exactly when the number $\mathbf{x}$—expressed in b-adic notation—is a logical axiom, is recursive (indeed with some effort one can prove that it is *primitive* recursive). Recall that we have also *stipulated* that the set of special axioms of arithmetic is decidable. That is, $SP(\mathbf{x})$ is recursive, where the relation holds exactly when the number $\mathbf{x}$, expressed in b-adic notation, is a special axiom.

It is a stipulation rather than an observation because we want to allow the widest variety of "practical" axiomatizations of arithmetic over the alphabet (1) of p. 265. If we restrict attention to the particular Peano axiomatization, then one can actually prove (with some effort) that the set of special axioms *is* in fact primitive recursive (cf. [53]).

Then

$$inProof(\mathbf{y}, \mathbf{x}) \equiv Seq(\mathbf{y}) \wedge (\forall \mathbf{i})_{<lh(\mathbf{y})} \big(SP((\mathbf{y})_{\mathbf{i}}) \vee (\mathbf{y})_{\mathbf{i}} \in \Lambda_1 \vee$$
$$(\exists \mathbf{j}, \mathbf{k})_{<\mathbf{i}}(MP((\mathbf{y})_{\mathbf{j}}, (\mathbf{y})_{\mathbf{k}}, (\mathbf{y})_{\mathbf{i}}) \vee MP((\mathbf{y})_{\mathbf{k}}, (\mathbf{y})_{\mathbf{j}}, (\mathbf{y})_{\mathbf{i}})) \big) \ \square$$

A.4.6 Lemma. *The set of* all *theorems of any axiomatization of arithmetic satisfying the stipulations (1)–(5) above (p. 264) is semi-computable.*

Proof. Let AR be the set of all theorems of an axiomatization that is as stated above. By the preceding lemma, $\mathbf{x} \in AR \equiv (\exists \mathbf{y})inProof(\mathbf{y}, \mathbf{x})$. $\square$

A.4.7 Remark. Thus (as we remarked long ago, p. 7; see also footnote 8 on that page), for any axiomatization of arithmetic that satisfies the stipulations (1)–(5), there is a computer program that without any input, and if it is allowed to run forever, will print (in coded form, as numbers) all its theorems. All the program has to do is to use as a subprogram a URM that computes a primitive recursive f that has AR as its range (cf. A.3.91). This computer program will behave simply as follows: **for** $i = 0, 1, 2, 3, \ldots$ **print** $f(i)$. $\square$

Let us call *complete arithmetic* (CA) the set of *all closed formulae* over our language of arithmetic ((1) on p. 265)—i.e., formulae with no free variables, also called *sentences*—that are true in $\mathfrak{N}$:

$$CA = \{\text{closed } A : \models_{\mathfrak{N}} A\}$$

We make some preliminary observations before we state and prove Gödel's first incompleteness theorem.

A.4.8 Exercise. We have noted that the "informal number n is captured in the language of arithmetic by the term

$$\underbrace{SS\ldots S}_{\text{length } n}0$$

which we abbreviate as $\widetilde{n}$". Make this statement precise by proving, using the table on p. 264 and simple induction on $n \geq 1$, that $(S^n 0)^{\mathfrak{N}} = n$, where I used the abbreviation "S^n" for the string $SS\ldots S$ of length n. $\quad\square$

Following on the above and noting that $n^{\mathfrak{N}} = n$ (A.1.5(5)), we have $(\widetilde{n})^{\mathfrak{N}} = n^{\mathfrak{N}}$, having written $\widetilde{n}$ for $S^n 0$. That is (A.1.6),

$$\models_{\mathfrak{N}} \widetilde{n} = n \tag{1}$$

Since (by first-order soundness and **Ax6**) $\models \widetilde{n} = n \rightarrow (A[\![\widetilde{n}]\!] \equiv A[\![n]\!])$[37] for any formula A over the language of arithmetic, we obtain, in particular, $\models_{\mathfrak{N}} \widetilde{n} = n \rightarrow (A[\![\widetilde{n}]\!] \equiv A[\![n]\!])$ and hence

$$\models_{\mathfrak{N}} A[\![\widetilde{n}]\!] \equiv A[\![n]\!] \tag{2}$$

By iterating this a finite number of times we can prove things such as $\models_{\mathfrak{N}}$ $A[\![\widetilde{n}, m, \widetilde{k}, j]\!] \equiv A[\![n, m, k, j]\!]$, mixing at will "formal numbers" $\widetilde{a}$ with imported constants $i, j, k, \ldots$ in the left hand side.

Pause. This is neat. Could we then go back and argue A.1.20 as follows?

We are given an interpretation $\mathfrak{D} = (D, M)$, a D-formula A, and a D-term t such that $t^{\mathfrak{D}} = i$ for some $i \in D$. We want

$$\left(A[\mathbf{x} := t]\right)^{\mathfrak{D}} = \left(A[\mathbf{x} := i]\right)^{\mathfrak{D}} \tag{3}$$

Since $\models t = i \rightarrow (A[\mathbf{x} := t] \equiv A[\mathbf{x} := i])$ as above, we have in particular $\models_{\mathfrak{D}} t = i \rightarrow (A[\mathbf{x} := t] \equiv A[\mathbf{x} := i])$. Taking also $t^{\mathfrak{D}} = i^{\mathfrak{D}}$ into account—that is, $(t = i)^{\mathfrak{D}} = \mathbf{t}$—we get $\models_{\mathfrak{D}} (A[\mathbf{x} := t] \equiv A[\mathbf{x} := i])$. Or, in different notation, we have (3) above. As for $\left(s[\mathbf{x} := t]\right)^{\mathfrak{D}} = \left(s[\mathbf{x} := i]\right)^{\mathfrak{D}}$ we can use 7.0.8 rather than **Ax6**. Right?

We will need one last preparatory item:

A.4.9 Definition. (Correctness; [48]) An axiomatization of arithmetic is termed "correct" precisely when the *standard interpretation* $\mathfrak{N}$ is a model of the set of the special axioms. $\quad\square$

Correctness is not the same as soundness. All first-order theories satisfy the soundness theorem (A.1.21). However, there are consistent axiomatizations of arithmetic that are *not* correct ([48, 53]).

---

[37] The notation "$[\![\ldots]\!]$" was introduced in Definition A.1.14.

A.4.10 Theorem. (Gödel's First Incompleteness Theorem) Any *axiomatic system for arithmetic that satisfies* (1)–(5) *of p. 264 and, moreover, is correct* must *be* incomplete *in the sense that its set of theorems cannot contain the set* CA: *There will be* true sentences *of arithmetic that the system* cannot prove.

Proof. We will show that if Gödel's theorem fails, then we can build a URM that solves the halting problem—a known impossibility (A.3.88).

Let us then have a set of (special) axioms over the language of arithmetic—an axiomatization—that satisfies (1)–(5) and that contains as theorems all of the set CA. Thus, if Θ is the set of theorems of the axiomatization,

$$CA \subseteq \Theta \tag{$*$}$$

We note that Θ is semi-computable (A.4.6); thus, for some recursive $Q(\mathbf{y}, \mathbf{x})$,

$$\mathbf{x} \in \Theta \equiv (\exists \mathbf{y})Q(\mathbf{y}, \mathbf{x}) \tag{$**$}$$

We will accept here a fact proved in the next section: The relation $\phi_x(x) \uparrow$ is definable in $\mathfrak{N}$ (cf. A.1.14). More specifically, we can find a formula of arithmetic, A, of *one free variable* $\mathbf{x}$ such that, for all $i \in \mathbb{N}$, $\phi_i(i) \uparrow \; \equiv \; \models_{\mathfrak{N}} A [\![i]\!]$. In view of the remarks preceding the theorem, we can rewrite the last equivalence as

$$\text{for all } i \in \mathbb{N}, \phi_i(i) \uparrow \; \equiv \; \models_{\mathfrak{N}} A [\![\widetilde{i}]\!] \tag{$***$}$$

This is extremely useful, since $A [\![\widetilde{i}]\!]$ is over the alphabet (1) of p. 265 and thus we can code each true "statement" of the form "$\phi_i(i) \uparrow$" by a natural number (whose 35-adic notation is) $A [\![\widetilde{i}]\!]$.

Here then is how we can solve the halting problem. We are given a number i and are asked to compute an answer to the question "$i \in K$?"—that is, "$\phi_i(i) \downarrow$?"

(a) We compute the number that the b-adic string $A [\![\widetilde{i}]\!]$ represents; say it is b.

(b) Using some fixed verifier for ($**$) and a URM that computes the function $\lambda \mathbf{x}.d((\mu \mathbf{z})T(\mathbf{x}, \mathbf{x}, \mathbf{z}))$ we start simultaneously computing the answers to "$b \in \Theta$?" and "$\phi_i(i) = ?$"

(c) By ($*$), if $\phi_i(i) \uparrow$ is true, then $A [\![\widetilde{i}]\!] \in \Theta$. Conversely, if $A [\![\widetilde{i}]\!] \in \Theta$, then as correctness guarantees that the special axioms are true in $\mathfrak{N}$ and as soundness guarantees that proofs propagate truth, we have that $\models_{\mathfrak{N}} A [\![\widetilde{i}]\!]$, that is, $\phi_i(i) \uparrow$ is true. Thus, if the first process halts, then we stop everything and proclaim "$i \notin K$" (i.e., "$\phi_i(i) \uparrow$").

(d) If the second process halts, then we stop everything and proclaim "$i \in K$" (i.e., "$\phi_i(i) \downarrow$").

Thus, ($*$) is untenable.

The hand-waving can be (mostly) eliminated simply: Let us call $\lambda i.f(i)$ the function that on input i computes the number whose 35-adic notation is $A [\![\widetilde{i}]\!]$. Once we have pinned down the formula A (in the next section), we can see that finding the number

$$A[\mathbf{x} := \underbrace{SS\ldots S}_{\text{length } i} 0]$$

tedious as it may be, is rather computationally straightforward—if you do not agree see also the following remark! In the absence of details we can readily accept that $f \in \mathcal{R}$ (and more work can show that actually $f \in \mathcal{PR}$; see remark below). But then, setting $g = \lambda i.(\mu\mathbf{y})\big(Q(\mathbf{y}, f(i)) \vee T(i, i, \mathbf{y})\big)$, we see that, first, $g \in \mathcal{R}$ since it certainly is in $\mathcal{P}$, and is total (why?) Second,

$$i \in K \equiv T(i, i, g(i)) \qquad\qquad \Box$$

Note the emphasized *any* in the theorem. It draws attention to the fact that we have not *fixed* any *particular* correct and "reasonable" theory—whatever we said holds for all such theories. *Any*, moreover, implies that every theory that the theorem speaks of misses an infinite chunk of CA. Indeed, if it misses only a finite chunk, $\mathcal{S}$, then adding all the formulae of $\mathcal{S}$ as special axioms we still get a recursive set of special axioms (closure properties, and the fact that finite sets are recursive) and one that is still correct (why?).

But the set of theorems of this new axiom set covers all of CA (why?) contrary to Gödel's theorem.

A.4.11 Remark. (Tarski's Trick) We know ("one point rule", 3, p. 175) that $\vdash (\exists\mathbf{x})(\mathbf{x} = \widetilde{i} \wedge A) \equiv A[\![\widetilde{i}]\!]$ and hence $\models (\exists\mathbf{x})(\mathbf{x} = \widetilde{i} \wedge A) \equiv A[\![\widetilde{i}]\!]$.

Thus, in the proof above, we might as well input to our verifier the number represented in 35-adic by "$(\exists\mathbf{x})(\mathbf{x} = \widetilde{i} \wedge A)$", rather than inputting the one represented by "$A[\![\widetilde{i}]\!]$". But why would we want to deal with a more complex counterpart of $A[\![\widetilde{i}]\!]$? Because it is easier to compute the number represented by the more complex formula as long as we have computed once and for all the *constant* that has as 35-adic notation the formula A! The complex formula isolates $\widetilde{i}$ in just one place, to the left of A. The "simpler" formula $A[\![\widetilde{i}]\!]$, on the other hand, may have $\widetilde{i}$ occur in several places, making it very hard to reuse (the number) A.

Here is the calculation in detail:

First, we want a primitive recursive function "*tilde*" such that, for all $n \geq 0$, $tilde(n)$ is the number whose 35-adic representation is $S^n 0$. *We understand $S^0 0$ as a verbose way to say "0".* Since the number so represented by $S^n 0$ is $12 + 33(35 + 35^2 + \cdots + 35^n)$, we have that

$$tilde(n) = 12 + \left\lfloor \frac{1155(35^n - 1)}{34} \right\rfloor$$

Let next f be as in the above proof, except that it refers to $(\exists\mathbf{x})(\mathbf{x} = \widetilde{i} \wedge A)$ rather than to $A[\![\widetilde{i}]\!]$. To fix notation, we take the formal variable $\mathbf{x}$ (which is unspecified in the notation "$\mathbf{x}$") to be x (lightface!) of the alphabet (1). Let a be the (constant!)

number represented (always in 35-adic) by the string " $\wedge A)$ " and c be the number represented by the string " $(\exists x)(x = $ ". Then, for all $i \in \mathbb{N}$,

$$f(i) = c * tilde(i) * a$$

See also the definition of concatenation, "$*$", on p. 268. Thus f is primitive recursive as promised.   □

A.4.12 Exercise. Prove that every finite set is primitive recursive.   □

A.4.13 Corollary. *The set* CA *is* not *semi-recursive.*

Proof. Assume that CA is semi-recursive; thus for some recursive relation Q[38] we have $\mathbf{x} \in CA \equiv (\exists \mathbf{y})Q(\mathbf{y}, \mathbf{x})$. We use a trivial variation on the proof of A.4.10 and still solve the halting problem! This time we change (b) to say: "Using some fixed verifier for $\mathbf{x} \in CA$ and a URM for the function $\lambda \mathbf{x}.d\big((\mu \mathbf{z})T(\mathbf{x}, \mathbf{x}, \mathbf{z})\big)$ we start simultaneously computing $b \in CA$ and $\phi_i(i)$." The subprocess (c) now simply reads, "Thus, if the verifier halts, then we stop everything and proclaim that $i \notin K$". All else is the same.   □

A.4.14 Corollary. *The set* CA *is* not *recursive.*

Proof. By A.3.85.   □

A.4.14 is a weak form of Church's theorem. The latter actually says that the set of all *theorems* of any axiomatization of arithmetic as described in Gödel's theorem is not recursive ([3, 4]). Nevertheless, the corollary is useful in pointing out that the "trivial solution" toward obtaining as theorems of arithmetic all of CA is wrong: This solution would be to take all of CA as the set of special axioms, but then the set of special axioms would cease being practical (recursive).

A.4.15 Remark. The original version of Gödel's first incompleteness theorem ([16]) was stated exclusively in syntactic terms: *In any recursive and ω-consistent extension of Peano arithmetic there will be* undecidable sentences *in the language of arithmetic, that is, closed formulae A such that neither A, nor $\neg A$ are provable.*

ω-consistency of an axiomatic number theory is the property that for no formula A of one free variable $\mathbf{x}$ is it possible to have all of $\neg A\,[\![\,\widetilde{n}\,]\!]$—for $n \in \mathbb{N}$—and $(\exists \mathbf{x})A$ provable. This condition implies, but is not implied by, consistency. Rosser ([42]) strengthened the incompleteness theorem to read: *In any recursive and consistent extension of Peano arithmetic there will be undecidable sentences.*

The reader will note that in our semantic version correctness took the role of consistency (indeed, every axiom set that has a model—here $\mathfrak{N}$—is necessarily consistent; cf. A.1.22). Any true formula $A\,[\![\,\widetilde{i}\,]\!]$ (A as in the proof of A.4.10) that

---

[38]This is a different "Q" than the one in the proof of Gödel's theorem, of course, but I use the same letter here so that the changes needed to adapt said proof to the one for the corollary are minimal.

is unprovable—and the essence of the proof is that not all such *true* sentences are provable—is undecidable in the sense of Gödel: Indeed, its negation is false, thus it cannot be provable either by correctness and soundness.

Gödel's original proof relied on a syntactic version of the *liar's paradox*, the latter being the paradoxical (semantic) utterance "I am lying".[39] Gödel formulated within Peano arithmetic the self-referential statement, "I am not a theorem", and showed it to be an undecidable sentence.

The phenomenon of self-reference is excellently explored in Hofstadter's *Gödel, Escher, Bach* ([23]). A related gripping (fictional) account of a brilliant reclusive mathematician's efforts to settle "Goldbach's Conjecture", and how he was shocked to learn of Gödel's incompleteness theorems, is Doxiadis's *Uncle Petros and Goldbach's Conjecture* ([12]). □

A.4.16 Exercise. If a set of special axioms Γ over the language of arithmetic is ω-consistent, then it is also consistent. □

A.4.1 Supplement: $\phi_x(x) \uparrow$ Is First-Order Definable in $\mathfrak{N}$

We will continue using boldface for variables in the metatheory to avoid confusion with the formal x, y, etc., listed in (1) on p. 265. Let us define inductively the set of *arithmetical* relations of the metatheory.[40]

A.4.17 Definition. The set of *arithmetical relations* is the smallest set of relations that:

(A) Contains the "initial" relations $\mathbf{z} = \mathbf{x} + \mathbf{y}$, $\mathbf{z} = \mathbf{x} \cdot \mathbf{y}$, and $\mathbf{z} = ex(\mathbf{x}, \mathbf{y})$ (cf. A.3.34)—where $\mathbf{x}, \mathbf{y}, \mathbf{z}$ are distinct variables in the metatheory

and, moreover,

(B) If $Q(\vec{\mathbf{x}})$ and $P(\vec{\mathbf{y}})$ are in the set, then so are $\neg Q(\vec{\mathbf{x}})$ and $Q(\vec{\mathbf{x}}) \vee P(\vec{\mathbf{y}})$.

(C) If $R(\mathbf{y}, \vec{\mathbf{x}})$ is in the set, then so is $(\forall \mathbf{y}) R(\mathbf{y}, \vec{\mathbf{x}})$.

(D) If $Q(\vec{\mathbf{x}})$ is in the set, then so are all its *explicit transformations*.

Explicit transformations ([47, 1]) are exactly the following: substitution of any constant into a variable, expansion of the variables-list by "don't care" variables (arguments), permutation of variables, identification of variables— that is, Grzegorczyk operations (ii)–(iv) (cf. A.3.26), albeit applied to relations. □

---

[39]Epimenides of Crete actually proclaimed a slightly different statement: "All Cretans are liars."

[40]The arithmetical relations have a lot of tolerance for variations in their definition: Sometimes as much as all of $\mathcal{R}_*$ is taken as the "initial" arithmetical relations. Sometimes as little as $\mathbf{z} = \mathbf{x} + \mathbf{y}$ and $\mathbf{z} = \mathbf{x} \cdot \mathbf{y}$. For technical convenience we have added the graph of exponentiation rather than choosing the most minimalist approach.

Clearly the set of arithmetical relations is closed under the remaining Boolean connectives and $(\exists \mathbf{y})$.

A.4.18 Lemma. *Every arithmetical relation is first-order definable in $\mathfrak{N}$ over a language, L, of arithmetic that contains beyond the functions $S, +,$ and $\times$ a function symbol for exponentiation.*

Proof. To keep the alphabet fixed to that of (1) on p. 265 the exponentiation function symbol (of arity 2) will be taken to be "(x)" whose natural semantics in $\mathfrak{N}$ are: $(x)^{\mathfrak{N}} = \lambda \mathbf{x}\mathbf{y}.ex(\mathbf{x}, \mathbf{y})$. We remind ourselves of the table on p. 264 that states the semantics of the remaining nonlogical symbols, and proceed by induction along the cases of Definition A.4.17. The basis contains three cases, $\mathbf{z} = \mathbf{x} + \mathbf{y}$ and $\mathbf{z} = \mathbf{x} \cdot \mathbf{y}$ and $\mathbf{z} = ex(\mathbf{x}, \mathbf{y})$.

Thus, writing the term "$+xy$" in the friendlier infix notation, we have that $z = x + y$[41] defines (cf. A.1.14) $\mathbf{z} = \mathbf{x} + \mathbf{y}$, since for any a, b, c in $\mathbb{N}$,

$$(a = b + c \,^{42})^{\mathfrak{N}} = \mathbf{t} \underset{\text{cf. A.1.6}}{\text{iff}} (a^{\mathfrak{N}} = b^{\mathfrak{N}} +^{\mathfrak{N}} c^{\mathfrak{N}}) = \mathbf{t}$$
$$\text{iff} \quad a = b + c$$

An entirely analogous case can be made for $\mathbf{z} = \mathbf{x} \cdot \mathbf{y}$ and $\mathbf{z} = ex(\mathbf{x}, \mathbf{y})$. For example (using the formal exponentiation "(x)" in infix notation),

$$(a = b \,(x)\, c)^{\mathfrak{N}} = \mathbf{t} \,\text{iff}\, (a^{\mathfrak{N}} = (x)^{\mathfrak{N}}(b^{\mathfrak{N}}, c^{\mathfrak{N}})) = \mathbf{t}$$
$$\text{iff}\, a = ex(b, c)$$

We leave it to the reader to verify that if $R(\vec{\mathbf{x}})$ and $Q(\vec{\mathbf{y}})$ are defined by the formulae A and B respectively, then $\neg R(\vec{\mathbf{x}})$ and $R(\vec{\mathbf{x}}) \vee Q(\vec{\mathbf{y}})$ are defined by $\neg A$ and $A \vee B$ respectively.

Next, we show that $(\forall \mathbf{y}) R(\mathbf{y}, \vec{\mathbf{x}}_r)$ is defined by $(\forall y) A$, if $R(\mathbf{y}, \vec{\mathbf{x}}_r)$ is defined by A. So, let without loss of generality $y, x_1, \ldots, x_r$ be all the free variables of A. We have for any $c, b_1, \ldots, b_r$ in $\mathbb{N}$ that

$$R(c, b_1, \ldots, b_r) \text{ iff } \left(A [\![c, b_1, \ldots, b_r]\!] \right)^{\mathfrak{N}} = \mathbf{t} \tag{1}$$

Now, we fix $b_1, \ldots, b_r$ in $\mathbb{N}$. $(\forall \mathbf{y}) R(\mathbf{y}, b_1, \ldots, b_r)$ holds iff for all $c \in \mathbb{N}$ we have that $R(c, b_1, \ldots, b_r)$ holds. By (1) this is equivalent to "for all $c \in \mathbb{N}$ we have $\left(A [\![c, b_1, \ldots, b_r]\!] \right)^{\mathfrak{N}} = \mathbf{t}$".

By A.1.6 the latter says precisely $\left((\forall x) A [\![b_1, \ldots, b_r]\!] \right)^{\mathfrak{N}} = \mathbf{t}$.

We conclude by looking into explicit transformations. Let then $Q(\mathbf{y}, \vec{\mathbf{x}}_r)$ be defined by the formula (over L) A. Then, for any fixed $i \in \mathbb{N}$, $Q(i, \vec{\mathbf{x}}_r)$ is clearly defined by $A [\![i]\!]$, since for all $a, b_1, \ldots, b_r$ we have $Q(a, b_1, \ldots, b_r)$ iff

---

[41] I used the specific formal variables x, y, z here. Any other set of distinct formal variables will also work. For example, $x_{101} = x_{222} + x_{303}$ also defines $\mathbf{z} = \mathbf{x} + \mathbf{y}$.

[42] In brackets we have a formal formula over the language $L(\mathbb{N})$: $(z = x + y)[z := a][x := b][y := c]$.

$\models_{\mathfrak{N}} A [\![a, b_1, \ldots, b_r]\!]$ and thus $Q(i, b_1, \ldots, b_r)$ iff $\models_{\mathfrak{N}} A [\![i, b_1, \ldots, b_r]\!]$. While $A [\![i]\!]$ is not over the language of arithmetic, we may use instead $A [\![\tilde{i}]\!]$, which is (cf. remarks following A.4.8). The case of identifying or permuting variables being trivial, we conclude by looking at the case of adding one "don't care" variable (extensible to any fixed number by a trivial induction). So let A, over L, define $Q(\vec{\mathbf{x}}_r)$ and let $\mathbf{z}$ be a new metamathematical variable. I take z, a variable different from all the free variables of A (which, without loss of generality, are $x_1, \ldots, x_r$) and argue that $A \wedge z = z$ defines the relation $R = \lambda \mathbf{z} \vec{\mathbf{x}}_r . Q(\vec{\mathbf{x}}_r)$:

We have, on one hand, for all $b_1, \ldots, b_r$: $Q(b_1, \ldots, b_r)$ iff $\left(A [\![b_1, \ldots, b_r]\!] \right)^{\mathfrak{N}}$ $= \mathbf{t}$. On the other hand, for all $c, b_1, \ldots, b_r$, $Q(b_1, \ldots, b_r) \equiv R(c, b_1, \ldots, b_r)$ and $\left(A [\![b_1, \ldots, b_r]\!] \right)^{\mathfrak{N}} = {}^{43} \left(A [\![b_1, \ldots, b_r]\!] \right)^{\mathfrak{N}} \wedge (c = c)^{\mathfrak{N}}$ hold. $\qquad \square$

To show that $\phi_{\mathbf{x}}(\mathbf{x}) \uparrow$ is first-order definable in $\mathfrak{N}$ over L it suffices, because of the lemma, to prove that it is arithmetical. In turn, since $\phi_{\mathbf{x}}(\mathbf{x}) \uparrow \equiv \neg(\exists \mathbf{y}) T(\mathbf{x}, \mathbf{x}, \mathbf{y})$, it suffices to prove that the Kleene predicate is arithmetical. It will so follow if we can prove that for every function $f \in \mathcal{PR}$ its graph is arithmetical, for then if χ_T is the characteristic function of T, we will have that $\chi_T(\mathbf{x}, \mathbf{y}, \mathbf{z}) = \mathbf{w}$—and therefore $\chi_T(\mathbf{x}, \mathbf{y}, \mathbf{z}) = 0$ by explicit transformation—is arithmetical.[44]

A.4.19 Lemma. *The following relations are arithmetical.*
(1) $\mathbf{x} = 0$ *(and hence* $\mathbf{x} \neq 0$*)*
(2) $\mathbf{x} \leq \mathbf{y}$ *(and hence* $\mathbf{x} < \mathbf{y}$*)*
(3) $\mathbf{z} = \mathbf{x} \dotdiv \mathbf{y}$
(4) $\mathbf{x} \mid \mathbf{y}$
(5) $Pr(\mathbf{x})$
(6) $Seq(\mathbf{z})$
(7) $Next(\mathbf{x}, \mathbf{y})$ *(meaning* $\mathbf{x} < \mathbf{y}$ *are consecutive primes)*
(8) $pow(\mathbf{z}, \mathbf{x}, \mathbf{y})$ *(meaning* $\mathbf{x} > 1$ *and* $ex(\mathbf{x}, \mathbf{y})$ *is the highest power of* $\mathbf{x}$ *dividing* $\mathbf{z}$*)*
(9) $\Omega(\mathbf{z})$ *(meaning* $\mathbf{z}$ *has the form* $p_0 p_1^2 p_2^3 \cdots p_n^{n+1}$ *for some* $\mathbf{n}$*)*
(10) $\mathbf{y} = p_{\mathbf{n}}$
(11) $\mathbf{z} = \exp(\mathbf{x}, \mathbf{y})$ *(cf. A.3.60)*

Note. We need not worry about bounding our quantifications, for it is not our purpose to show these relations in $\mathcal{PR}_*$. Indeed we know from earlier work that they are in this set. This time we simply want to show that they are arithmetical.

Proof.
(1) $\mathbf{x} = 0$ (and hence $\mathbf{x} \neq 0$): $\mathbf{x} = 0$ is an explicit transform of $\mathbf{x} = \mathbf{y} + \mathbf{z}$; $\mathbf{x} \neq 0$ is obtained by negation.
(2) $\mathbf{x} \leq \mathbf{y}$ (and hence $\mathbf{x} < \mathbf{y}$): This is equivalent to $(\exists \mathbf{z})(\mathbf{x} + \mathbf{z} = \mathbf{y})$.
(3) $\mathbf{z} = \mathbf{x} \dotdiv \mathbf{y}$: This is equivalent to $\mathbf{z} = 0 \wedge \mathbf{x} < \mathbf{y} \vee \mathbf{x} = \mathbf{z} + \mathbf{y}$.

---

[43]Equality on the set $\{\mathbf{t}, \mathbf{f}\}$.
[44]Gödel proved all this without the need to have exponentiation. However, adopting this operation makes things considerably easier and, as mentioned earlier (footnote 40 on p. 275), it does not change the set of arithmetical relations.

(4) $\mathbf{x} \mid \mathbf{y}$: This is equivalent to $(\exists \mathbf{z})\mathbf{y} = \mathbf{xz}$ (I am using "implied multiplication" throughout: "$\mathbf{xy}$" rather than "$\mathbf{x} \times \mathbf{y}$" or "$\mathbf{x} \cdot \mathbf{y}$").

(5) $Pr(\mathbf{x})$: This is equivalent to $\mathbf{x} > 1 \wedge (\forall \mathbf{y})(\mathbf{y} \mid \mathbf{x} \to \mathbf{y} = 1 \vee \mathbf{y} = \mathbf{x})$.

(6) $Seq(\mathbf{z})$: This is equivalent to $\mathbf{z} > 1 \wedge (\forall \mathbf{x})(\forall \mathbf{y})(Pr(\mathbf{x}) \wedge Pr(\mathbf{y}) \wedge \mathbf{x} < \mathbf{y} \wedge \mathbf{y} \mid \mathbf{z} \to \mathbf{x} \mid \mathbf{z})$.

(7) $Next(\mathbf{x}, \mathbf{y})$: This is equivalent to $Pr(\mathbf{x}) \wedge Pr(\mathbf{y}) \wedge \mathbf{x} < \mathbf{y} \wedge \neg(\exists \mathbf{z})(Pr(\mathbf{z}) \wedge \mathbf{x} < \mathbf{z} \wedge \mathbf{z} < \mathbf{y})$.

(8) $pow(\mathbf{z}, \mathbf{x}, \mathbf{y})$: This is equivalent to $\mathbf{x} > 1 \wedge ex(\mathbf{x}, \mathbf{y}) \mid \mathbf{z} \wedge \neg ex(\mathbf{x}, \mathbf{y} + 1) \mid \mathbf{z}$[45]

(9) $\Omega(\mathbf{z})$: This is equivalent to $Seq(\mathbf{z}) \wedge \neg 4 \mid \mathbf{z} \wedge (\forall \mathbf{x})(\forall \mathbf{y})(Next(\mathbf{x}, \mathbf{y}) \wedge \mathbf{y} \mid \mathbf{z} \to (\exists \mathbf{w})(pow(\mathbf{z}, \mathbf{x}, \mathbf{w}) \wedge pow(\mathbf{z}, \mathbf{y}, \mathbf{w} + 1)))$.

(10) $\mathbf{y} = p_{\mathbf{n}}$: This is equivalent to $(\exists \mathbf{z})(\Omega(\mathbf{z}) \wedge pow(\mathbf{z}, \mathbf{y}, \mathbf{n} + 1))$.

(11) $\mathbf{z} = \exp(\mathbf{x}, \mathbf{y})$: This is equivalent to $(\exists \mathbf{w})(pow(\mathbf{y}, \mathbf{w}, \mathbf{z}) \wedge \mathbf{w} = p_{\mathbf{x}})$. $\square$

We can now prove the following theorem that concludes the business of this sub-section.

A.4.20 Theorem. *For every $f \in \mathcal{PR}$, its graph $\mathbf{y} = f(\vec{\mathbf{x}}_n)$ is arithmetical.*

Proof. We do induction on $\mathcal{PR}$ (cf. A.3.24):

(1) *Basis.*   There are three graphs to work with here: $\mathbf{y} = \mathbf{x} + 1$, $\mathbf{y} = 0$ and $\mathbf{y} = \mathbf{x}$ (or, fancily, $\mathbf{y} = \mathbf{x}_i$; or more fancily, $\mathbf{y} = U_i^n(\vec{\mathbf{x}}_n)$). They all are explicit transforms of $\mathbf{y} = \mathbf{x} + \mathbf{z}$.

(2) *Composition.*   Say, the property is true for the graphs of $f, g_1, \ldots, g_n$. This is the induction hypothesis (I.H.). How about $\mathbf{y} = f(g_1(\vec{\mathbf{x}}_m), g_2(\vec{\mathbf{x}}_m), \ldots, g_n(\vec{\mathbf{x}}_m))$? Well, this graph is equivalent to (by repeated application of the informal 3 on p. 175)

$$(\exists \mathbf{u}_1) \cdots (\exists \mathbf{u}_n)(\mathbf{y} = f(\vec{\mathbf{u}}_n) \wedge \mathbf{u}_1 = g_1(\vec{\mathbf{x}}_m) \wedge \cdots \wedge \mathbf{u}_n = g_n(\vec{\mathbf{x}}_m))$$

and we are done by the I.H.

(3) *Primitive recursion.*   This is the part that benefits from the work put into A.4.19. Here's why: Assume (I.H.) that the graphs of h and g are arithmetical, and let f be given for all $x, \vec{y}$ by

$$f(0, \vec{\mathbf{y}}) = h(\vec{\mathbf{y}})$$
$$f(\mathbf{x} + 1, \vec{\mathbf{y}}) = g(\mathbf{x}, \vec{\mathbf{y}}, f(\mathbf{x}, \vec{\mathbf{y}}))$$

Now, to state $\mathbf{z} = f(\mathbf{x}, \vec{\mathbf{y}})$ is equivalent to stating

$$(\exists \mathbf{m}_0)(\exists \mathbf{m}_1) \cdots (\exists \mathbf{m}_{\mathbf{x}})\Big(\mathbf{m}_0 = h(\vec{\mathbf{y}}) \wedge \mathbf{z} = \mathbf{m}_{\mathbf{x}} \wedge$$
$$(\forall \mathbf{w})(\mathbf{w} < \mathbf{x} \to \mathbf{m}_{\mathbf{w}+1} = g(\mathbf{w}, \vec{\mathbf{y}}, \mathbf{m}_{\mathbf{w}}))\Big) \tag{i}$$

---

[45]Note that ex is that of A.3.34, and $ex(\mathbf{x}, \mathbf{y} + 1) \mid \mathbf{z} \equiv (\exists \mathbf{u})(\mathbf{u} = \mathbf{y} + 1 \wedge ex(\mathbf{x}, \mathbf{u}) \mid \mathbf{z})$, applying 3 on p. 175 within the metatheory.

The trouble with the "relation" (i) above is that it is not a relation at all,[46] because it has a variable-length prefix: $(\exists \mathbf{m}_0)(\exists \mathbf{m}_1)\cdots(\exists \mathbf{m}_\mathbf{x})$. We invoke coding to salvage the argument. Let us use a single number,[47]

$$\mathbf{m} = p_0^{\mathbf{m}_0} p_1^{\mathbf{m}_1} \cdots p_\mathbf{x}^{\mathbf{m}_\mathbf{x}}$$

to represent all the $\mathbf{m}_i$, for $i = 0, \dots, \mathbf{x}$. Clearly,

$$\mathbf{m}_i = \exp(i, \mathbf{m}), \text{ for } i = 0, \dots, \mathbf{x}$$

We can now rewrite (i) as

$$(\exists \mathbf{m})\Big(\exp(0, \mathbf{m}) = h(\vec{\mathbf{y}}) \wedge \mathbf{z} = \exp(\mathbf{x}, \mathbf{m}) \wedge$$
$$(\forall \mathbf{w})\big(\mathbf{w} < \mathbf{x} \to \exp(\mathbf{w}+1, \mathbf{m}) = g(\mathbf{w}, \vec{\mathbf{y}}, \exp(\mathbf{w}, \mathbf{m}))\big)\Big) \tag{ii}$$

The above is arithmetical because of the I.H. Some parts of it are more complicated than others. For example, the part

$$\exp(\mathbf{w}+1, \mathbf{m}) = g(\mathbf{w}, \vec{\mathbf{y}}, \exp(\mathbf{w}, \mathbf{m}))$$

is equivalent to

$$(\exists \mathbf{u})(\exists \mathbf{v})\big(\mathbf{u} = \exp(\mathbf{w}+1, \mathbf{m}) \wedge \mathbf{v} = \exp(\mathbf{w}, \mathbf{m}) \wedge \mathbf{u} = g(\mathbf{w}, \vec{\mathbf{y}}, \mathbf{v})\big)$$

The above is arithmetical by the I.H. and the preceding lemma. This completes the proof. □

---

[46]*Relation* is the informal counterpart of *formula*, of course.
[47]We do not need the fancier coding $\langle \mathbf{m}_0, \dots, \mathbf{m}_\mathbf{x} \rangle$ here, since the length, $\mathbf{x}$, of the sequence is known.

References

1. J. Bennett. *On Spectra*. PhD thesis, Princeton University, 1962.

2. N. Bourbaki. *Éléments de Mathématique; Théorie des Ensembles*. Hermann, Paris, 1966.

3. Alonzo Church. A note on the Entscheidungsproblem. *J. Symbolic Logic*, 1:40–41, 101–102, 1936.

4. Alonzo Church. An unsolvable problem of elementary number theory. *Amer. Journal of Math.*, 58:345–363, 1936. (Also in Davis [7, 89–107]).

5. S. Cook. The complexity of theorem-proving procedures. In *Proceedings, 3rd ACM Symposium on Theory of Computing*, pp. 151–158, 1971.

6. M. Davis. *Computability and Unsolvability*. McGraw-Hill, New York, 1958.

7. M. Davis. *The Undecidable*. Raven Press, Hewlett, NY, 1965.

8. R. Dedekind. *Was sind und was sollen die Zahlen?* Vieweg, Braunschweig, 1888. (In English translation by W.W. Beman [9]).

9. R. Dedekind. *Essays on the Theory of Numbers*. Dover Publications, New York, 1963. (First English edition translated by W.W. Beman and published by Open Court Publishing, 1901).

10. Edsger W. Dijkstra. Go to statement considered harmful. *Communications of the ACM*, 11(3):147–148, 1968.

11. Edsger W. Dijkstra and Carel S. Scholten. *Predicate Calculus and Program Semantics*. Springer-Verlag, New York, 1990.

12. Apostolos Doxiadis. *Uncle Petros and Goldbach's Conjecture.* Faber and Faber, London, 2000.

13. Herbert B. Enderton. *A Mathematical Introduction to Logic.* Academic Press, New York, 1972.

14. S. Feferman and A. Levy. Independence results in set theory by Cohen's method II. *Notices of the Amer. Math. Soc.*, 10:592, 1963. (Abstract).

15. K. Gödel. Die Vollständigkeit der Axiome des logischen Funktionen-kalküls. *Monatshefte für Mathematik und Physik*, 37:349–360, 1930.

16. K. Gödel. Über formal unentscheidbare Sätze der Principia Mathematica und verwandter Systeme I. *Monatshefte für Math. und Physik*, 38:173–198, 1931. (Also in English in Davis [7, 5–38]).

17. David Gries and Fred B. Schneider. *A Logical Approach to Discrete Math.* Springer-Verlag, New York, 1994.

18. A. Grzegorczyk. Some classes of recursive functions. *Rozprawy Matematyczne*, 4:1–45, 1953.

19. J. Hartmanis. *Feasible Computations and Provable Complexity Properties.* SIAM, Philadelphia, 1978. (Volume 30 in the *CBMS-NSF Regional Conference Series in Applied Mathematics*).

20. Leon Henkin. The completeness of the first-order functional calculus. *J. Symbolic Logic*, 14:159–166, 1949.

21. H. Hermes. *Introduction to Mathematical Logic.* Springer-Verlag, New York, 1973.

22. D. Hilbert and P. Bernays. *Grundlagen der Mathematik I and II.* Springer-Verlag, New York, 1968.

23. Douglas R. Hofstadter. *Gödel, Escher, Bach: An Eternal Golden Braid.* Basic Books, New York, 1979.

24. John E. Hopcroft, Rajeev Motwani, and Jeffrey D. Ullman. *Introduction to Automata Theory, Languages and Computation.* Addison-Wesley, Boston, 3rd edition, 2007.

25. L. Kalmár. An argument against the plausibility of Church's thesis. In *Constructivity in Mathematics, Proc. of the Colloquium*, Amsterdam, pp. 72–80, 1957.

26. S.C. Kleene. General recursive functions of natural numbers. *Math. Annalen*, 112:727–742, 1936.

27. S.C. Kleene. Recursive predicates and quantifiers. *Transactions of the Amer. Math. Soc.*, 53:41–73, 1943. (Also in Davis [7, 255–287]).

28. Donald E. Knuth. Structured programming with go to statements. *ACM Computing Surveys*, 6(4):261–301, 1974.

29. Christopher C. Leary. *A Friendly Introduction to Mathematical Logic.* Prentice Hall, Englewood Cliffs, NJ, 2000.

30. William J. LeVeque. *Topics in Number Theory*, volumes I and II. Addison-Wesley, Reading, MA, 1956.

31. Harry R. Lewis and Christos H. Papadimitriou. *Elements of the Theory of Computation.* Prentice Hall, Englewood Cliffs, NJ, 1998.

32. V. Lifschitz. On calculational proofs. *Annals of Pure and Applied Logic*, 113:207–224, 2002.

33. Yu. I. Manin. *A Course in Mathematical Logic*. Springer-Verlag, New York, 1977.

34. A. A. Markov. Theory of algorithms. *Transl. Amer. Math. Soc.*, 2(15), 1960.

35. Elliott Mendelson. *Introduction to Mathematical Logic*. Wadsworth & Brooks, Monterey, CA, 3rd edition, 1987.

36. Rózsa Péter. *Recursive Functions*. Academic Press, New York, 1967.

37. Emil L. Post. Finite combinatory processes. *J. Symbolic Logic*, 1:103–105, 1936.

38. Emil L. Post. Recursively enumerable sets of positive integers and their decision problems. *Bull. Amer. Math. Soc.*, 50:284–316, 1944.

39. W. V. Quine. Concatenation as a basis for arithmetic. *J. Symbolic Logic*, 11:105–114, 1946.

40. J.A. Robinson. A machine oriented logic based on the resolution principle. *JACM*, 12(1):23–41, 1965.

41. H. Rogers. *Theory of Recursive Functions and Effective Computability*. McGraw-Hill, New York, 1967.

42. J. Barkley Rosser. Extensions of some theorems of Gödel and Church. *J. Symbolic Logic*, 1:87–91, 1936.

43. K. Schütte. *Proof Theory*. Springer-Verlag, New York, 1977.

44. J. C. Shepherdson and H. E. Sturgis. Computability of recursive functions. *JACM*, 10:217–255, 1963.

45. Joseph R. Shoenfield. *Mathematical Logic*. Addison-Wesley, Reading, MA, 1967.

46. M. Sipser. *Introduction to the Theory of Computation*. PWS Publishing, Boston, 1997.

47. R. M. Smullyan. *Theory of Formal Systems*. Number 47 in *Annals of Math. Studies*. Princeton University Press, Princeton, 1961.

48. R. M. Smullyan. *Gödel's Incompleteness Theorems*. Oxford University Press, Oxford, 1992.

49. G. Tourlakis. *Computability*. Reston Publishing, Reston, VA, 1984.

50. G. Tourlakis. A basic formal equational predicate logic—Part I. *BSL*, 29(1–2):43–56, 2000.

51. G. Tourlakis. A basic formal equational predicate logic—Part II. *BSL*, 29(3):75–87, 2000.

52. G. Tourlakis. On the soundness and completeness of equational predicate logics. *J. Logic Computat.*, 11(4):623–653, 2001.

53. G. Tourlakis. *Lectures in Logic and Set Theory, Volume 1: Mathematical Logic*. Cambridge University Press, Cambridge, 2003.

54. G. Tourlakis. *Lectures in Logic and Set Theory, Volume 2: Set Theory*. Cambridge University Press, Cambridge, 2003.

55. Alan M. Turing. On computable numbers, with an application to the Entscheidungsproblem. *Proc. London Math Soc.*, 2(42, 43):230–265, 544–546, 1936, 1937. (Also in Davis [7, 115–154].).

56. A. N. Whitehead and B. Russell. *Principia Mathematica*, volume 2. Cambridge University Press, Cambridge, 1912.

57. R. L. Wilder. *Introduction to the Foundations of Mathematics*. Wiley, New York, 1963.

INDEX